WELFARE RANCHING

*The Subsidized Destruction
of the American West*

Edited by George Wuerthner and Mollie Matteson

ISLAND PRESS

WASHINGTON • COVELO • LONDON

FOREWORD

Douglas R. Tompkins

The undertaking of a book of this type is always an immense project. I am now on my sixth such project, with three more on the drawing board. I beg no sympathy from anyone as to why I should submit to such punishment as a book of this complexity inevitably exacts. It requires a tireless and dedicated team, experts on the issue, and activists committed to public policy change that would begin the long process of restoration. If I were to congratulate myself on any aspect of this book, it would be for the choice of George Wuerthner as editor, photographer, and writer. George is one of the most committed and involved activists on this issue. Without him, this book simply would not have come into being. Perhaps I may have catalyzed this project, but George made it happen, along with his wife, Mollie Matteson, and the great design layouts of Roberto Carra.

Livestock grazing in the arid West is as outmoded as is whaling in today's oceans. It is a thing of the past, a "tradition" whose practitioners are still immersed in a livelihood in which ecological reality has yet to sink in. Americans, still in love with their frontier history and the western livestock culture, have yet to come to grips with the hopeless future of grazing cattle and sheep in the West (and arid places worldwide). The western landscapes, open and vast; the cowboy image and myths; a lifestyle of rugged individuals—all have conspired to perpetuate the idea that ranching on these lands can somehow go on forever. Such times are over. It is time to begin the long, slow process of recovery, of rewilding a landscape that should never have been domesticated in the first place. That these are public lands being severely degraded by private interests—what amounts to little more than a handful of millionaires, corporations, and hobby cowboys—makes it all the more timely to call for an end to the misguided grazing permit system. It is a pernicious and ultimately landscape-degenerating and extinction-provoking system put in place because the western ranching industry used political power to avail itself of the public grasses. The general public is basically unaware of how this access by private, for-profit businesses, be they small or large, abuses public lands for a pittance of a fee, and that a vast and expensive bureaucracy is required to manage this industry. The costs of the attempt (although unsuccessful and arbitrary in its methodology) to manage private grazing are far greater than the income from fees. In short, all U.S. citizens who pay taxes are essentially subsidizing grazing permit holders to trash the public lands. This is an absurd situation, akin to that of logging on public (national and state) forests, in which logging fees don't cover the costs of federal or state management agencies.

What this book shows, and shows with no tricks of photography or superficial science, is that wherever cattle are grazing at the public trough, we discover often irreversible ecological damage. Fauna of all kinds are either extinct,

Opposite: Wildflowers along Elk Creek, an area grazed by wildlife but not by livestock, Yellowstone National Park, Wyoming.

viii–ix: Livestock-free grasslands, Buenos Aires National Wildlife Refuge, Arizona.

extirpated, or endangered; riparian zones are trampled and degraded; introductions of exotic grasses and foiled mitigation attempts abound. The landscape is altered slowly over time in such a way that the change is difficult to comprehend. Media and commercial advertising keep time-capsule imagery alive in the public mind; hardly anyone goes out there to make the comparison with how the West *used to be*. In this book we present images and scenery to show how it was, how it is, and how it could be. The contrasts are staggering; the possibilities for a restored *wild* West are tantalizing. There are already examples (mainly in our national parks, most—but not all—of which have no grazing of domestic livestock) of what ecological integrity means to living landscapes. Domestic livestock have no place in those landscapes, and the landscape's health is solid testimony to what happens if complex ecosystems are left to themselves. The abundance of wildlife is staggering; the delight to our citizens is abundantly clear as reintroductions of wolves, bears, bison, and pronghorn antelope grasp the imagination of all who see them. Some parts of our landscapes simply need to be wild for biological, spiritual, and emotional reasons.

Basic human ethics suggests that we must not humanize every square yard of the planet, that the human economy can flourish and humanness become more profound when we protect and promote the wild world and learn to share the planet with all her creatures. Human economic activities either improperly conducted or managed, or simply inappropriate for the location and landscape, must be recognized and halted. Domestic livestock grazing in the arid western United States, *especially on public lands, which contribute less than 3 percent of U.S. meat production*, is unnecessary and wasteful. There is no right way to do the wrong thing. It is simply time to throw in the ranching towel and face the ecological reality, the only true reality we have.

Read this book; do the necessary work to be up-to-date on this issue; become an activist (or a better one) by knowing your subject matter. Join all of us already on the path toward healing a century and a half of reckless ranching in the arid West. Our economy will gain, not lose, if we retire grazing permits on public lands. We will have a vibrant and living western landscape, the revival of extirpated species, and beautiful testimony to true human values—making biodiversity the foremost value, from which should flow our social, economic, and cultural systems.

You will learn here what is being proposed to seek a just and equitable solution for the ranchers themselves, as well as for the land and all Americans. The ranchers' plight is of great concern and must be dealt with in good faith and in recognition of economic reality, not solely ecological reality. It will not be a simple and easy transition for everyone, as each permit holder has a different configuration of lands and business interests, but it can be worked out fairly and equitably if all sides are committed to a socially and ecologically just end. To that end, this book and the campaign it champions are totally devoted. May it be a win-win for all concerned.

WHAT IS A

300 MILLI

T STAKE?

The 300 million acres of
public land grazed by livestock in the West
equal the area of three Californias.

ON ACRES

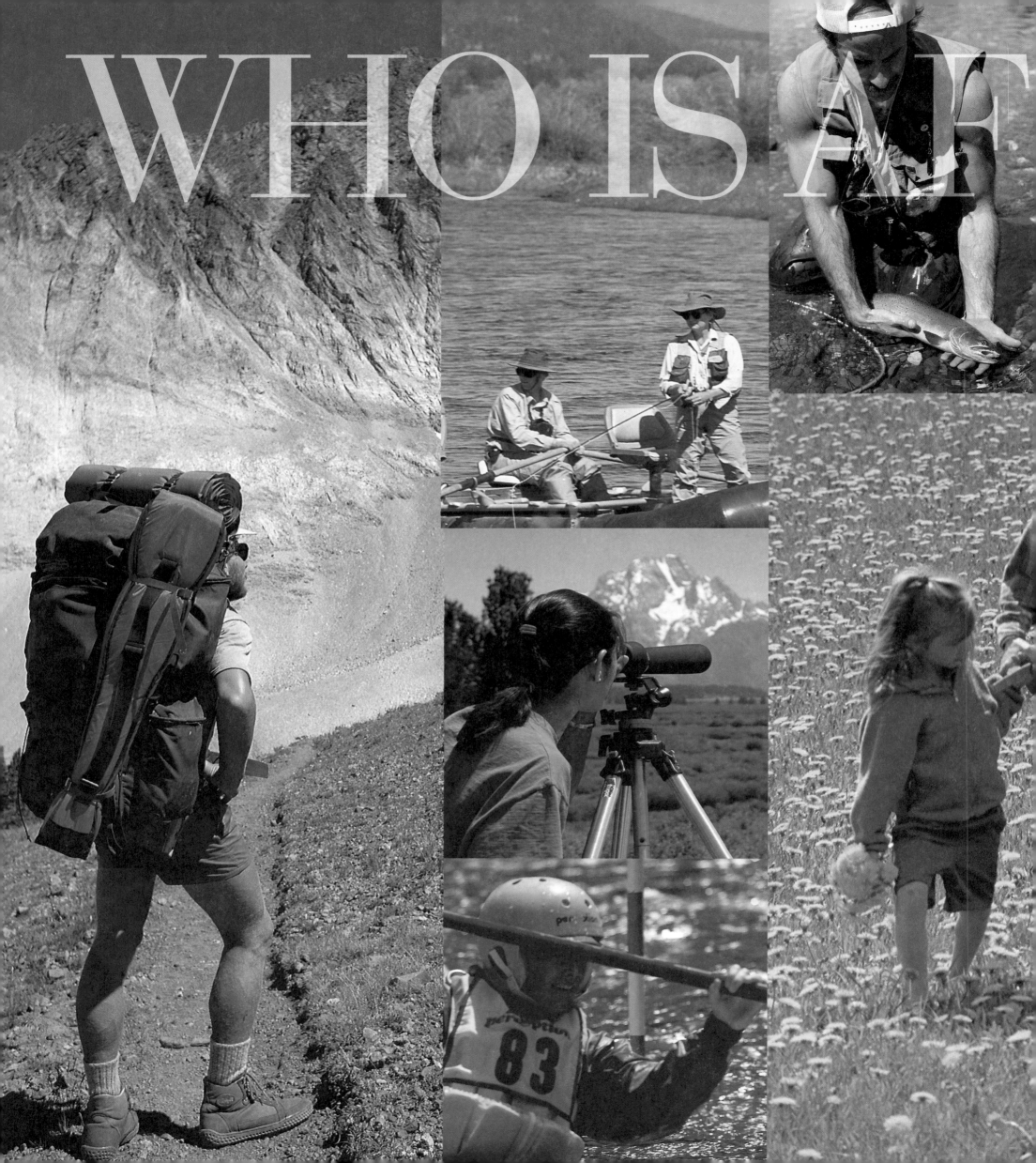

WHO IS AF

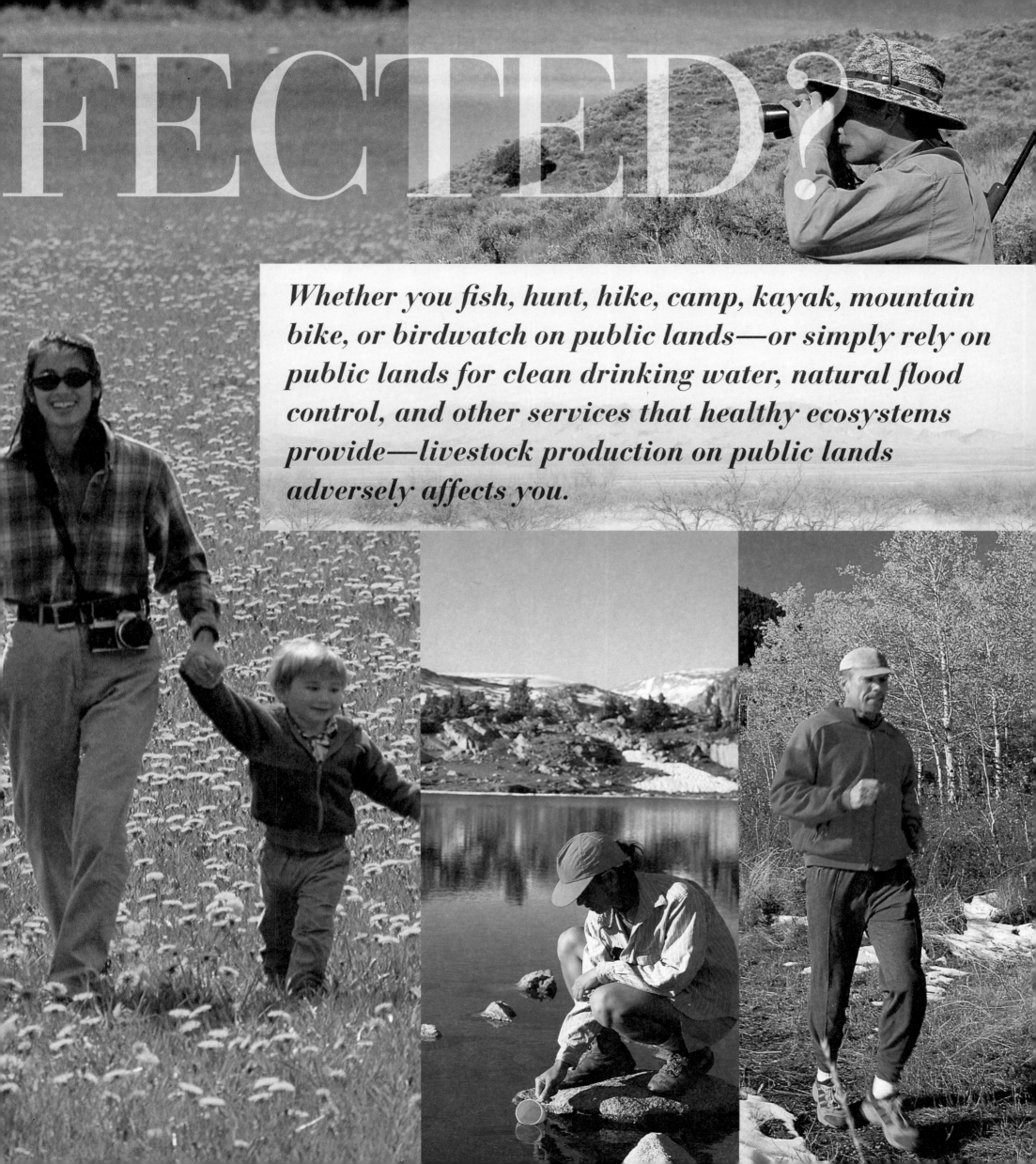

FECTED?

Whether you fish, hunt, hike, camp, kayak, mountain bike, or birdwatch on public lands—or simply rely on public lands for clean drinking water, natural flood control, and other services that healthy ecosystems provide—livestock production on public lands adversely affects you.

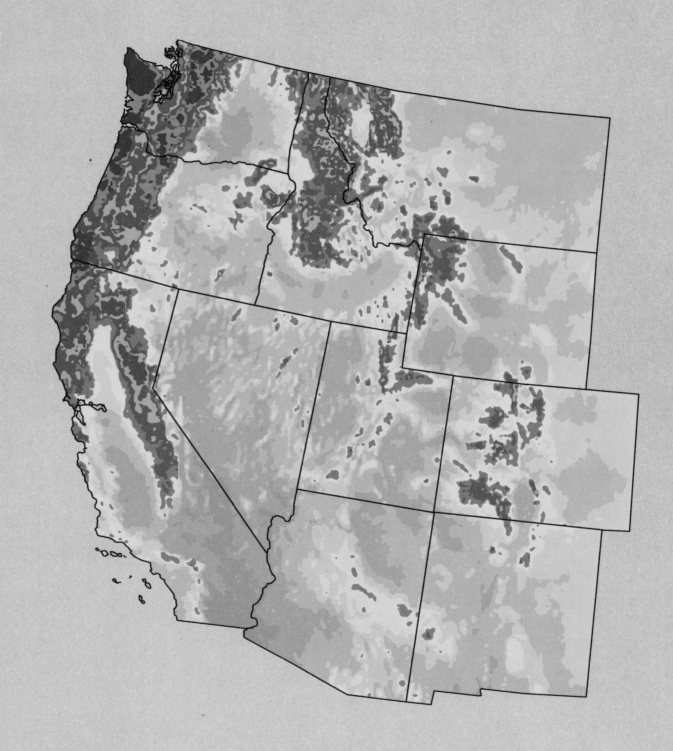

AVERAGE ANNUAL PRECIPITATION

WESTERN UNITED STATES

Period: 1961–1990
Units: inches

Legend (Inches per Year)

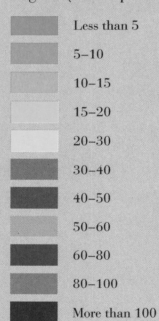

Less than 5

5–10

10–15

15–20

20–30

30–40

40–50

50–60

60–80

80–100

More than 100

INTRODUCTION

Three hundred million acres. That is what is at stake. In round figures, some 300 million acres of public lands—federal, state, and county—are currently leased for livestock production. This figure includes some 90 percent of all Bureau of Land Management holdings, 69 percent of the lands managed by the U.S. Forest Service, plus national wildlife refuges, national parks, and other nature preserves. The combined area is as large as the entire eastern seaboard from Maine to Florida, with Missouri thrown in! And 300 million acres is what potentially could be restored if public lands livestock production were eliminated. Nowhere else in the United States is there such potential for large-scale ecosystem restoration at so little cost—and ultimately affecting so few people—as in the termination of domestic livestock production on our public lands.

Although the impacts associated with livestock production vary from region to region, and even from ranch to ranch, there is overwhelming evidence that livestock production has impoverished the West's biological capital. This damage is not confined to the public lands. Most of the private lands in the West are devoted to livestock production in one way or another, and suffer equally from environmental degradation. Public values—such as clean water and healthy, abundant wildlife populations—are diminished by poor land use practices in the private sector. When discussing livestock production and the threat it poses to the natural heritage we share as a society, this book focuses primarily on public lands grazing, but we must also consider a more complete picture—an industry operating in both the public and private arenas, extracting private profit while depleting public resources and jeopardizing natural systems.

Subsidies

Whether on public or private lands, the western livestock industry is subsidized in multiple ways. First, there is the abundance of federal and state funding that props up the industry, including below-market grazing fees, emergency feed programs, low-interest federal farm loans, and many other taxpayer-funded programs.

Even more important are the environmental costs—most of them not counted in any way, and certainly not absorbed by livestock producers. These costs include soil erosion; degraded water quality and the costs of cleanup; the spread of exotic weeds and the subsequent reduction in plant community productivity; and the costs of saving species endangered by livestock production.

Finally there are the social costs resulting from beef consumption. Though the subject is beyond the scope of this book, a heavy meat diet contributes to numerous health problems that society pays for directly and indirectly, from higher costs to reduced life expectancy.

The Cost of Aridity

Though this book strives to make the case that public lands livestock production should cease, we hasten to say that ranchers are not bad people. They are pursuing what they consider to be an honorable occupation, striving to provide products they believe America needs and wants. Nevertheless, as we hope readers will agree, the costs of running this industry in the arid West are simply too high. Ranchers are struggling against insuperable geographic limits. It is our view that western ranching cannot now, nor ever will, be made ecologically benign, sustainable, or profitable because of the scarcity of the most enduring and powerful ingredient for all life—water.

The West is a land of steep and rocky mountains, deep canyons, and vast, open spaces. It is also a land of aridity. The major focus of this book is that dry realm between the Sierra Nevada and Cascade Ranges on the west and the tallgrass prairie on the east. Not only is this area home to all of the United States's official deserts—including the Mojave, Sonoran, Great Basin, and Chihuahuan Deserts and the Colorado Plateau—it is also home to semiarid grasslands, such as the Palouse prairie and the Great Plains. And in addition to their lack of precipitation, these areas also have highly variable patterns of precipitation. To paraphrase the great western historian Walter Prescott Webb, the prevailing climate of the West is drought punctuated by occasional periods of wetness.

There is a direct correlation between moisture and plant productivity. The more moisture an area receives, the more biomass the land can produce. The West is not nearly as productive as the more humid, wetter parts of the country. Nor are these dry lands nearly as tolerant of abuse as more moist environments. Damage occurs more easily. And recovery takes longer—if it occurs at all.

The rugged terrain and arid conditions impose limitations on what ranchers can do with domestic livestock, and on how much they can spend to mitigate the negative impacts associated with livestock production. To have enough forage for a viable cattle operation, ranchers must own huge spreads—or have access to large areas of cheap public grazing land. Indeed, in some states like Nevada, it may require 250 acres or more to support one cow for a single year. Meanwhile, in a place with a relatively moist and warm environment, such as Mississippi or Missouri, a rancher can sustain a cow year-round on a single acre. Western ranchers can compete with livestock operators in higher-rainfall regions only because western operators have had almost unlimited access to free or cheap forage.

Grazing livestock over huge expanses brings huge expenses. It costs more for fencing, because the amounts of materials and hours of labor needed are greater. It costs more to get cattle or sheep to market, because ranchers have to hire "semi" trucks and drive hundreds of miles to transport stock. It costs more time simply to drive out and check on livestock, as well as to monitor the well-being of the range. It costs more money to put in irrigation systems and build reservoirs. It costs a great deal to guard livestock from predators, or even to hire someone to kill predators. Ranchers have squeaked by, in part, because they've managed to avoid paying many of these costs, transferring them instead to the taxpayer or to the land itself. These days, the livestock industry is being asked more frequently to pay the real costs. The public—and perhaps some ranchers as well—is discovering that the meager amount of meat produced on western rangelands cannot justify the costs of operation.

Another consequence of aridity is that cows, widely distributed over the landscape to get adequate forage, are harder to protect from predators than livestock grazing the back forty and being called in to shelter in a barn every night. It is not feasible nightly to round up, or even patrol, an entire herd of cows on the typical western grazing allotment. The ranchers' solution has been simple, and devastating: get rid of the predators. Thus, wolves were wiped out in the West, as were grizzly bears—except for a few, isolated strongholds; mountain lions continue to be persecuted; coyotes are killed by the tens of thousands every year; and even smaller predators, such as foxes and eagles, are frequently resented and sometimes shot.

Because of aridity, western livestock operations are nearly all dependent on irrigated pasture and hay production. Such irrigation results in the dewatering of rivers and the pumping of groundwater away from natural springs and seeps. The livestock industry's thirst for water also requires the construction of storage reservoirs that fragment river systems and change water flow regimes.

In moist landscapes, plants can recover their reserves after grazing if given a rest of a few months. In the arid West, rest in the middle of a drought is not granting any recovery. To make matters worse, during drought periods ranchers must graze even more acreage to make up for the land's lower productivity. It is a situation that nearly guarantees a cycle of land degeneration.

Aridity also affects how livestock use the land. Cattle evolved in the moist woodlands of Eurasia. As a result of their natural tendency to congregate near water sources and dense vegetation—in the West, primarily found in streamside areas and around seeps and springs—cattle do an inordinate amount of damage by trampling vegetation and soils, stripping plant material, breaking down stream banks with their hooves, and fouling water with their wastes.

The Costs of Mitigation

Some livestock proponents argue that the negative effects of livestock can be mitigated. For example, to keep livestock from damaging riparian zones (those thin green bands of water-dependent vegetation found along streams and springs), proponents advocate fencing. This proposal, however, has numerous problems, not least of which is that there are literally hundreds of thousands of miles of riparian area in the West. Fencing even a small portion of the total would be extremely costly. Furthermore, fencing of riparian areas still leaves many small seeps and headwater streams subject to the pounding of cattle hooves and the stripping of vegetation by bovine vacuums. Plus, fences become barriers to the free movement of wildlife.

Other supposed solutions, such as pumping water from seeps or springs to water tanks or troughs, create other problems. For example, rings of nearly bare ground usually appear around water developments as entire herds of livestock descend on them. These sacrifice zones become compacted, with many native plants driven out, to be replaced by exotics and tough, unpalatable plants. As for the removal of the water itself, it requires only the simplest logic to realize that with less water in a spring or stream, there is less habitat for water-dependent native species. Even if no cow goes near a stream, the diversion of water to stock tanks and the like means survival becomes trickier for trout, for willows, for frogs, for water ouzels.

Finally, who pays for this mitigation? Ranchers, already operating under marginal economics, cannot afford to pay for mitigation measures. So, then, the taxpayers do. And why should they bear the costs?

No matter whose pocketbook the money is coming out of, it doesn't really make sense to invest in making the arid West a better cow pasture. With so many other areas—outside of this dry, rugged part of the country—where it is possible to raise livestock without such massive manipulation, infrastructure, and cash outlay, the question of why anyone should participate in or support such foolishness looms larger than ever.

Livestock's Contribution to Biological Impoverishment

As in any work that takes a broad overview of a topic, there will always be exceptions to the points being made. Livestock proponents like to point to a few exceptional ranching operations, then try to portray them as feasible, sustainable, and environmentally benign, if not representative of how the industry as a whole currently conducts itself. We respond by stating that no ranching in the West is environmentally benign, but even if there are a few exceptional operations, they don't invalidate the general rule: that livestock production in the arid West has contributed to major biological impoverishment.

Livestock production, by its very nature, is a domestication of the landscape. It requires using the bulk of water, forage, and space for the benefit of one or two domestic animals—at the expense of native creatures. Although this is characteristic of agriculture everywhere, the expropriation of resources for the raising of livestock is particularly egregious in the arid West because natural productivity is limited and highly variable. The majority of the West is directly or indirectly influenced by livestock production, either as rangeland, as cultivated land or pasture growing feed for livestock, or as delimited reserves of nature where naturally migrating wildlife are persecuted the instant they step outside the boundaries people have imposed on them. If you add in the hundreds of millions of acres of farmland in the Midwest devoted to the production of livestock forage, it's clear that the total physical and ecological footprint of livestock production is enormous.

A Choice

The issue of western livestock production is largely about the wise use of resources. Contrary to the prevailing myth of the West, the majority of cattle are not raised here, but east of the hundredth meridian. Missouri grows more beef than Montana. Louisiana is a bigger cattle producer than Wyoming—the "Cowboy State." And tiny Vermont produces more beef than all the public lands in Nevada. Given the small percentage of meat produced off the vast western range and the tremendous costs to native ecosystems as well as to tax-payers, who indirectly and directly subsidize the western livestock industry, any amount of commercial livestock production here is difficult to justify.

The elimination of livestock grazing on public lands in the West would be of very little consequence to the overall meat supply of the nation. As is discussed later in this book, the number of people whose livelihoods would be adversely affected by ending grazing on public lands is remarkably low, and even then, probably a good many ranchers on public lands would have the option of turn-ing solely to private lands to continue their operations. Without the crutch of cheap public lands forage, some ranchers might turn to innovative ventures on their private spreads. This is already happening in parts of the West. For example, although the history of dude ranching goes back nearly as far as that of western ranching itself, these days more and more ranches are offer-ing opportunities for outdoor recreation, as well as nature observation and environmental education. The hard fact for some is that an outside income—selling insurance, working as a schoolteacher, and so forth—is already a necessity, whether a public lands grazing permit is tied to the ranch or not. The number of permittees who would face financial ruin for the sole reason that their public lands grazing had ended is likely very small—and far lower than the number displaced by the typical corporate downsizing move.

Still, it is possible for our society to show compassion and generosity to public lands ranchers while at the same time acting to protect and restore the ecolog-ical integrity of western landscapes. Keep in mind that society legally owes public lands ranchers nothing. Grazing on the federal lands has always been a privilege, not a right, and permits have always been subject—theoretically—to revocation if environmental damage is deemed significant by the managing agency. However, the reality is that for over a century, the public lands have been neglected and abused. Government officials have been thwarted from standing up to politically powerful ranchers (if that were even an inclination in the first place). And the interaction of the banking industry and the live-stock industry has resulted in real monetary value being attached to public lands grazing permits that are associated with specific ranch properties.

Thus, a variety of proposals for phasing out livestock grazing on public lands are now being developed and discussed. Some are explained near the end of this book. Most involve trading retirement of public lands grazing permits for some type of one-time transition payment to permittees—whether funds come from federal sources or from private organizations.

In light of the marginal nature of western ranching, the ever-growing econom-ic pressures on the western livestock industry, and the burgeoning desire of the public to see the public lands serve truly broad, collective interests—including that of the natural world itself—we hope that ranchers will begin to see that the time has come to close this chapter in the pageant of the West. We hope they will see that a new chapter on the western public lands is unfolding, and that they have the power to participate positively in how the story pro-ceeds. Or they can resist and deny—as has been the case for too long—and then forces beyond their control will likely wrench matters out of their hands in the end, anyway. We ourselves do not know whether such a scenario would lead to more beneficent care of the land. Perhaps narrow, though different, interests would prevail again, taking the resources of the West for self-serving purposes. It is, after all, a familiar western drama.

We do know, however, that there is today great opportunity to salvage and restore a West of true wildness, for all people and for all the species that inhab-it this land. There is yet time to renew a West of clear-flowing rivers rippling with trout, a West of hills undulating with large and elegant herds of elk and bison, bighorn sheep and pronghorn antelope. There is still a chance. We can call back the mighty grizzly, the darting black-footed ferret, the loping wolf, the gregarious prairie dog. Three hundred million acres of public land in the West is one of our best hopes—perhaps the last, very best hope—for setting aside a portion of our continent that not only protects a magnificent land-scape, not only harbors a remarkably rich diversity of plants and animals, but also honors a land of the imagination and the human spirit.

Before America was a land of settlers, of pioneers, before it was dreamed of as a land of empire and conquest, it was another kind of place. It was home, and it was sacred landscape, at one and the same time. Here resided creatures and powers to be revered and respected. Although much has changed in the last half-millennium, some things have not. Our human longing for a world beau-tiful and big and beyond our measuring lives on. The western lands held in trust for all of us still offer the chance to know, to experience, an immeasur-able and mysterious yet dazzlingly real world.

These lands are under siege, and much is lost and irretrievable. This book is our call to alarm. Yet, ultimately, we work for what is still here, for what may yet be again. The arid West is a land of limits; this we have said, and what follows in this book should make that abundantly clear. Yet limitations can produce innovation; limitations can drive creativity, in human societies as in nature. Our society sought to make the West over, to make it into a place we carried collectively in our minds, from ancient memories and cul-tural myths. Now, as our society presses against the ecological limits of the West, it is time to create a new story, one that better matches the physical boundaries of this place. Just exactly how that new story should go is not yet clear, though we are suggesting its outlines.

Our foremost recommendation, the plotline we see most clearly for the new story of the West, is to end the wasteful, destructive, tragic abuse of our public lands by the livestock industry. It is a very tall order; to some, it no doubt sounds extreme. But, there simply is too much at stake—some 300 million acres of land at stake—to settle for some weaker, less ambitious option. That has been tried—the history of grazing reform is distressingly redundant—and today, while more species are in more dire straits than ever and some places are on the verge of ecological collapse, livestock proponents are still fid-dling with grazing schedules and stocking rates, building this water develop-ment or that fence, looking for that elusive, perfect management scheme that will, at last, make cattle and sheep benign beasts in this irredeemably parched land. How many more chances, at how much cost, should the public allow ranchers and public range managers? How many more species do we care to see become endangered, or extinct, before we, the owners of the public land, say it is time to give up on trying to develop the kinder, gentler cow and instead focus on fixing the damage that has been done, and on putting our western landscapes back together?

How much more time before the public reclaims the western public lands? We, of course, hope that it is very little time at all.

HOW

IT WAS...

HOW

IT IS...

HOW

IT CAN BE

THIS LAND IS

YOUR LAND

PART I

The public lands of the United States are a hallmark of our democracy and harbor some of the greatest resources of our nation. Federally managed lands—owned by all Americans—total 623 million acres, or more than 25 percent of the U.S. land base. There are four major federal land agencies—the Bureau of Land Management (BLM), the U.S. Forest Service (USFS), the National Park Service (NPS), and the U.S. Fish and Wildlife Service (USFWS). State agencies and other government departments oversee millions of acres of additional public land.

The vast majority of the federal public lands are in the western United States, where they serve as sources of clean water, recreation, scenic beauty, and inspiration. The public lands are wildlife habitat and in many cases provide the only remaining suitable environments for jeopardized species. On the large blocks of acreage provided by the public lands, restoration and maintenance of landscape-scale ecological processes—such as wildfires—are feasible and desirable. Elsewhere, the prerogatives of commercial enterprise and other human needs usually dominate.

Unfortunately, resource exploitation of various kinds has driven public lands management for many decades. Mining, logging, oil and gas drilling, and even farming have occurred and continue to occur on public lands. But the most widespread commercial use of western public lands is livestock production. Nearly all public lands that have any forage potential for livestock are leased for grazing. This includes 90 percent of BLM lands, 69 percent of USFS lands, and a surprising number of wildlife refuges and national parks. This land—your public land—is frequently managed as if it were a private feedlot rather than the common heritage of all Americans.

Next time you go out to visit your public lands and encounter a fence you must cross, a gate you must open, a campground fouled with cow manure, a trout stream trampled by cows, a hay meadow rather than a natural wetland, weeds instead of native grasses, cattle and sheep instead of prairie dogs, remember, this is your land. Do you like what you see?

xvi–xvii: St. Mary's Valley, Glacier National Park, Montana.

xviii–xix: Cows on Bureau of Land Management land, Trout Creek Mountains, Oregon.

xx–xxi: Elk along the Gibbon River, Yellowstone National Park, Wyoming.

xxii–1: Bull elk.

Opposite: Livestock-free grasslands, Arches National Park, Utah.

This land is your land, this land is my land
From California to the New York Island
From the redwood forest to the Gulf Stream waters
This land was made for you and me.
—Woody Guthrie

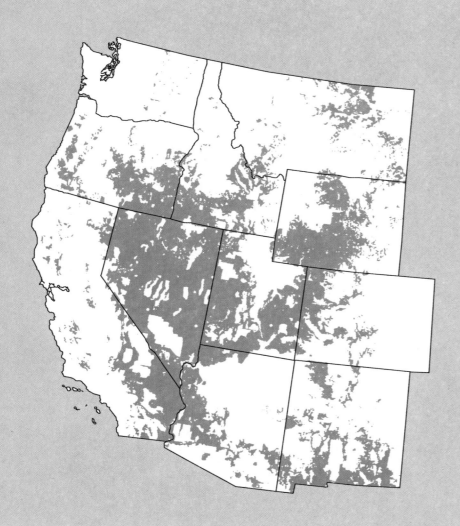

**BUREAU OF LAND MANAGEMENT
LANDS**

 BLM lands

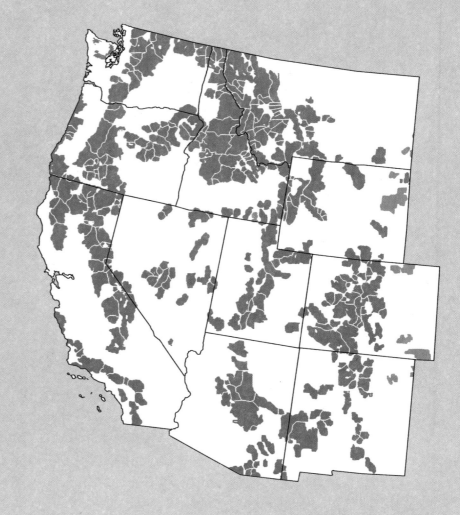

U.S. FOREST SERVICE LANDS

 National Forest Lands

National Grasslands

PUBLIC LANDS RANCHING

BY THE NUMBERS

Unless otherwise noted, the following statistics are for the eleven western states where the bulk of public lands grazing occurs: Arizona, California, Colorado, Idaho, Montana, New Mexico, Nevada, Oregon, Utah, Washington, and Wyoming. Total U.S. acreage: 2.3 billion. Total acreage in the lower forty-eight states: 1.9 billion. Total acreage in the eleven western states: 750 million. Acreage in the West under Bureau of Land Management (BLM) management: 177 million. Acreage of BLM land leased for grazing: 163 million. Percentage of BLM land leased for grazing: 92%. Acreage in the West under U.S. Forest Service (USFS) management: 141 million. Acreage of USFS land leased for livestock production: 97 million. Percentage of USFS land in the West leased for livestock production: 69%. Total BLM and USFS acreage leased for livestock production: 260 million. Total acreage of other federal lands (National Park Service and U.S. Fish and Wildlife Service) leased for livestock production: 5 million. Total federal acres leased for livestock production: 265 million. Total state land leased for livestock production: 36 million. Total acreage of other public lands—county, city, etc.—leased for livestock production: 5 million. TOTAL ACREAGE OF WESTERN PUBLIC LANDS LEASED FOR LIVESTOCK PRODUCTION: 307 million. Total acres of private lands in the West used for livestock production: 184 million. Total acres of Indian reservations in the West used for livestock production: 35 million. TOTAL ACREAGE OF THE WEST DIRECTLY USED FOR LIVESTOCK GRAZING: 525 million. (This is slightly more than one-quarter of the U.S. land area, excluding Alaska.) The above figures do not include the tens of millions of acres of cropland and pasture used to grow livestock feed and fodder in the West, nor the hundreds of millions of acres in the Midwest that also grow livestock feed, some of which ends up fed to "western" livestock. Total number of U.S. cattle producers: 1.6 million. Number of cattle producers in the West: 180,000. Percentage of cattle producers in the East: 89%. Percentage of cattle producers in the West: 11%. Number of permittees on western federal lands: 22,000. Percentage of U.S. cattle producers ranching on western BLM and USFS lands: 1.4%. Percentage of U.S. cattle producers ranching on all western public lands: 1.9%. Percentage of U.S. population ranching on western public lands: 0.012%. Percentage of U.S. livestock feed (crops, pasture, range forage) supplied by BLM and USFS lands: 2%. Percentage of U.S. livestock feed supplied by other public lands: 1%. Percentage of U.S. livestock feed supplied by all public lands: 3%. Percentage of U.S. livestock feed supplied by western private lands: 18%. Percentage of U.S. livestock feed supplied by all private lands: 97%. Percentage of western livestock feed supplied by all western public lands (federal, state, county, etc.): 18%. Percentage of BLM forage controlled by top twenty BLM permittees: 9.3%. Percentage of all BLM permittees represented by top twenty BLM permittees: 0.1%. Acreage of western public lands grazing allotments controlled by J. R. Simplot, the richest person in Idaho: 2 million. Other large public lands grazing permittees include Metropolitan Life Company, Agri-Beef Company, Nevada First Corporation, and the Packard family, of Hewlett-Packard fame.

RANCHING MYTHS

MYTH
Ranchers Are Good Stewards of the Land

TRUTH

Cattle at water trough, Bureau of Land Management lands, eastern Oregon.

More than 410 million acres of U.S. rangelands—public and private—are in unsatisfactory ecological condition, according to an estimate by the Natural Resources Conservation Service. This is an area four times the size of California, or 21 percent of the continental United States, and nearly all of it is in the West. These lands are severely damaged, with at least 50 percent of the desirable plant species eliminated, high erosion and weed invasion rates, and riparian areas unable to function normally.

Although public lands usually get more attention from the media, statistics compiled by the Natural Resources Conservation Service indicate that more total acres and a higher percentage of private lands in the West are in unsatisfactory condition as compared with public rangelands. This is particularly egregious in that private lands tend to be more productive and better watered than public lands—hence more resilient to livestock abuses.

In truth, ranchers are fighting an impossible battle against the natural limitations of the landscape. The West is not only an arid region but one in which annual precipitation varies widely. The amount of precipitation that falls in a year is directly reflected in the amount of grass production, meaning that forage quantity varies widely from year to year as well. This makes it very difficult for ranchers to maintain a stable business operation while also managing herds so as not to damage the land.

To be a good steward, ideally one not only must have a sense of responsibility and concern for the land—as many ranchers do—but also must treat the land in a way that conserves its fertility, productivity, diversity, and beauty for the future. Yet by raising domestic animals that demand large quantities of water and forage in a place that is dry, and by favoring slow-moving, heavy, and relatively defenseless livestock in terrain that is rugged, vast, and inhabited by native predators, ranchers have put themselves in a position of constant warfare with the land. They funnel most of the grass into their own animals, at the expense of the wild herbivores. They divert water from rivers to grow hay and other crops to feed cows, leaving fish and other aquatic life with hot, shallow trickles. They allow their cattle to graze and trample riparian areas—habitat on which 75 to 80 percent of all wild animal species in the West depend—polluting waterways with manure and adding excessive sediments to the water as they denude the land. And although "beauty is in the eye of the beholder," it's arguable whether most people would prefer a place where the grass is chewed down to stubs and the ground is littered with cow pies, over a grassland of tall and waving stems, dotted with wildflowers.

Windmill and water storage tank, Arizona.

MYTH

Rangeland Conditions Are Improving

TRUTH

Spring trampled by cattle, Humboldt National Forest, Nevada.

Rangelands were so severely overgrazed in the late nineteenth and early twentieth centuries that most places just couldn't get any worse. Since then, there has been limited improvement, mostly because of a steep reduction in domestic sheep numbers. Yet it would be wrong to imply that our rangelands are seeing significant advances toward biological sustainability. Hundreds of millions of acres are still in an ecologically degraded condition. For example, according to statistics compiled by the Society for Range Management, 15 percent of Bureau of Land Management (BLM) lands are improving in ecological condition and function. However, 14 percent of BLM lands are continuing to decline. And although the vast majority of BLM holdings are rated "stable," a high proportion of the acreage in this category is in such poor shape that it cannot get much worse. Livestock proponents like to say that the majority of western rangelands are "stable and improving." Yet by combining the large percentage of "stable" lands with the smaller percentage of "improving" lands, what livestock advocates have done is to disguise the reality that most of these public lands are ecological disaster zones.

Most improvement on public lands has been on the uplands (areas upslope of valley bottoms and streams), because of the decreasing numbers of livestock there, while the devastation of biologically critical riparian areas continues. In fact, according to a 1990 Environmental Protection Agency report, riparian areas are in the "worst condition in history." And, as a 1989 General Accounting Office report found, livestock are the major source of riparian degradation on public lands in the West. It is possible for livestock proponents to claim that the range condition of a particular allotment is improving even while the riparian zones within it are worsening, because of the way official range assessments average all parts of an allotment together.

In most cases, improvement on an allotment is a consequence of lowered stock density or a shortened grazing season. In effect, fewer livestock means better range condition, and in nearly all instances, termination of all livestock grazing would result in much more rapid rangeland recovery.

MYTH
Livestock Benefit Wildlife

TRUTH

Bighorn sheep ram.

Hundreds of species across the West are in danger of extinction, primarily because of livestock production. Species as varied as the Bruneau Hot Springsnail, the southwestern willow flycatcher, and the Bonneville cutthroat trout are in jeopardy as a consequence of habitat loss or degradation due to livestock grazing and its associated activities. No other human activity in the West is as responsible for the decline or loss of species as is livestock production.

Predator and pest control has extirpated many species, from wolves to prairie dogs. Dewatering of rivers for irrigation has contributed to the decline of many aquatic species, including many native trout. Livestock trampling of riparian areas, wet meadows, seeps, and springs has harmed habitat for a great variety of creatures, from songbirds to frogs. Livestock consumption of grass and other vegetation decreases hiding cover for many animals, making them more vulnerable to predators. Disease transmission from livestock to wildlife, as has frequently occurred with domestic and bighorn sheep, can diminish or eliminate certain wild animal populations. Most forage on public rangelands is allotted to livestock, leaving little food for native species to consume.

A few species have increased with the spread of livestock production. Yet, just as one could demonstrate that rats and pigeons flourish in the city and thereby incorrectly assert that wildlife benefit from urbanization, so too is it false to point to the proliferation of deer, Canada geese, cowbirds, and a few other opportunists and suggest that livestock production enhances conditions for wildlife in general.

Several big-game species, such as elk, pronghorn antelope, and bighorn sheep, have increased from early twentieth-century lows, when market and subsistence hunting nearly drove them to extinction. However, the rise in the numbers of these species is a consequence of intensive game management—such as adoption of strict hunting seasons, reintroductions, and habitat acquisition—rather than any inherent compatibility with livestock. Indeed, many of these big-game animals are still limited by having to compete for forage, water, and space with domestic livestock.

Livestock advocates suggest that water developments, such as troughs and stock ponds, benefit wildlife. While some wild animals undoubtedly use them, these facilities tend to lack adequate surrounding vegetation for hiding cover, nesting habitat, foraging, and other wildlife needs. Thus, these structures are almost useless to most wild species, and they exist at the expense of natural seeps, springs, and streams that would support far more native creatures if left intact.

MYTH

Public Lands Grazing Supports the Family Rancher

TRUTH

Cowboys herding cattle.

Public lands grazing subsidies, like most agricultural subsidies, disproportionately benefit large landholders. In a 1992 Government Accounting Office profile of Bureau of Land Management (BLM) permittees, the largest 500 permittees, out of nearly 20,000 total, controlled 36 percent of the public lands forage. Just 16 percent of all permittees controlled 76.2 percent of the AUMs (animal unit months—one AUM being the amount of forage required by a cow-calf pair for a month) available on BLM lands. Most of these permittees were big corporations or very wealthy individuals. The smallest 2,000 permittees controlled less than 0.13 percent of BLM forage.

This inequality is a result of the process for assigning public lands allotments. Access to permits requires ownership of private base operations. Since wealthy ranchers own more land, and thus more base property, they wind up with more federal lands allotments.

In addition, few ranchers depend entirely on their public lands allotments to meet all their forage needs. Although the percentage varies from operation to operation and state to state, most ranchers fulfill the majority of their annual forage needs from private lands. Only the largest operations actually use public lands for a significant amount of their livestock's forage. If the public lands were to become unavailable to these large ranches, most of their operators could reasonably afford alternatives for grazing their stock.

Alternatively, most smaller ranches today represent status or lifestyle choices for their owners—the vast majority of ranchers who use public lands. Most western ranches do not depend exclusively on livestock for their income, or for even an important fraction of their income. Growing and selling livestock is usually a break-even enterprise at best. Jobs in town or other business ventures are what allow families to maintain their status and appearance as "ranchers"—not running cattle or sheep on the range. If these ranchers chose to give up, or were forced to relinquish, their public lands allotments, most would adjust through reducing their herd size to match their private holdings, or through leasing the private grazing lands of other landowners. Family ranchers might also continue to diversify their income—as many are already doing—either with new enterprises on the ranch (for example, guest ranches, and guided fishing and hunting), or with other work off the ranch.

MYTH

Cattle Have Replaced the Bison

TRUTH

Bull bison.

Although cattle and bison have a common evolutionary ancestor, so do the polar bear and black bear. Yet we would not suggest that these two bears can inhabit the same type of landscape or that they are ecological analogues of one another. Cattle evolved in moist Eurasian woodlands and are poorly adapted to arid regions. In comparison with bison, cattle use more water, spend more time in riparian areas, and are less mobile. They are poorly adapted to dry western rangelands—one reason why livestock grazing has been so detrimental to these ecosystems.

Bison feed in one place for a few days, then move on, whereas cattle tend to "camp out" in the same location for weeks, overgrazing the landscape in the process. Bison survive on available, native forage. Cattle require extra feed to survive northern winters, which typically means hay production and accompanying dewatering of streams. Cattle are poorly adapted to dealing with predators, being rather slow and unintelligent. Bison retain their wild instincts for avoiding and fending off wolves, grizzlies, and other carnivores.

Wild bison functioned within ecosystems in ways that livestock do not. Their bodies served as food for predators and were scavenged by ravens, coyotes, and magpies. What was left of their carcasses decomposed and was returned to the soil. Bison were a part of, and contributed to, a great diversity of life. Livestock, on the other hand, represent a large net loss of energy and biomass to an ecosystem, as their bodies are removed for human consumption elsewhere.

Despite the simplistic claim that cows merely replace bison, it's not just bison that have been replaced by this exotic, domesticated species. On most rangelands today, cattle are the *only* major herbivore. Yet in the days before livestock, an entire suite of species fed on the grassland plants, from grasshoppers and sage grouse to prairie dogs and pronghorn. Substituting a single species—with different dietary preferences—for this diverse group of herbivores results in overuse of some plant species and grants competitive advantage to others. These other plants are often invasive and less palatable to many native herbivores.

MYTH

Rangelands Must Be Grazed to Stay Healthy

TRUTH

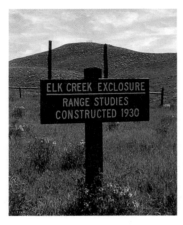

Exclosure, for studying range where livestock are excluded, Boise National Forest, Idaho.

Over much of the area that is now public land in the West, native plant communities evolved largely in the absence of grazing herd animals. Between the Sierra Nevada–Cascade crest and the Rocky Mountains lies the arid Intermountain West, composed of areas such as the Great Basin, the Palouse prairie, and the deserts of the Southwest, where bison were mostly absent and even herds of pronghorn antelope, bighorn sheep, elk, and other herbivores tended to be small and widely distributed. Consequently, the plant species of this region are not adapted to continual heavy grazing and trampling, as occurs with domestic livestock.

Yet some livestock proponents argue that although no large herds of grazing or browsing animals occurred in the Intermountain West in historic times, during the last Ice Age great numbers of wild horses, mastodons, giant sloths, and other herbivores roamed these lands. Thus, livestock advocates claim, cattle are merely filling a niche left empty since the extinction of these Pleistocene mammals. The problem, however, is that climatic conditions were very different during the Ice Age—precipitation was higher, for example—and plant communities were much different in composition, as well as generally more productive than today. Cattle are not filling some long-vacant ecological niche but are, in fact, exotic animals that have dramatically altered the native plant communities of the arid West.

Even where large herds of bison, elk, and pronghorn were common, such as on the Great Plains, plants do not need to be grazed. Rather, many Great Plains grasses *tolerate* grazing by compensating for losses in leaf and stem materials through additional growth. However, when plants move carbohydrates up from their roots to produce new leaves, root growth may slow, and seed production may be inhibited. Only plants with unlimited access to water and nutrients and with no competition (conditions found only in a growth chamber) can withstand repeated cropping without harm. In nature, plants repeatedly munched by livestock suffer from diminished root mass—a potentially lethal situation for the plant during a drought. Of course, drought occurs commonly in the West, including the Great Plains.

MYTH

Ranching Is the Foundation of Rural Economies

TRUTH

Cows in stream, southern Arizona.

Many livestock supporters attempt to portray public lands livestock production as an essential element of rural economies. It's easy to see the fallacy in this argument if you think about the numbers involved. For example, in Nevada there are fewer than 800 public lands grazing permittees. And in the entire state less than 2,000 people are engaged full-time as farmers or ranchers. One casino in Las Vegas employs more people than work in agriculture in all of Nevada. Although other states may have higher numbers of people involved in ranching, livestock production is proportionally a small part of the economic picture in all western states.

Ranching and associated activities provide very few jobs. Furthermore, most ranch operations, except the very biggest, are not highly profitable. Both of these truths help explain the rather interesting finding of one University of Arizona study: that instead of rural towns being dependent on the livestock industry for their economic survival, the reverse was true. Ranch families depend on nearby towns and cities to provide full- or part-time jobs that help keep the ranch financially afloat. Without family income from such positions as schoolteachers, local civil servants, store clerks, salespeople, and so forth, ranch ownership would be impossible. The vast majority of people who call themselves ranchers enjoy the lifestyle and the prestige, but they are not choosing a lucrative pursuit (as indeed many will complain!). Therefore, it can be argued that, financially, rural towns would likely survive without ranchers, but most ranchers would be hard-pressed to survive without the towns.

As ranching is relatively unimportant in local economies, it is even less important on state and regional scales. According to the Department of the Interior's 1994 *Rangeland Reform Environmental Impact Statement,* the elimination of all public lands livestock grazing would result in a loss of 18,300 jobs in agriculture and related industries across the entire West, or approximately 0.1 percent of the West's total employment. Natural resource economist Thomas Power has calculated that all ranching in the West, on both public and private lands, accounts for less than 0.5 percent of all income received by western residents.

MYTH
It's Either Ranching or Subdivisions

TRUTH

Ranch for sale.

Livestock advocates try to silence critics by saying that reducing or eliminating livestock from public lands will lead to subdivisions. Yet, supporting the livestock industry—even increasing its subsidies—will not stop the parceling out of ranchland into housing tracts.

Ranching in the West has always depended on the ready availability of large acreages of land. Western ranchers have competed with stock growers in more productive regions of the country by using more space, and by getting the forage on that land cheaply. However, when land prices rise, western ranchers lose their one advantage. Wetter, milder areas produce more cattle per acre than western rangelands, without as many of the costs and challenges, such as predators, scarcity of water sources, and the need for miles of fencing.

Subdivision is also market driven. But a supply of millions of acres of land for sale (as is the case in the Great Plains) does not alone draw developers. To be attractive to developers, and to eventual buyers of residential lots and homes, land must offer a favorable mix of amenities: proximity to jobs, outdoor recreation, arts and culture, good schools, a pleasant climate, and beautiful scenery. These are the qualities that stimulate subdivision.

Sprawl has gobbled up farmland in California's Central Valley and the Los Angeles Basin, despite the fact that these are some of the most valuable agricultural lands in the world. There is no way marginal ranches and rangelands in the West can compete when a high demand for housing occurs in an area.

The threat of subdivisions needs to be put in perspective. Ultimately, population growth is the problem. In the meantime, livestock production has a physical footprint far greater than urban and suburban areas. In California—the most populous western state—less than 5 percent of the land area is devoted to cities, towns, and subdivisions. Agriculture—farming and ranching—dominates more than 70 percent of the state's acreage. In other western states, the fraction of land occupied by housing and urban/suburban development is even smaller.

Fortunately, there are at least three proven ways to protect open space, wildlife habitat, and other environmental values on private lands: zoning, conservation easements, and outright fee purchase. If the same amount of money we currently throw away on subsidies to the livestock industry were devoted to protecting and buying up wildlife habitat instead, the land would be far better off.

MYTH

Good Livestock Production and Ecosystem Preservation Can Coexist

TRUTH

Cow-blasted riparian area, Elk Creek, Beaverhead National Forest, Montana.

Perhaps the biggest fallacy perpetrated by the livestock industry is the idea that if we would only reform or modify management practices, there would be room both for livestock and for fully functional ecosystems, native wildlife, clean water, and so on. Unfortunately, even to approach meaningful reform, more intensive management is needed, and such management adds considerably to the costs of operation. More fencing, more water development, more employees to ride the range: whatever the suggested solution, it always requires more money. Given the low productivity of the western landscape, the marginal nature of most western livestock operations, and the growing global competition in meat production, any increase in operational costs cannot be justified or absorbed. If the production of meat as a commodity is the goal, then an equal investment of money in a moister, more productive stock-growing region—such as the Midwest or the eastern United States—would produce far greater returns.

Even if mitigation were economically feasible, we would still be allotting a large percentage of our landscape and resources—including space, water, and forage—to livestock. If grass is going into the belly of a cow, there's that much less grass available to feed wild creatures, from grasshoppers to elk. If water is being drained from a river to grow hay, there's that much less water to support fish, snails, and a host of other life forms. The mere presence of livestock diminishes the native biodiversity of our public lands.

The choice is really between using the public lands to subsidize a private industry or devoting them to ecological protection and preserving the natural heritage of all Americans. On private lands, native species face an uncertain future. It would be a prudent and reasonable goal to make preservation of biological diversity and ecosystem function the primary goal on public lands. To suggest that we know how to conduct logging, livestock grazing, or other large-scale, resource-consumptive uses while sustaining native biodiversity is to perpetuate the greatest myth of all.

THE IRON

The livestock industry, government agencies, universities, elected politicians, banks: these five elements constitute the Iron Pentagon, a fortresslike nexus of power and influence that keeps public lands livestock grazing safe from scrutiny, criticism, or reform. Each institution or group has a vested interest in maintaining the hegemony of the livestock industry in the West. All have their reasons for supporting the continued subsidization and environmental degradation associated with livestock production. Yet the bias in favor of livestock is mostly invisible to the uninformed and is merely accepted as the status quo.

Government agencies are often seen as "neutral," mediating between greater public interests and narrow special interests. However, when it comes to public lands ranching, budgets and staff as presently configured depend on the maintenance of livestock grazing programs. Support for and ties to the livestock industry can be a political necessity for bureaucrats to gain promotions, or even to hold onto existing positions. For example, being a range conservationist, a position equivalent to that of a forester in the Forest Service, has often been a Bureau of Land Management (BLM) prerequisite for becoming a district manager or state director. As a consequence of such interdependence, agency officials are predisposed to accommodate livestock interests.

Furthermore, government workers hear from ranchers continually in their day-to-day dealings, whereas the average citizen is not often encountered, and certainly not in the form of a long-term, ongoing relationship. Officials receive far more input from the ranching industry, generally speaking, than they do from the public at large.

Many federal employees have to live in small, rural towns where hostile attitudes toward government are often prevalent. Unless they are willing to risk social isolation and ostracism in such places, government workers at the local level tend to avoid taking adversarial positions with community members. All of these factors skew government decisions in favor of the livestock industry at the expense of the land, wildlife, taxpayers, and all American citizens.

State employees are even more muted in their criticism of the livestock industry. State legislatures in the West, for a variety of reasons discussed below, are frequently dominated by extractive interests. Through control of funding by conservative legislatures, and through the political appointment of state agency directors, the pressure on state employees to avoid controversy is strong. This is particularly evident in fish and game departments. Officials in these agencies are loath to criticize the ranching industry, in part because a significant amount of hunting and fishing opportunity occurs on private lands. Antagonizing livestock interests can result in fewer acres available to hunters and fishers—an outcome that state fish and game agencies are not likely to favor.

University research is also biased in favor of the agriculture industry. The problem is particularly difficult because of the relatively small number of large, private research universities in the West. Most well-known western universities are publicly funded. Many are land grant institutions, originally financed through the sale of public lands and established with an explicit mission to support agriculture. Therefore, range and animal science departments in western universities tend to see their primary goal as assisting the livestock industry. Researchers focus on ways to make livestock operations more efficient and more productive. At times they may also study range techniques less apt to damage the land, but rarely is there research into the numerous ways livestock production alters natural systems or into how much livestock production costs taxpayers, directly or indirectly. The assumption of nearly all university range scientists is that livestock grazing can and should continue.

The Iron Pentagon *(clockwise, from the top)*: the livestock industry, government agencies, universities, elected politicians, banks.

16

Even academic programs with no direct connection to agriculture can be silenced by the politics of ranching. Because so many western universities are public institutions, dependent on financial support from state legislatures, they cannot afford to disregard influential interests that control the state purse strings. As legislative bodies in the West disproportionately represent agricultural interests, universities tend to discourage criticism of the livestock industry from within any branch of the academy. The fear of losing funding is justified, as there are numerous examples of western legislatures threatening to cut funding for state universities or individual departments merely because they wish to silence a single professor or student. Academic freedom at such universities is more imaginary than real.

As has already been noted, members of the livestock industry are well represented among western politicians, at all levels of government. From county commissioners to state legislators to members of Congress, a disproportionate number of ranchers, and people with strong ties to ranching, hold political office. These people work in various ways to maintain the power of ranchers over public resources.

How is it that ranchers have so much political clout? Several factors skew politics in their favor. First, most ranchers are longtime residents of their communities. They have strong vested interests in the outcome of decisions on taxes, subsidies, government regulations, and the like. Compared with the average American family that moves every three to five years, ranch owners have much to gain over the long term from participating in and influencing local and state politics. In addition, ranching families are often well known in their communities because of their long residency, sometimes going back several generations; their status as large landowners; and, typically, their relative affluence. When ranchers run for office, they often get instant name recognition. This is a decided advantage in an election campaign.

The way elected positions are structured also tends to favor those in agriculture. For example, in many parts of the rural West, offices such as county commissioner are part-time positions. The low salaries rule out participation by people without other sources of income and without flexible schedules. Most citizens with regular, full-time jobs cannot afford to hold public positions, particularly those positions that demand holding office hours during the regular working day. Yet ranchers—being self-employed, having the ability to juggle ranch chores around other commitments, and frequently having the financial latitude to engage in off-ranch pursuits—are able to hold office with less sacrifice than such work would require of others.

The same forces tend to limit participation in many state political offices. Most western legislatures meet only once every two years, and then only for two to three months during the winter. The pay is low and certainly not enough to support an individual year-round. There are few people with full-time jobs who can leave for several months and expect to get their jobs back after a legislative session is over. The fact that most legislatures meet in winter also favors ranchers, as well as farmers, for this is usually their slack period.

Probably the least-known power broker in the Iron Pentagon configuration is the banking industry. In many parts of the West, financial institutions have loaned money to ranchers based on the presumed "value" of their grazing permits. The difference between the very low, subsidized fees ranchers pay to the government and the prices equivalent grazing opportunities could command in the private arena is what banks recognize as a "value," against which ranchers can borrow. Even though livestock grazing on public lands is legally a privilege—not a right—banks have been allowed to loan to ranchers on the basis of their public lands allotments. Worse yet, government officials not only tend to ignore this misuse of public resources, they often feel compelled to continue grazing allotments and high stocking levels—that is, a high number of cattle relative to the amount of forage available—because otherwise ranchers may lose the supposed "value" they hold in public lands permits. Banks, of course, don't wish to see ranchers default on their loans, so they apply enormous pressure on bureaucrats to maintain the status quo with regard to grazing permits and to keep fees low.

CULTURAL
HISTORICA

AND
L ROOTS

The Grasp of the Cowboy on Contemporary Consciousness

Marlboro

PART II

Today, "myth" is commonly understood as something opposed to reality—a fallacy. Such a definition applies to the use of the term in the preceding section of this book. However, a myth is also something much grander: a story, ostensibly tied to historic events but functionally an explanation or expression of a people's worldview. A myth reveals what a society believes about itself, its origins, its proper relationship to nature, and the manner in which individuals should behave. In the United States, there is probably no greater myth than that of the cowboy.

Myths can be pervasive and inescapable, and their powerful influence may not even be recognized by most people. To challenge them can be dangerous and, at the very least, may draw a great deal of skepticism. Thus, to take up the matter of the damage done to public lands by the cowboy's cow is no simple project of laying out facts and statistics. Nor is it enough to employ the direct, nonverbal power of photographs, though we do so abundantly throughout this book.

Before there can be an honest discussion of what has happened to the native species and ecological systems of the West because of the influence of livestock production, we need to confront the cowboy myth. In this section, Christopher Manes and George Wuerthner dissect the roots of the cowboy myth, including its relationship to cultural beliefs about meat, manhood, leadership, and nature. Thomas Fleischner relates the manner in which cattle came to dominate the landscapes of the American West, and how stockmen—in no small part because of the high degree of societal deference paid them—have wielded enormous control over western land use policy during the last century and a half. T. H. Watkins takes a closer look at the laws and policies regarding public lands ranching that developed during the Great Depression and Dust Bowl era. Relief provided to ranchers during that dark, desperate time was rapidly institutionalized into a system of subsidies that stands today. "Welfare rancher" is a term at least as valid as "welfare mother," if not vastly more so, yet public lands livestock producers enjoy much greater success at opposing welfare reform. The resilience of the American myth of the cowboy has much to do with this success. As Andy Kerr and Mark Salvo explain, even in national parks and designated wilderness areas—set aside for their aesthetic and conservation values—ranchers' economic interests have prevailed over the public interest. Livestock grazing continues in some of America's most treasured natural landscapes.

Finally, Edward Abbey offers his own sort of antimyth. His barbed humor and outrageous rhetoric may offend—"sacred cows" of any sort were always the target of Abbey's sharp mind. What lies at the center of his raucous and unmannerly language, however, are his own fiercely felt loyalties: to wildness, to beauty, to truth. In the long run, the Marlboro Man hasn't got a chance.

18–19: Marlboro cigarette ads capitalize on and help perpetuate the American cowboy mythology.

Opposite: The cultural lineage of the modern-day rodeo can be traced back to the bull-worshipping ceremonies of Mesopotamia and other ancient Western civilizations.

IN THE BEGINNING: COW

Christopher Manes

Cultural memory of cattle worship can be traced from the dawn of Western civilization
through the mythology and rituals of ancient Greece, Rome, and Scandinavia, among others, and on
into European settlement of North America. Deeply rooted notions of masculinity, power and
wealth, freedom and honor stem from this millennia-old "cow mythos"
and pervade even contemporary American culture.

Christopher Manes *is an attorney and author of* Green Rage: Radical
Environmentalism and the Unmaking of Civilization *(1990) and* Other
Creations: Rediscovering the Spirituality of Animals *(1997). He is a student
of ancient literature—a former Fulbright Scholar to Iceland, studying Old
Norse literature—and is a Ph.D. candidate in medieval literature at the
University of Oregon. He resides near Palm Springs, California.*

In Western cultures, cattle have long been symbolic of broader ideas
about nature. At different times and in different places, these views
have included: nature (and cattle) as tame and tranquil; nature as
dangerous and requiring human control; and nature as dumb and
without feeling—a resource for human use.

In the beginning, in myth after myth, from Mesopotamia to Scandinavia to
Texas, is the genus *Bos*. As Paul Shepard points out, the mythology of the
cow—sprung from the rootless, acquisitive life of herdsmen on horseback—
haunts us still, so that the "modern consumer in the supermarket may have
received his browsing instinct and tendency to perpetually take and move on
from the equestrian drift of the pastoral mind across open country."[1] We
domesticated cattle; but in return the ruminant and the myths it engendered
contributed to our becoming the ecologically feckless people we are today.

This is in part why cattle are still allowed to trample the public lands of the
American West, despite all the environmental and economic data indicting
the policy.[2] The continued use of wildlands for cattle grazing is a tribute not
only to the political clout of the ranching industry but to the power and persistence of the cow mythos. Like the cover of a Harlequin romance novel, a
history of colorful fantasies about longhorns and cowboys, rugged individual-
ists and bucolic beasts overlies the commercial misuse of public lands by
ranchers. Unravel the mythos, take away the rustic imagery, and behold: the
rancher, the icon of the West, wears no clothes.

Once you begin to look for cows in western myth, they appear everywhere,
from the Bible to Gilgamesh to the Old Norse *Elder Edda*. Our narratives of
cattle almost always circle around themes of hierarchy, aggression, and
subordination of nature.[3] This is no coincidence. Popular belief sees cattle
keeping as an ecologically benign form of primitive culture, akin to hunter-
gathering. In fact, nomadic pastoralism most likely developed out of
agriculture and ultimately supplied the warrior class that made the first
agrarian states possible.[4]

Perhaps something like this happened: When early farmers began to prosper
at animal husbandry, forage near cultivated lands became scarce. Large ungulates like cattle had to be driven farther and farther afield to graze, eventually
leading to sojourns of days or weeks. This difficult task fell increasingly on
young males, who for the first time in their lives found themselves beyond the
authority of the village. With the introduction of the horse, herding became
an economy unto itself, no longer tied to the farm, and the independent,
proud, immature males who pursued the life of a nomad developed their own

rituals, values, heroes, and gods, with a military bent and a rapacious streak as broad as modern Wall Street.[5]

The age-old rivalry between farmers and ranchers was born, first memorialized in the story of Cain, a tiller of the soil, slaying Abel, a sheepherder. But if ranchers and farmers were often at odds, they also lived in economic symbiosis, alternatively raiding and trading. And when the great agricultural states of the Middle East and Mediterranean emerged, the mounted herdsmen, with their martial ethos and mercenary values, were perfectly cast for a role as the warrior class.

A distant memory of this process seems to linger in the Greek myth of Europa, where Zeus, in the form of a beautiful white bull, abducts the Phoenician princess Europa and carries her off to Crete to found the Minoan civilization. Her brother Cadmus sails to Greece in search of her but learns instead of his own exalted destiny from the Delphic Oracle, who instructs him to follow a certain cow until it grows weary and rests, and on that spot to build a city. Cadmus obeys the soothsayer and founds Thebes, which begins the mythic cycle that leads to Oedipus and his complex. European civilization, according to this myth, traces its origins to the lust of a divine steer.

Similarly, in the Bible, bovines are consistently identified with aggressive agricultural states—and their idolatry—in contrast to peaceful, sheep-raising Israel. The golden calf forged by faithless Jews in the camp of Moses brands them as followers of Egyptian paganism. And in the remarkable apologia Jehovah offers to the prophet Jonah about divine mercy, the spiritually blind Ninevites become virtually synonymous with their kine: "And should not I spare Nineveh, that great city, wherein are more than sixscore thousand persons that cannot discern between their right hand and their left hand; and also much cattle?" (Jonah 4:10).

Scandinavian mythology offers another variation on this theme, linking cattle not just to the civilized world but to the cosmic order. In the Norse creation myth, the birth of the gods themselves is attributed to a cosmic cow, Audhumbla, who licked a block of ice until the father of the gods emerged to thaw out.

Unlike agrarian myths, with their focus on gratitude for the Earth's gifts, the cosmology of cattle tends to emphasize deal making and the accumulation of wealth.[6] Whether trading cows for brides as dowry, or sacrificing heifers to inveigle divine favor, the mythology of herdsmen reflects a society based on exchange, not gift giving, on acquisition, not generosity. In prehistoric Europe, cattle began to function as currency, dividing the rich from the poor, the warrior from the peasant, beyond any such distinctions found among early small-scale agrarians. Thus our words *chattel* and *capital* derive directly from *cattle*. And the word *fee* is cognate with the Germanic *fexu*, meaning "cattle," with the ultimate Indo-European root being *pek*, also meaning "cattle" and also the origin of the word *pecuniary*. Cattle equaled wealth and power—just as chattel/capital/fees still do—and the more cattle, the more power and wealth.

From this equation and the deracinated lifestyle of nomadic herdsmen came the ethos of competition, accumulation, and "might makes right," characterizing much of what we call civilization, including our own. Cattle rustling—with the bloody conflicts it entailed—was a point of honor among herdsmen, "a noble activity protected by the warrior god and sanctioned by myth."[7] Pick up any early Irish epic and you will be hard pressed to find a Celtic hero who is not in some way connected with cattle raids, the main canvas on which ancient Eire depicted manhood (significantly, a large percentage of American cowboys were Irish immigrants). In the *Odyssey*, the crew of Odysseus, true to Bronze Age mores, steal the Cattle of the Sun and have a barbecue. Hindu myth presents the story of Visistha and Visvamitra, again about a quarrel over a stolen cow. Change the cattle to wealth or power, and you have most of our modern heroic archetypes, from Faust to Sergio Leone's Man with No Name. "The hero, the warrior and the cowboy," concludes Shepard, "are almost inextricable."[8]

As a virtually all-male pursuit, cattle herding produced a mythos that exalted the masculine virtues of warrior prowess, guile, aggression. The result was that beef inevitably became associated with maleness and male sexuality. Meat as an aphrodisiac, as a male possession to be bartered for sex, may go back to primordial times.[9] But the relationship between beef and masculinity reached new heights with the rise of cattle culture. It is no accident that the myth of Europa involves abduction and rape by a divine bull. Out of the stew of pastoralism's erotic fascination with bovines come a number of sexually charged narratives, such as the myth of the Minotaur. Half bull, half man, the Minotaur is the offspring of Cretan Queen Pasiphaë's unnatural desire for a steer. The Cretans sacrifice children to the monster (who apparently uses them for his sexual pleasure) until the arrival of Theseus, a world-class philanderer and conqueror of the Amazons, a race of female warriors. Theseus slays the Minotaur, a gesture that seems to symbolize his own mastery of its sexual powers, to be unleashed instead on his girlfriends and man-hating Amazons alike.

The link between meat and manhood lives on today. The word *beef* has become a virtual morpheme for *male*, as in the use of *beefcake* to describe erotic male photography, as opposed to its female counterpart, *cheesecake*—a mere dairy product. Up until the nineteenth century it was in fact considered somewhat gauche for a woman to eat steak in public.

And if beef equals maleness, it also connotes a particularly masculine form of political power. For almost every Western country, but especially America with its cowboy traditions, steak has overtones of nationalism and national power, while vegetarianism suggests the passivity of Gandhi and the mooniness of Jerry Brown. We don't even call an establishment that specializes in beef a restaurant, but rather a "steakhouse"—as if such places embodied traditional values not found in newfangled eateries.

Ronald Reagan, sporting a cowboy hat, would take off to his "ranch" in California; Bill Clinton and Jimmy Carter both drew attention to their southern hunting roots; and Theodore Roosevelt's political persona and big-game hunting jacket were all but indistinguishable. George W. Bush, who also owns an obligatory pseudo-ranch, wore cowboy boots to his inaugural ball, which was described by an Associated Press writer in these gushing, meaty terms: "Bush's fellow Texans were putting on the party of the night, with 9,000 people chomping on 7,000 pounds of beef brisket, 6,000 pounds of smoked ham, and 60,000 pieces of jumbo shrimp."[10] The scene uses cuisine to make the case for Republican Party macht (though unintentionally, it may have suggested to some a positive side to mad cow disease).

"Where's the beef?"—a phrase from a hamburger commercial—was used as a campaign slogan by Walter Mondale during the 1984 presidential race to imply a lack of substance in opposing candidate Reagan's political promises. The remarkable crossover of this slogan from the realm of ground beef to presidential politics could only take place because a prior symbolic order existed in which cattle and power were one—the pastoralism that gave us so many of our social, ethical, and mythic structures.[11]

The apotheosis of the cow in the modern world is out of place and pernicious. But we should remember that the *Bos* of ancient times was not the denatured, medicated bossy we see today. When the Norse thought of cattle, they had in the back of their minds the fierce, wild aurochs *(Bos primigenius)*, the ancestor of the modern cow, that stomped around the fringes of Europe before going extinct in the seventeenth century, a formidable animal having more in common with a Cape buffalo than a Holstein. To kill an aurochs was a warrior rite of passage for ancient Teutons. The famous Viking drinking horns were originally torn from one of these rampaging protocattle, assuming it didn't gore the Viking to death first. And for the Greeks, bovines meant the powerful, piebald Minoan bull, the subject of "bull dancing" depicted on Cretan frescoes, a dangerous sport that involved turning a somersault over one of these horned killers' back as it charged.

Modern cattle first came to North America around A.D. 1000 on the knorrs of the Norse, and died out with the ill-fated colonies of Vineland. One wonders what course this continent's environmental history would have taken had these animals gone feral, reverted to the ancestral aurochs, and thundered up and down the eastern seaboard to greet the next wave of European colonists in the sixteenth century. Perhaps the pilgrims would have been less satisfied with their own domesticated stock, as the wild emblems of a not-altogether-alien mythos warmed them to the wildness of the new country, and—who knows?—bull dancing. What ecstatic sermons might Cotton Mather have written had he known the joy of leaping over the horns of a Cretan bull, his clerical buckles all askew?

But that was not to be. The colonists who came to America transplanted not only their domesticated cattle but the ethos of cow culture, which thrived profusely once it reached the open grazing lands of the West. Surely one of the great parallel cultural developments in history is how America's cowboys—with their cattle rustling and range wars with farmers, their prowess at weapons and horsemanship, and their general lawlessness—reprised the herding societies of early Europe and the Middle East. A Bronze Age Hittite would have felt quite at home in a saloon in postbellum New Mexico. This was possible because so much of our society's mythic imagery derives from those earlier ruthless, cow-loving vagabonds.

"Men die, cattle die," sighs the grim Norse book of wisdom, *The Havamal*, lamenting the brevity of this life. Myths also perish. Aurochs go extinct. Bull dancing ends. Cow-chosen Thebes is put to the sword, and cow-spared Nineveh sinks into the sand. We can best honor the legacy of the aurochs and the hoofprints they left in our cultural history by freeing cattle from the shabby rhetoric of the ranching industry. We can best respect the raging bulls of the past by getting the subsidized cows of modernity off our wild lands.

The way of life among mobile herdsmen is essentially a culture of hierarchy, theft, rebellious sons, and competitive use of the earth. The North American ranchers, the gauchos of the Pampas, the Somali and Mongol cattlemen, the Australian cowboy, and other bearers of the bovine idiom continue the mindset of ancient, mounted cattlekeepers, their ideology and ecological effects, be they Incan or Aryan.

—Paul Shepard, *The Others: How Animals Made Us Human*, 1996

BEEF, COWBOYS, THE WEST
American Icons

George Wuerthner

Powerful symbols and cultural beliefs shape the public lands grazing debate, but their pervasive
influence is largely unexamined. American attitudes about beef as food, cowboys, and the western
frontier are strongly intertwined with notions of virility, freedom, physical competence, and
moral integrity. Modern ranchers, by extension, are seen as defenders of open space
against the creep of urbanization and all its social as well as ecological ills;
the ecological damage wrought by livestock is thus obscured.

In the ongoing and often rancorous debate over public lands grazing in the United States, environmentalists, public land management agencies, hunting and fishing groups, taxpayer advocacy organizations, and others attempting to regulate, modify, or restrict public lands livestock production have encountered stiff political and social resistance to change, despite the numerous well-documented ecological, economic, and health costs.[1] This resistance can be attributed, in large part, to the significant symbolic value of beef in American culture, combined with the powerful iconic status of the cowboy in American culture and mythology.

This cultural context of public lands livestock policy goes mostly unacknowledged, though it greatly influences the direction of scientific research, the management of public lands by federal and state agencies, and the range of potential solutions considered in grazing conflicts. Without addressing the cultural components of the public lands livestock grazing issue, science and "public process" alone will not lead to a satisfactory resolution.

The reasons for the entrenchment of beef and the cowboy in the American mythos can be traced to three main factors:

1. The American dietary predilection for beef consumption has its origins in Europe, where cattle had religious, economic, and social significance.
2. The European cattle culture complex was transferred to North America and with westward settlement became deeply enmeshed in the American frontier experience.
3. The cowboy came to embody valued American traits associated with the frontier, such as individualism, virility, and competency. The cowboy *is* the idealized American, a national symbol and therefore reflective of American cultural values.

The modern-day rancher, such as this one in southwest Montana, often rides a four-wheeler—rather than a horse—to herd his cattle.

Organizations and individuals attempting to "rein in" cowboys are not just challenging a minor industry or special interest; rather, they are challenging how America thinks about itself. Without an understanding of the cultural roots of America's love affair with cattle, cowboys, and beef, the scientific, rationalist approach to livestock grazing controversies will continue to fall short, and public perceptions of the role of ranching in the American West will remain unchanged. Indeed, the cowboy and western cattle culture are an American tradition that Americans may wish to preserve regardless of its costs.

Ancient Connections

The relationship between cattle and humans is ancient. Cattle were domesticated about 7,000 years ago in Mesopotamia, where they were used primarily for religious sacrifices.[2] After sacrifice, or as part of the religious ceremony itself, the animal would be consumed. In the pre-Christian era, cattle cults spread throughout the Middle East, North Africa, and elsewhere.[3] For many early cultures, cattle—and bulls in particular—represented both virility and fertility, particularly among agriculturists. In ancient Egypt, the sky goddess of fertility, Hathor, was often depicted as a cow.[4] Early Greek worshippers of Dionysus, the god of life and fertility, saw the bull as an embodiment of their god and would sacrifice the animal in their frenzied rituals. Devouring the raw flesh of the bull was believed to confer godlike qualities and blessed the land with bovine blood.[5]

Beef as Status Symbol and Health Food

Eventually, however, the religious symbolism of cattle as a source of fertility was secularized. Throughout Europe, cattle came to stand for wealth. According to scholar Paul Shepard, meat became part of the "language of social obligation and kinship" in ancient cultures and societies. The word *meat* is derived from the Old German *Gemate*, often shortened to *Mate*, or one with whom food is shared. Meat consumption figured prominently in human social interactions from the earliest times.[6]

The domestication of livestock further cemented this tradition. Originally, the preferred domestic animal in northern Europe was the pig. Cattle gradually rose to greater prominence as food over a period of centuries as the region's extensive forests were cleared, creating additional pasturage for bovines. By the fifteenth and sixteenth centuries A.D., cattle herding was the primary use of lands unsuitable for crop production, particularly in the higher regions of the Alps, the highlands of the British Isles, and parts of Scandinavia.[7] In these regions—too cold, steep, or infertile for grains, fruit, or other food crops—cattle were raised.

By the 1800s, expanding industrialization in Europe, particularly England, created a growing class of affluent mercantilists and others eager to spend their wealth on such luxuries as meat. Scotland and Ireland, or what has been labeled the "Celtic fringe," increased production of cattle for export to England's growing population, particularly the middle and upper classes.[8] While the aristocracy of Europe consumed plate after plate of meat, the poor were lucky to get enough grains to eat.[9]

Many of the early colonists to North America, particularly in New England, were from the British Isles. Once in America, meat consumption among these people increased dramatically. Land was cheap or free for the taking, and even the poorest backwoodsman could afford a few cows. Beef, which in England had been a luxury largely reserved for the upper classes, became a standard part of the colonists' diet. Both wild game and domestic animals were consumed, and in far greater amounts than all but the wealthy of Europe had enjoyed. In effect, even the poorest American pioneer became an aristocrat, at least in diet, playing directly into the American ideal of the liberated, independent yeoman farmer and the self-determining man of the frontier.

This trend continued throughout the nineteenth century, so that even members of the working class in America's growing industrialized cities had far more to eat, including meat, than had the average person in Europe. Immigrants were amazed at the abundance of meat. Meat consumption became a status symbol among the immigrant masses in the late 1800s, as powerful a mark of "making it" as ownership of a television became among Americans in the 1950s.[10]

Meat consumption was also stratified by race and gender. In the wake of Darwin's theory of evolution, it became popular to ascribe English domination of the world to the perceived superiority of "civilized" races and upper classes, whose diet was an expression of their sophistication. Meat consumption was also considered necessary for "thinking" people. For example, nineteenth-century physician George Beard prescribed meat for men who suffered from "nervous exhaustion." According to Beard, "brain-workers" required more meat than "savage" or "lower" classes of society.[11] Indeed, Beard saw meat consumption as a civilizing influence:

> In proportion as man grows sensitive through civilization or through disease, he should diminish the quantity of cereals and fruits, which are far below him on the scale of evolution, and increase the quantity of animal food, which is nearly related to him in the scale of evolution, and therefore more easily assimilated.[12]

Beard, who lived during the height of England's colonial era, rationalized the conquest of other peoples as a rightful consequence of intellectually superior meat eaters:

The rice-eating Hindu and Chinese and the potato-eating Irish peasant are kept in subjection by the well-fed English. Of the various causes that contributed to the defeat of Napoleon at Waterloo, one of the chief was that for the first time he was brought face to face with a nation of beef-eaters.[13]

In general, meat consumption was considered a masculine tradition. Among working-class families in nineteenth-century England, most meat was reserved for adult males, with women and children getting little or none.[14] During World War II, American soldiers were fed extra rations of meat to make them superior warriors.[15] And even today, in any small-town café you can find the meat-laden "rancher's breakfast" or "trucker's breakfast"—designed for "real" men who do manly work.

There is even a hierarchy of meats in American culture, with red meat being the truest, most potent flesh. According to folklore, red meat traditionally confers such admired qualities as aggression, strength, passion, and sexuality.[16] Red meat is associated with males and maleness, while white, "bloodless" meats like chicken are associated with feminine qualities. Meat eating is associated with virility. A vegetarian is effeminate and becomes soft and passive like a "fruit."

Further illustration of the ways in which we perceive the power of meat and the passivity of vegetables can be found in common phrases and sayings: We get to the "meat" of the situation. We serve "hearty" beef stews. "Where's the beef?" implies that what is known or present is less than full or complete. We deal with "meaty" (read "substantial, challenging, real") questions. To be in a coma or brain dead is to be a "vegetable." To be a "couch potato" is also to "vegetate" (lack energy, initiative). These ideas were carried forward in a recent ad campaign of the American Beef Council with the slogan "Beef—real food for real people." Researcher and social critic Jeremy Rifkin suggests that "the identification of raw meat with power, male dominance, and privilege is among the oldest and most archaic cultural symbols still visible in contemporary civilization."[17] As we shall shortly see, these are among the traits commonly ascribed to the American cowboy.

Rancher Versus Cowboy

At this juncture, there's an important distinction to make. Although the terms *rancher* and *cowboy* are often used interchangeably, even by ranchers and cowboys themselves, there are real economic and class differences.[18] Despite the romantic notion of the cowboy as the independent, steadfast, resourceful American male, the real cowboys of yesteryear were, by and large, young, illiterate farm boys who were treated as little more than expendable hired hands.[19]

And though the stereotypic cowboy is usually portrayed as a white male, many hands were black or Mexican.[20] Indeed, the term *cowboy* most likely came about on Texas rangelands, which were largely controlled by wealthy Anglo-Americans from the antebellum South, whose black slaves tending cattle were often referred to as "boys." Cowboys were poorly paid laborers, doing a mostly dirty, boring, unglamorous job.[21]

The rancher, on the other hand, though he might dress the same as his cowboys and ride a horse, was the "cattle king" and was often born of wealth. The roots of the western livestock industry lie in Mexico, where haciendas, or large estates, were awarded to conquistadors or other members of the upper class by the Spanish crown as a means of encouraging colonization.[22] Riding a horse and land ownership were privileges originally restricted to a small group of elite males of European ancestry. However, as the Mexican cattle industry developed, these restrictive laws began to break down in the face of economic necessity. Hacienda owners found they required teams of skilled horsemen to manage open-range livestock, and the most available labor sources were the populations of local Indians, mestizos, and others of the Mexican lower classes. Given the status attached to horseback riding and horse ownership, for many poor Mexicans, becoming a cowhand, or *vaquero*, was a means of achieving upward mobility and rank.[23]

A similar stratification between cowboy and rancher existed in the western United States.[24] Many of the larger cattle outfits in the West, particularly in the northern plains of Wyoming and Montana, were started by English remittance men—wealthy second or third sons of aristocrats who had no chance of inheriting the royal estate.[25] These men came to the western frontier to carve out their own empire. Although there were plenty of small-time ranchers who by hard work or by the judicious use of a rope and branding iron were able to acquire their own modest cattle ranch, most of the larger ranches in the West were started and owned by men who were already wealthy—whether transplants from elsewhere in the United States or from Europe.[26]

Today, large corporations and wealthy individual ranchers dominate the livestock industry. For instance, according to a 1992 General Accounting Office report, 11 percent of all Bureau of Land Management permittees are corporations, another 8 percent are partnerships, and another 5 percent are in the "other" category. Indeed, the ten largest public lands permit holders are all corporations or billionaires. The report also found that the largest 2,000 (10 percent) of all ranch operations grazing public lands control 74 percent of the public lands forage. The other 90 percent of all ranch operations share the remaining 26 percent of public forage.[27] And even this group of "small" ranchers has considerably more wealth than the average American. The popular notion that the rancher is just a poor, working stiff trying to make ends meet has little to do with reality.[28]

The Mythological Cowboy and the Taming of the West

The history of the American West makes sense when placed within the much larger history of European colonialism.[29] To understand the place of the cowboy, one must understand the culture out of which he was born. The values of Europe, particularly the British Isles, where cattle were a symbol of power and wealth, transferred easily to America. If the American West had been colonized by the Japanese or some other non-beef-eating group, it is doubtful that cattle would now be grazing in Nevada or Wyoming. The European cattle culture predisposed the American frontier society for continued western expansion, and influenced how it would be accomplished.

Open-range cattle ranching is the least intensive form of commercial agriculture; thus, it is relegated to the most remote, least populated zones. Moister, more productive lands are used for activities of greater monetary value and return.[30] In northern Europe, the cattle herding systems were concentrated in the British Isles and Scandinavia, occupying the windswept, rainy, rocky lands. As historian Terry Jordan has noted, Old World cattle folk have always been peripheral to civilization. In North America, the open-range cattle cul-

ture developed in the West on the lands that could not be farmed or used for other exploitative purposes.[31] In the American West, the recent spread of subdivisions, ski resorts, and other land use changes on former ranchland represents a further marginalization of cattle herding culture.[32] It is a pattern that has been ongoing for thousands of years.

Though cattle ranching seldom survives long in areas of increasing urbanization, it has a long tradition as the vanguard of colonialism. As Jordan suggests, "Cattle-ranching provided an innovative land-use strategy that facilitated the advance of the Euroamerican settlement frontier at the expense of the native peoples."[33] One could add that it also came at the expense of native fauna and flora.[34]

Ranching is the first step in the domestication of the landscape, and in the taming of wilderness (which I define as undeveloped and unmanaged lands—not necessarily unpeopled). The subjugation of the West and the wilderness was a Christian burden, and the cowboy, like the missionary, was doing "God's duty" in civilizing the land.[35] At an even deeper level, the cattle culture is based on a worldview that sees nature as requiring control, and those who do the controlling as powerful people. The most direct expression of this view is in the way the animals themselves are manipulated and controlled to serve the ends of their "masters." Hearkening back to the rites of ancient cattle-worshipping and cattle-sacrificing cultures, this control of cattle is ritualized in such events as the Spanish bullfight and the American rodeo.

The modern rodeo is a descendant of the Iberian tradition, and of the ancient cattle sacrifice rites.[36] It represents the taming of uncontrolled, wild nature. The lone cowboy, by virtue of his strength, toughness, and cunning, outwits and outendures a wild bronco or bull. Each eight-second ride is reaffirmation of human control over the dark forces of the universe.[37] This is why the idea of preserving wildlands and biodiversity is exactly at cross-purposes to the survival of the cowboy. Not only does livestock production require manipulation of the landscape, the larger myth of the cowboy assumes that nature should and must be directed and managed. Wilderness, the organisms that dwell there, and evolutionary processes such as wildfire, weather, and predation are jeopardized, directly or indirectly, by such attempts to turn the land to strictly human purposes.

The cowboy is closely associated with the frontier.[38] The much-quoted western historian Frederick Jackson Turner argued that the frontier conveyed a particularly American experience and produced a superior individual. Turner asserted in his book *The Frontier in American History* that

> to the frontier the American intellect owes its striking characteristics. That coarseness and strength combined with acuteness and inquisitiveness; that practical inventive turn of mind, quick to find expedients; that masterful grasp of material things, lacking in the artistic, but powerful to effect great ends; that restless, nervous energy; that dominant individualism, working for good and for evil, and with all that buoyancy and exuberance which comes with freedom—these are traits of the frontier.[39]

And into this frontier rode the cowboy, taming nature, and bringing order.

The West was the grand backdrop the cowboy needed to achieve enduring, truly mythical proportions. The West presented a problem—the Great

American Desert. The nation required a figure bigger than life to take on a landscape with almost limitless horizons.[40] The cowboy found a "wasteland" and made it productive. Heroic humans were taming epic landscapes by virtue of Manifest Destiny. The West is thus a land of possibility and promise and also a place that requires someone with special virtues and strengths to make it useful and fruitful.[41]

While controlling nature, the western hero is in turn a product of the frontier, an expression of this untamed land. Owen Wister's novel *The Virginian*, published at the beginning of the twentieth century, epitomized Turner's western character.[42] The Virginian was, according to scholars Joe Frantz and Julian Choate,

> a man of honesty and virility, healthy, deeply loving, strong, shrewd, fun-loving, and gentle—containing all the personal qualities inherent in the heroes of any race or nation, from daring in battle to gallantry in love, and always with tremendous reserve. He has everything except the advantage of good breeding, which can be overcome by the love and example of a good woman.[43]

The Virginian is the everyman of the frontier—a little rough on the outside, but decent, honest, and hard working. He is, in short, an American as Americans think of themselves. His myriad images represent a national metaphor for individualism, strength, work, action, achievement, and courage. Against the backdrop of the frontier, the cowboy is democracy at work. He is an individual conquering nature and himself, and making both himself and the land better for it.

The cowboy is a romantic figure unlike any other in the parade of frontier characters. This romanticism can be seen in everything from art to politics. Even little kids play "cowboys," not miners, loggers, mercantilists, or railroad coolies. America has "cowboy poets" and "cowboy art," but no "miner poets" or "miner art." We have "cowboy music" and "cowboy movies," but no tradition of "logger songs" or a genre of "logger movies." And when a real man lights up a Marlboro cigarette, he doesn't tend to sick people, teach children, or even cut trees or dig rocks from the ground; he rides a horse and chases cows across a vast and open land. Interestingly, the Marlboro ads epitomize how the image of the West and the image of the cowboy have been collapsed together, so that one cannot be envisioned without the other.[44]

Politicians for decades have exploited the popularity of the cowboy myth. From Teddy Roosevelt to Ronald Reagan to George W. Bush, politicians have taken on the image of the cowboy to promote themselves as self-assured, self-reliant, action-oriented men. As a modern-day cowboy, Reagan carried out his antigovernment, anti-Communist agenda. Would a cowboy back down from the Russians? Would a cowboy surrender to OSHA regulations and fret about getting hurt on the job? A real cowboy gets up, dusts off his jeans, and limps off, even if his leg is broken, and accepts it all as part of the job.

More recently, the rancher-cowboy has come to be regarded not so much as a tamer of the wilderness but as a defender of the frontier from the evils of creeping urbanization.[45] Instead of fighting rustlers and Indians, today's mythical cowboy is out making his stand against subdivisions, urban degeneracy, and the questionable value system of people who do not get their hands dirty in making a living. The cowboy stands for an older and better way of life. The cowboy is keeper of the purity of the landscape and society. Though the assumption that ranching can prevent subdivisions is widely held, it is a dubious notion and does not account for the very real, negative impacts of livestock production on the land and on taxpayers.[46] The real reason the rancher is seen as a protector of the environment has more to do with our cultural biases and the history of the cowboy image than it does with the reality of contemporary trends in economics, population, and land use.

Final Thoughts

The rancher-cowboy comes from a long tradition of power and status. As a producer of beef, a food with considerable symbolic value to Americans, as well as being the personification of American values of individualism, personal integrity, strength, and male competency, the cowboy is firmly ensconced in the nation's iconography. Thus, attempts to eliminate or merely reform livestock production will not be successful until the symbolism of "meat" and "cowboy" is carefully deconstructed, and the premise of controlling nature—inherent in the livestock industry—is challenged. Ultimately, the public's understanding of and reverence for the American cowboy will need to shift, or at least be tempered by historical and contemporary realities, if real change in public policy is to occur.

The teaching mythology we grew up with in the American West is a pastoral story of agricultural ownership. The story begins with a vast innocent continent, natural and almost magically alive, capable of inspiring us to reverence and awe, and yet savage, a wilderness. A good rural people come from the East, and they take the land from its native inhabitants, tame it for agricultural purposes, bringing civilization: a notion of how to live embodied in law. The story is as old as invading armies, and at heart it is a racist, sexist, imperialist mythology of conquest; a rationale for violence—against other people and against nature.

—William Kittredge, *Owning It All*, 1987

Riparian area trampled by cattle, Blacksmith Fork, Wasatch-Cache National Forest, Utah.

LAND HELD HOSTAGE
A History of Livestock and Politics

Thomas L. Fleischner, Ph.D.

Since the early sixteenth century, the West has been used for livestock grazing. Yet winter die-offs, drought, and overgrazing have long plagued western ranching. On public lands, federal regulation of livestock grazing was slow to develop. The political influence of stock growers has enabled them to largely resist policy shifts that would allow ecological recovery. Today, increasing scientific knowledge and general appreciation of arid lands may finally effect real grazing reform.

Thomas L. Fleischner, Ph.D., *is professor of environmental studies at Prescott College in Arizona, and the author of numerous articles on conservation biology, natural history, and environmental policy. He chaired the Public Lands Grazing Committee of the Society for Conservation Biology and currently serves on the board of governors of that organization.*

European livestock came to North America only a year after the first European settlers. On Columbus's second voyage, in 1493, he brought horses, sheep, and cattle from Spain and the Canary Islands to the Caribbean island of Santo Domingo. From there, Spaniards shuttled them to Cuba, and in 1521, Cortés and Gregorio Villalobos brought them to the mainland of Mexico. Cattle raising caught on instantly in the new colony. Within a decade, scores of ranches occupied the plains and valleys near Mexico City, and a registry for brands was established. By the end of the sixteenth century, these big immigrant herbivores were munching placidly throughout Mexico.[1]

Cattle first arrived in what is now the United States in 1540.[2] Coronado, in his quest for the fabled Seven Cities of Cibola, traveled northward through present-day Arizona, New Mexico, and Colorado, and as far northward and eastward as Kansas.[3] Although a few strays probably escaped from Coronado's immense herds, the period of continuous livestock grazing in the future United States didn't begin until about 1700. As Padre Eusebio Kino, a zealous Jesuit missionary, traveled widely throughout what is now northwestern Mexico and the southwestern United States, he established twenty-four missions and preached two gospels: Jesus Christ and livestock husbandry. Kino gave livestock to the Indians, hoping to draw them into the convenience of European pastoral ways. It is hard to say which conversion—to Christianity or to herding large Eurasian mammals—ultimately had a more dramatic impact on the face of North America.[4]

Missions that had been established in Texas, New Mexico, and Arizona between 1670 and 1690 became livestock centers soon after 1700. During the mission period, 40,000 to 50,000 sheep and 10,000 to 20,000 cattle were brought to Texas. By 1834, twenty-one missions in California thronged with 423,000 cattle; 61,600 horses; and 321,500 sheep, goats, and swine. Mission San Luis Rey alone was home to 80,000 cattle; 10,000 horses; and 100,000 sheep, goats, and swine.[5] Meanwhile, all the trappings we associate with the quintessential American icon—the cowboy—had been invented back in Mexico.[6]

In the early nineteenth century, livestock populations began to swell in the American West. By the start of the Civil War, Texas swarmed with more than

A stock pond is just one of many alterations ranchers make to the landscape to better accommodate their cattle, to the detriment of native wildlife.

3.5 million head of cattle.[7] Conditions were right for a market frenzy: cattle were abundant, consumers in the north and east were eager, and the new railroads crossing the continent made previously inconceivable linkages between the Texas range and the dinner plates of eastern townspeople. To take advantage of this opportunity, the long cattle drive—another tradition from Mexico—was adopted.[8] Several general routes developed, often converging at river crossings and railroad depots. In the next twenty years, more than 5 million cattle were driven northward from the ranges of Texas.[9]

What made this possible? The advent of the railroads, surely. But the glory and riches of the great cattle drives were equally dependent on the health of the immense grasslands that stretched between Texas and the waiting rails. Those grasslands, though, were vulnerable: the 20 million hooves that trod toward the railroads quickly degraded vegetation and soils.

Such explosive economic activity did not occur in a vacuum. Self-promotion ran rampant. The Cheyenne, Wyoming, newspaper, for example, ran sixty-three articles in two years touting the state as a cattleman's paradise.[10] Exploitation of rangelands accelerated wildly as dreamers speculated fortunes on the new gold: livestock. As in all get-rich-quick booms, rationality and long-term concern were flung aside. Stockmen judged their success by the number of head of livestock, rather than any standard of sustainability. By 1880, Utah was home to almost 100,000 cattle; New Mexico, $1/3$ million; Texas, more than 4 million. Between 1870 and 1886, the number of cattle in what are now the seventeen western states[11] leaped from 7.9 to 21.6 million head.[12]

Inevitably, the bottom fell out. Even in the best of times, the semiarid West couldn't withstand grazing pressures like these. A recent study at Capitol Reef National Park, Utah, analyzed past vegetation change and concluded that "the most severe vegetation changes of the last 5,400 years occurred during the past 200 years. The nature and timing of these changes suggest that they were primarily caused by 19th-century open-land sheep and cattle ranching."[13]

Two key features of the West's geography had begun to affect the burgeoning cattle industry: it was dry, and its climate was unpredictable. Euro-Americans chose to ignore the first fact and hadn't been here long enough to perceive the second. The next couple of years provided a harsh lesson. The severe winter of 1885–1886 caused massive die-offs of cattle throughout the West. As much as 85 percent of herds perished in some regions. The spring thaw simply teased, as it led into a serious summer drought. Hot and dry weather, especially in the northern plains, caused cattle and their forage plants to suffer further losses. Then, to add insult to injury, the ensuing winter arrived direct from the Arctic. Blizzard followed blizzard, and the losses mounted higher still. Paper fortunes collapsed from Cheyenne to London. Bones littered the ground from the prairies of the north to the deserts of the south.[14]

A Reluctant Path Toward Policy

By the 1890s, cattle had grazed in the West for three and a half centuries, but regulation of grazing practices was nonexistent. Yet even during the most reckless times, a few ranchers recognized the need for maintaining a dependable forage supply. The ruinous years after 1885 compelled a few more to face reality. Some even began to clamor for government interven-

tion to help control abuse of the common resource. The last two decades of the nineteenth century saw a massive decimation of forests as huge corporations sped to convert forests to dollars. People both inside and outside government grew alarmed. Public discourse on the treatment of American forests served as a harbinger of debates soon to come to rangelands. Rangelands—more hidden from the public eye—were continuing to deteriorate, and stockmen grew more worried. But resistance to federal involvement in range issues also persisted. As early as 1878, John Wesley Powell, the original explorer of the Colorado River, had urged in his *Report on the Lands of the Arid Region of the United States* that federal land policy must be fundamentally different in the West than in the moist East.[15] But his ideas ran against the grain. During the grass boom of the 1870s and 1880s, his proposals were rejected.

Even so, stockmen's enlightened self-interest began to awaken to the need for regulation. As larger and larger chunks of capital were invested in ranch operations, ranchers grew uneasy about the vagueness of their legal basis to graze on federal land. In 1884, the National Cattle Growers Convention passed a resolution favoring a federal leasing program. In 1898, the American National Livestock Association passed resolutions asking that public lands be given protection from overgrazing. Still, these pleas went unheeded. Three years later, the American Cattle Growers endorsed a congressional bill that would impose lease fees; the bill withered.

In 1901, Teddy Roosevelt, an avid big game hunter and conservationist, swept into the White House with a promise of progressive reform. Roosevelt had spent considerable time hunting in the West and recognized that the range was in trouble. In 1903, he assembled a group of experts to advise him on issues of rangeland deterioration and the possible need for government intervention. The report of the Public Land Commission was presented to Congress in 1906. It stated:

> The general lack of control in the use of public grazing lands has resulted, naturally and inevitably, in overgrazing and the ruin of millions of acres of otherwise valuable grazing territory. Lands useful for grazing are losing their only capacity for productiveness, as, of course, they must when no legal control is exercised.[16]

Included in the report were the results of a survey that gathered the opinions of 1,400 stockmen from throughout the West. An impressive 78 percent favored some sort of government control of grazing.[17] Yet the commission's report was ignored by Congress.

Why was there such stout resistance to implementing range regulation? Some congressmen invoked lofty ideals of freedom and democracy in their opposition to regulation. Some were convinced that range ills would be cured more effectively by providing individual citizens with private homesteads. Until the mid-1920s, turf wars between the Department of Agriculture and the Department of the Interior also worked against action, as both bureaucracies claimed to be the logical overseer of the range. Some westerners objected to "government meddling" just on principle. "States' righters" insisted that public lands should be given to the states. That many states didn't want federal lands was, apparently, an irrelevant detail. Also inhibiting any move toward a rangeland policy was the plain fact that the general public was less interested in grasslands and deserts than in forests.

By 1920, Congress had snubbed several major initiatives for reforming grazing policy.[18] After America entered World War I in 1917, the needs of hungry soldiers increased the demand for beef—and decreased public concern with the health of rangelands. Livestock numbers swelled to their highest peaks since the crash of the 1880s, even though none of the ecological problems of that era had been rectified.

During the gleeful profligacy of the Roaring Twenties, conservation issues receded even further from the public eye. Behind closed doors, debate on grazing policy focused on the pros and cons of a leasing system for the public domain, and which department might administer it—Interior or Agriculture.[19] In spite of Chief Forester Gifford Pinchot's utilitarian jump start, the Forest Service was unpopular with many ranchers because it had had the gall to charge fees for what had always come free with the scenery—the right to graze. The Forest Service found itself caught between the proverbial rock and a hard place. The nation had wracked up a huge debt during the war, and Congress was sniffing everywhere for cash. The House committees on agriculture and forestry, especially, pushed the Forest Service to raise grazing fees.

Stockmen across the West set their political friends into motion. The result was that while some congressmen were wagging their fingers at the Forest Service for charging stockmen too little, others, from the West, were denouncing the agency for having the audacity to charge anything at all.[20] Despite the stalemate, the hard times of the Great Depression—economic, ecological, and governmental—argued against continued complacency. The time was overdue for the federal government to acknowledge that livestock grazing on its lands demanded action.

In 1932, a bill that would have established federal oversight of livestock grazing was endorsed by both the Interior and Agriculture Departments and approved by the House of Representatives, but it shriveled in the Senate. Nonetheless, it was the first time a grazing bill had ever passed either house of Congress. Edward Taylor, a conservative congressman from the western slope of Colorado and a prominent voice for states' rights for a quarter of a century, had come to realize that transferring federal lands to states was politically unpalatable nationally, and might not help the range anyway. He reintroduced the failed bill and promoted it forcefully.

Many western congressmen attacked the bill as an assault on the American dream. Debate followed familiar grooves—grazing regulation was antidemocratic and anti-individual. Taylor responded:

> The West was built . . . largely upon the courage, privations, and frightfully hard work of the pioneer homesteaders. . . . But my dear sirs, if those hardy pioneers had had to go onto the kind of barren land that is contemplated within this bill, the West would still be a barren wilderness.[21]

Congress shuffled along, avoiding decision. But in the strain of the Depression and the fresh hope of the New Deal, public sentiment veered toward action. The Dust Bowl became harder and harder to ignore, as it literally dominated the atmosphere of the country.

Secretary of the Interior Harold Ickes upped the ante. He discovered that he had legal authority to create a federal reservation out of all otherwise undesignated federal lands—which included the vast majority of the public range.

The bill's progress toward passage suddenly sped up, and it was passed by the House of Representatives in April 1934. The next month, a ghostly presence descended on Washington and made senators take heed of the issue they had been endlessly debating. Some of the worst dust clouds in history blocked the prairie sun, then, buoyed by prevailing winds, blew a thousand miles and filtered down around the Capitol. The Senate approved the bill the following month, and President Roosevelt signed the Taylor Grazing Act into law on June 28, 1934.[22] The United States of America was now in the business of managing its rangelands.

The Taylor Grazing Act officially ended the giveaway policy for federal lands, which had been the cornerstone of national land policy since the Revolution. All federal lands that had not already been appropriated or given away were set aside as "the public domain." But the heart of the act was its resolve to manage livestock grazing on this newly designated land: "To stop injury to the public grazing lands by preventing overgrazing and soil deterioration, to provide for their orderly use, improvement, and development, to stabilize the livestock industry dependent upon the public range, and for other purposes." The act authorized the secretary of the interior to establish grazing districts on lands that were, in his opinion, valuable chiefly for grazing. Furthermore, the secretary was instructed to provide "for cooperation with local associations of stockmen, State land officials and official State agencies engaged in conservation . . . of wild life."

Ickes created a Division of Grazing within the Department of the Interior; five years later its name would change to the U.S. Grazing Service. Farrington Carpenter, a Colorado rancher who had studied law at Harvard and Princeton, was named as its first head. Carpenter immediately set up a series of meetings with ranchers throughout the West to discuss the new law. Committed to local control of grazing policy, he coined a phrase still heard today—"home rule on the range." He created grazing district advisory boards, composed exclusively of ranchers.

From the very first, local ranchers were given control over such issues as who should receive permits and how many animals the range could handle. In 1940, representatives from district advisory boards were organized into a National Advisory Board Council, which wielded tremendous influence nationwide. A member of that council later recollected that ranchers wrote the entire Federal Range Code at the council's first meeting, with government officials polite enough to offer to leave so as not to interfere.[23]

Meeting the two primary goals of the Taylor Grazing Act—stopping injury to rangelands and stabilizing the livestock industry—required two basic decisions: How many livestock could the range reasonably support? And whose stock should they be? To answer the first question, the government relied on the opinions of local advisory boards, since neither funding, personnel, nor inclination existed to conduct scientific studies of carrying capacity.

The second question—which stockmen would get to use the public domain—was especially touchy. The majority of previous users had to go; there simply wasn't enough land for all. Carpenter's interpretation of the Taylor Grazing Act clarified criteria for determining who should receive one of the initial grazing permits. First, priority was given to landowners and those with water rights. Second, priority was given to ranchers who owned enough land to partially sustain their livestock, but not so much land that they didn't need public forage. Still, there remained three times too many applicants. So Carpenter established another crucial criterion: preference would be given to those who had customarily used the land in the past—in a sense, codifying squatter's rights. Since it was often impossible to determine who, in fact, had been first in an area, the Department of the Interior set up a "priority period" of the five years immediately prior to the passage of the act.

Using these guideposts—land ownership, water rights, intermediate scale of ranch operation, and activity during the five-year priority period—the Division of Grazing largely succeeded at stabilizing the livestock industry in the West. Obtaining one of those precious few original permits was a windfall.[24] Those families that prevailed in the mid-1930s sweepstakes endure today as those with public land grazing "rights" (more accurately, privileges). The Taylor Grazing Act, and its subsequent clarifying interpretations, remains the most pivotal legislation concerning livestock grazing—"the watershed event" in grazing law, according to legal scholars.[25]

During the World War II years, the Grazing Service lost popularity among range users for the same reason the Forest Service had a couple of decades earlier— grazing fees. When the Grazing Service tried to raise grazing fees, ranchers and their political allies erupted. The Grazing Service found itself trapped in a crossfire of congressional factions, with the result that its funding was cut by almost 90 percent. Meanwhile, another Department of the Interior agency, the General Land Office, had been running out of work. For a century, it had been responsible for the "disposal" of federal land—recording homestead claims and land grants to railroads and states. But the Taylor Grazing Act had effectively closed the frontier, and so no more land was to be given away.

In 1946, the two sister agencies in the Department of the Interior were melded into a new one—the Bureau of Land Management (BLM). Congress created the BLM without providing any clear mandate on its mission. Two things were clear, though: the BLM's primary business was livestock grazing, and the new agency had inherited a shaky relationship with Congress.

The BLM faced a daunting set of challenges: ranchers wanted fees kept low, Congress wanted revenues, and it was expected to heal rangelands and stabilize the livestock industry. To make things worse, Congress trimmed the agency's budget: only eighty-six people remained to oversee 150 million acres of grazing land. The grazing district advisory boards, fearing a return to unregulated chaos, deflected funds they had been given for range improvements to pay salaries of BLM range employees. Thus, within the first couple of years of the BLM's history, its primary task—regulation of the range— was only being accomplished by the good graces of the very people it was supposed to be regulating.[26]

In 1948, Marion Clawson, who came from a Nevada ranch family and had a doctorate in economics from Harvard, was appointed director of the BLM. Clawson immediately set out to professionalize the agency. He shifted emphasis from Washington to field offices in the West and pushed the BLM toward the emerging concept of "multiple use"—the idea that public lands should be managed for more values than just livestock grazing. But Clawson's tenure as a strong leader for the BLM ran up against familiar obstacles. In 1953, newly elected President Eisenhower chose Douglas McKay, governor of Oregon and former Chevrolet salesman, as secretary of the interior. McKay boasted, "We're here in the saddle as an administration representing business and industry."

Giveaway McKay, as he soon became known, couldn't tolerate a BLM director like Clawson, who believed that government had an essential role in conserving rangelands. Within months of Eisenhower's inauguration, Marion Clawson was unemployed.[27]

Gradually, though, the BLM widened its view to encompass broader concerns. In the 1960s, the agency began to create "allotment management plans" for each grazing allotment, paying heed to values besides the production of beef. The Forest Service had officially become a "multiple-use" agency in 1960. Upon passage of the Multiple-Use Sustained-Yield Act, national forests were to be managed for five uses—timber, range, watershed, recreation, and fish and wildlife.[28] In 1964, Congress directed the BLM to follow the Forest Service down the multiple-use path. The Classification and Multiple Use Act (CMU), however, was only a temporary measure. In 1970, the CMU expired, and once again the BLM had no multiple-use mandate on the books.

Public lands management began to come of age, and natural resources laws tumbled out of Congress like pebbles down a chute. But most of this new environmental legislation passed rangelands by, like a cloud drifting over a mountain crest, dissipating in the dry air of the rain shadow. BLM lands went conspicuously unmentioned in the Wilderness Act. No new additions were made to BLM holdings to complement spectacular new national parks. Public concern over falling trees, chemicals in the air, and poisons in the water mounted visibly. But the simple munching of cattle on the range? The expanding environmental movement didn't think to comment on it. Four decades after its genesis, and a decade and a half after the Forest Service became a multiple-use agency, the Bureau of Land Management still hobbled along as well as it could without any clear mission statement. The agency continued to operate according to the nebulous guidelines of the Taylor Grazing Act and the unofficial traditions that sprang up in its wake. That was about to change.

In 1976, Congress passed the most significant law relating to BLM lands since the Taylor Grazing Act. The Federal Land Policy and Management Act (FLPMA) provided comprehensive direction to the BLM. The new law confirmed that the BLM was a multiple-use agency. Six uses were classified as appropriate: grazing, fish and wildlife, minerals, rights-of-way, recreation, and timber. Prior to FLPMA's passage, over three thousand federal laws pertained to BLM lands; the new law superseded more than three hundred of them. For all its sweeping revisions and synthesis, though, FLPMA lacked clarity. Bedeviled by political compromise, Congress, in effect, told the BLM to consider conflicting values, but didn't specify any standards or desired results.[29]

In Nevada, in 1978, three men who held BLM grazing permits, disgruntled with the environmental bent of Jimmy Carter's White House, put their heads together and developed a line of reasoning that seemed rock solid: federal lands in western states like Nevada and Utah, where Uncle Sam controlled so much of the land, should be handed over to the states. These three ranchers also happened to be legislators in the Nevada State Assembly, and within a year the assembly had passed a bill that called for state control of all BLM lands in Nevada. Quickly, they were imitated: Utah, Arizona, New Mexico, and Wyoming all passed similar bills. These self-styled revolutionaries called themselves "Sagebrush Rebels." In the presidential campaign that year, the Republican challenger, Ronald Reagan, grinned broadly as he said, "Count me in!" When he won in a landslide, Reagan installed well-known antiregulation champions into key roles: James Watt as secretary of the interior and Robert Burford, a rancher, as director of the BLM. Soon, what had been demands from the western fringe became official government policy.[30]

Resisting Change

A glance at the history of grazing politics reveals a remarkable consistency of controversies. Consider, for example, these two congressional opinions, uttered sixty years apart—Representative Ayres of Montana in 1933: "The trials and tribulations of the western people are only a magazine story to these bureaucrats" and Senator Simpson of Wyoming in 1993: "We are defending a Western life style in this Administration's war on the West."[31]

For several decades, it has been commonplace to hear some ranchers heartily complaining that the federal bureaucracy is impinging on their rights. Others have objected to academic scientists who don't live on their land telling them anything about it. West has often been pitted against East, but this mistrust and acrimony, like the prevailing winds, has generally flowed eastward, not westward. Congresspeople from range states have struggled to get government "off the backs" of their constituents, and have—if they get riled up enough—introduced legislation that would give federal land to western states. The same political scenario repeated itself in 1929, 1945, 1979, and 1994: Republicans took back control of Congress, and conservative congressmen began to flex their muscles by introducing bills to transfer millions of acres of federal land to their states.[32]

Were the Forest Service to announce that every single acre of forest was suitable for logging, the public would probably shriek. But the BLM has made an analogous decision for rangelands—grazing is still authorized on the vast majority of the BLM's 177 million acres in the lower forty-eight states[33]—and the public has responded with indifference. How has the ranching industry kept such a muscular grip on public policy and resisted new views of land use so successfully? The answers to this question include social, historical, literary, and—above all—political perspectives. Formidable cultural forces have worked to maintain the status quo of livestock grazing on public lands.

In his classic work *Politics and Grass*, Philip Foss analyzed problems with federal grazing policy. To begin with, he said, early federal homestead policies forced overgrazing by encouraging stockmen to stuff too many cattle on nubbins of land that were too small and too dry. Ecological abuse of the range became entrenched in tradition. Our political process itself worked to resist change, as everyday aspects of the American governmental system coincidentally served ranching interests. In the U.S. Senate, sparsely populated western states are represented disproportionately—every state, whether New York or Nevada, has but two voices. Thus, over a quarter of senators represent western grazing states. Western politicians tend to be unified in their defense of ranching interests, whereas eastern politicians know or care little about grazing issues. Besides, a senator from Ohio doesn't gain much political capital in Cleveland or Cincinnati by challenging ranchers, but a senator from Utah surely loses points if he or she doesn't defend grazing privileges.

Ranchers were among the first settlers in much of the West, and their control of land quickly translated into political influence among state politicians. In fact, in many cases, western politicians have been ranchers. In both houses of Congress, representatives from western states scramble to chair subcommit-

tees that oversee range policy. Thus, over the years, a small number of western politicians have exerted enormous control over federal grazing policy. Foss referred to this as a "special private government," rather than a "general public government." This special government dutifully serves the interests of one private group—ranchers—while holding true to rhetoric about the public good.[34]

Grazing district advisory boards are an example of what political scientists call a "captured" policy pattern: control of a policy by the very group such policy is supposed to regulate. In the 1930s, when the Taylor Grazing Act first codified grazing regulation, a view known to political scientists as "interest-group liberalism"—in which interest groups make policy for their own economic realms—held sway.[35] No other interest group has ever played this game of institutionalized self-interest as effectively as ranchers. During the legislative battles concerning the Taylor Grazing Act in the 1930s and FLPMA in the 1970s, ranchers succeeded at diffusing the laws' regulatory authority.

Legal scholars have described FLPMA as a "not very good law . . . [that] represent[s] congressional buck-passing," "internally inconsistent," and "schizophrenic."[36] Furthermore, judicial interpretation of this muddled law has discouraged federal land managers from applying it too enthusiastically. Legal analysts have noted that, because of lack of public interest, rangelands have avoided "many environmental laws and safeguards that have become common in other areas of modern public land law."[37]

The structure of the BLM itself contributes to a lackadaisical approach to grazing regulation. As part of Marion Clawson's initial reforms, the bureau is alone among federal land management agencies in being administered by a separate office in each state. Compared with other agencies, each state is largely a world unto itself. Like the local grazing advisory boards, this state-by-state structure of the BLM simplifies the expression of local political will.

More fundamentally, though, Americans have a love affair with the imagery and symbolism of cowboys. Within any culture, stories undergird attitudes and biases. In the American mind, the West is equated with cowboys.

In the 1970s and 1980s, a few observers in the media began to scrutinize range management more carefully. In 1972, *Readers Digest* published an article entitled "Nibbling Away at the West." A few years later, *Audubon* carried a piece called "The Eating of the West." By 1985, the sportsmen's magazine *Outdoor Life* had weighed in with an editorial that proclaimed "fish and wildlife's biggest enemy is the excessive livestock grazing being done on more than 200 million acres of rangelands managed by the Bureau of Land Management and the U.S. Forest Service."[38]

In the 1980s, an upsurge in public recognition of the recreational potential of BLM lands, especially in the Southwest, collided with a growing awareness that livestock were just about everywhere. Range management, which had been consistently ignored by the general public, began to be an issue of broader concern. Using rangelands for fun was a novel concept. The BLM was not funded for recreation until 1965. But as hikers, bikers, and environmentalists penetrated the once-exclusive domain of ranchers, they didn't always like what they saw. Bare ground and cow pies inspired disgruntled comments on trailhead registers: "Get the cows out!" "Welcome to Cowpie National Monument!" and "I found some grass—put more cows in, quick!" Bumperstickers with slogans like "Stop Welfare Ranching" began to appear on out-of-state cars at BLM trailheads. By the 1990s, contention was more abundant than forage on the western range. Droves of recreationists grew incensed that public lands were littered with feces and hoofprints. Environmentalists were fanning the flames of outrage at ecological abuse and economic subsidy. And ranchers circled their wagons in an understandably defensive stance.

Throughout the history of range management, the basic assumption that livestock grazing was the most sensible use of the arid West had gone unchallenged. The first sentence of the first textbook on range management was "The West is a land of livestock grazing."[39] But, beginning in the late 1970s, academic scientists and professional land managers began to probe deeper into grazing issues. A government symposium in 1977 concluded that grazing was "the single most important factor limiting wildlife production in the West."[40] An interagency committee in Oregon and Washington, composed of state and federal biologists, quietly released a report in 1979 concluding that livestock grazing was the most important factor degrading fish and other wildlife habitat in the eleven western states (Arizona, California, Colorado, Idaho, Montana, Nevada, New Mexico, Oregon, Utah, Washington, Wyoming).[41] Ecologists began to squawk more frequently about the awful consequences of cattle grazing in riparian areas.

Whereas range scientists had focused on forage, productivity, and effects on a few game species, more broad-based studies found that livestock grazing might be less benign in arid western ecosystems than previously thought. Biologists were especially concerned about livestock effects on western riparian areas, which are among the most productive habitats in North America, essential to many wildlife species for breeding, wintering, and migration. Cattle, too, prefer riparian areas for their shade, cooler temperatures, and

water. In addition, riparian zones have more abundant food. Not surprisingly, cattle spend a disproportionate amount of their time in riparian zones, where they inflict considerable damage. The Environmental Protection Agency concluded that riparian conditions throughout the West are now the worst in American history—livestock grazing is a primary reason.[42]

Scientists began to realize that these sorts of effects are remarkably widespread. In the eleven western states, 70 percent of the land area is grazed by livestock. In terms of the extent of area affected, other types of land use—logging, mining, crop agriculture—pale in comparison.[43] Significantly, in the 1990s, the normally staid scientific community began to clear its throat and speak with a firmer voice. The American Fisheries Society, the Society for Conservation Biology, and the Wildlife Society all endorsed formal position statements on the nefarious biological consequences of grazing. "Livestock grazing . . . has a host of negative ecological repercussions. . . . Much of the ecological integrity of a variety of North American habitats is at risk," intoned the conservation biologists. "The inherent productivity and ability to sustain fish, wildlife, and livestock is in jeopardy," echoed the wildlife biologists.[44]

Five hundred years after the first domestic cattle and sheep arrived in the Americas, and 450 years after they were brought to the western United States, domestic livestock are still nearly ubiquitous. Meanwhile, scores of native animal and plant species have been pushed to the margins. Some species are now extinct, owing in large part to the prerogatives of ranchers and the imperatives of an industry ill suited to the rugged, arid landscape. It remains to be seen whether future reforms, unlike those of the past, will make significant progress toward salvaging the native species, clean, free-flowing water, and diverse biological communities of the West.

Sunset and moonrise over sagebrush-covered Bureau of Land Management lands, Jarbidge area of southwest Idaho. The best use of the West's arid lands had long been thought to be livestock grazing. Today, many Americans—informed by ecological science and inspired by spacious western landscapes—wish to see native species instead of cows and sheep and believe the public lands are best utilized as sanctuaries of biological diversity and natural beauty.

AN EVIL IN THE SEASON
The Cattleman's Welfare System Begins

T. H. Watkins

The current system of federal subsidies to the western cattle industry was born out of the Great Depression and Dust Bowl era. The 1934 Taylor Grazing Act was intended to rescue, as well as to regulate, desperate stock growers. It entrenched local ranchers' control over federal lands management; granted permits erroneously regarded by many as akin to a property right; and established a system of cheap grazing fees. Today, most public lands grazed by livestock remain in an ecologically unsatisfactory condition.

T. H. Watkins, *editor of* Wilderness *magazine from 1982 to 1997, was the Wallace Stegner Distinguished Professor of Western American Studies at Montana State University at the time of his death in 2000. A prolific writer and esteemed American historian, his books include* The Redrock Chronicles *(2000) and* Righteous Pilgrim: The Life and Times of Harold L. Ickes, 1874–1952 *(1990).*

Whether you think of it as a little mom-and-pop cow-calf operation barely carving out a sustainable living or as a corporate monster raising cows for feedlots the size of Rhode Island, the western cattle industry has been riding the backs of taxpayers for nearly seventy years. At issue is the fee that the government of the United States charges ranchers to graze animals on federal grasslands, a fee so low compared with real market values that it amounts to a subsidy—one that is such a travesty, even economic conservatives can join with extreme environmentalists to agree it should be ended.

That subsidy is the gift of a history born in hard times, the hardest hard times that the West so far has ever suffered. The economic desolation of the Great Depression combined with a largely human-caused environmental disaster to inspire the cattle industry to give up its fondly held delusions of rugged individualism—just long enough to plead with the New Deal government for help. The New Dealers obliged, only to have their noblest hopes seized by the industry, which cheerfully transformed the government's well-meaning program into an engine of convenience. For the rest of the century, the livestock industry was assured it would receive federal benefits with a minimum of federal control—at a cost to the land we still do not know how to measure fully.

The Great Depression was at its deepest ebb: nearly 13 million people were out of work. Adding to the general misery, a terrible drought began in the lower Mississippi River Valley in 1930, and in 1931 the drought shifted westward. In the spring of that year, it was the upper Midwest that began to suffer; over the next three years the misery spread south and west into every state between the Front Range of the Rockies and the Ohio and Mississippi River Valleys. There was a brief respite for some states in 1932, but between June of 1933 and May of 1934 the plains states experienced the lowest rainfall on record, while from the Flathead Range in Montana to the San Juans in southern Colorado, winter snowfall in the Rockies ranged from one-third to one-half of normal—and in New Mexico the Sangre de Cristos received little more than a dusting.

With drought came heat, more heat than millions had ever experienced, even in the plains, where the most ordinary summer could be a torment. Everywhere in the agricultural checkerboard of the heartland, crops not

State-owned lands southwest of Tucson, Arizona. In the drier regions of the West, Dust Bowl–like landscapes are not some distant memory, but part of the current reality of livestock-grazed lands.

already eaten by grasshoppers shriveled under the withering blight of the sun, and livestock grew skeletal and frantic with thirst and hunger, in some places "dropping dead in their tracks from the heat." Journalist Meridel Le Sueur took an exploratory bus trip across Minnesota and into North Dakota, "trying not to look at the ribs of the horses and the cows, but you got so you couldn't see anything but ribs, like beached hulks on the prairie, the bones rising out of the skin. You began to see the thin farmer under his rags and his wife as lean as his cows."[1]

On September 9, 1934, reporter Lorena Hickok reported from eastern Wyoming to Harry Hopkins, head of the Federal Emergency Relief Administration in Washington, D.C.: "I saw range that looked as though it had been gone over with a safety razor."[2]

There was an "evil in this season," wrote James Agee in *Fortune*, the words accompanying a numbingly effective portfolio of drought photographs by Margaret Bourke-White. It was a time when much of the Northern Hemisphere was "little better than a turning hearth, glowing before the white continuous blast of the sun."[3]

Then there was the wind, as much a part of the geography of the plains as the buffalo grass through which it rippled. In a drought, the wind became an enemy collaborating with sun and heat to test the limits of human patience. "A high wind is an awful thing," Le Sueur wrote in describing what she experienced in 1934. "It wears you down, it nags at you day after day, it sounds like an invisible army, it fills you with terror as something invisible does."[4]

It was not just the wind itself that scraped at the nerves in this evil season but the burden it carried, the direct consequence of generations of disregard for what the land could and could not be expected to do. It was a lesson that should have been learned long before, of course, as early as 1864, when in *Man and Nature*, George Perkins Marsh had cited the nations of antiquity to demonstrate how entire empires could disintegrate once their land had been abused beyond redemption. In this country, Marsh's warnings had been validated during the ecologically ruinous 1880s. Out on the High Plains, millions of cattle, half-starved because the western range had been packed with animals and brutally overgrazed for years, perished during the winter of 1886–1887 in what was called the "Big Die-Up." When the spring winds swept the land clean of snow, carcasses dotted a landscape that Theodore Roosevelt, anticipating Lorena Hickok, described as "a mere barren waste; not a green thing could be seen; the dead grass eaten off till the country looked as if it had been shaved by a razor."[5]

In the 1930s, millions of acres on the Great Plains that had not fully recovered from the abuses of the 1880s still lay open to livestock use and intensive agriculture—and it was on these lands that much of the land-wrecking boom of the World War I years had just played out. The temptation had been considerable. As the terrible engine of war had begun grinding across the European landscape in August 1914, all but obliterating the ability of the warring nations to feed themselves, the demand for American-grown food had swollen to unprecedented levels. Henry C. Wallace, editor of *Wallace's Farmer* (and father of Franklin Roosevelt's secretary of agriculture), declared that the United States had a "moral responsibility to feed the hungry people of the world."[6] Farmers and ranchers cheerfully accepted the obligation, planting and harvesting more and more bushels of wheat and other grains, raising and shipping ever-increasing numbers of cattle and other livestock, while prices for all agricultural products rose in an exhilarating demonstration of the principles of supply and demand. When the United States entered the war in April 1917, setting up a Food Administration for War that put in place artificially high price supports to encourage even more production, farmers and ranchers could be forgiven for believing that it was impossible to lose money.

At the war's end, as a recovering Europe clamored for food, the levels of prices and production remained high well into 1920, even after the government removed its price supports. Then European demands for relatively expensive American food suddenly began to decline as cheaper wheat from Australia and Canada, and cheap beef and other meat products from Argentina and elsewhere, came in on suddenly submarine-free shipping lanes. An anti-inflation Federal Reserve announced an end to the easy credit of the war years. Prices sank every bit as dramatically as they had risen, the slide continuing throughout the rest of the decade, while farmers and ranchers desperately stepped up production, hoping to make up in volume for steadily declining prices.

By the 1930s, then, much of the western land had been broken and exposed by repeated plowing, leached of its nutrients by constant planting and replanting, grazed down to the dirt by cattle and sheep, its topsoil skinned off in sheets or gullied by water erosion during wet years. And it was on these lands that the sun had been doing some of its most devastating work during the drought years.

So now the wind: it came down on all that exposed and crippled land; scooped up hundreds of millions of tons of it as dust, then boiled it all up into choking clouds that rolled across entire states and at least twice—in May 1934 and March 1935—sailed so high into the jet stream that airborne earth from the Great Plains darkened eastern cities in the daytime and dusted the decks of transatlantic liners. However dramatic, the jet stream storms of 1934 and 1935 gave the East the barest taste of what had become a commonplace misery in the plains states. On March 15, 1935, after portions of Texas, Oklahoma, and Kansas already had experienced two weeks of intermittent dust storms, a big one swept down from southeastern Colorado and for the next several days shut much of Kansas down in a gloom of dust. That storm had no sooner abated when another piled across the southern plains from Oklahoma on March 24, this one destroying half the wheat crop in Kansas before sweeping up into Nebraska and killing virtually all of that state's wheat. Then came the sudden "black blizzard" of April 14, which concentrated most of its terrific energy in Kansas, stranding hundreds of travelers, burying and killing one child, and lasting so long that its gale-force winds (in some storms they reached 60 or 70 miles an hour) and light-obliterating dust inspired apocalyptic terror among many, particularly when they were accompanied by a drop in temperatures of nearly 50 degrees in a matter of hours. "This is ultimate darkness," one victim wrote in a daily log. "So must come the end of the world."[7]

"By the middle of last August," the October 1934 issue of *Fortune* declared, "a good third of our part of the continent was one wide crisp. The great map in the Washington office of Relief Administrator Harry Hopkins showed 1,400 counties in twenty-two states, of which 1,100 were counted as harmed beyond all help."[8]

"I have been so moved by the distressing effects of a widespread drouth," President Franklin D. Roosevelt remarked in a speech delivered while on a cross-country train tour in August 1934, "and at the same time so strengthened in my belief that science and cooperation can do much from now on to undo the mistakes that men have made in the past and to aid the good forces of nature and the good impulses of men instead of fighting against them."[9]

Generally speaking, the New Dealers would discover that it proved easier to harness the tools of science to aid the good forces of nature than to invoke the spirit of cooperation and build on the good impulses of human beings—whom Roosevelt and his people would find (then as always) maddeningly unpredictable and not easily squeezed into sociopolitical molds. Take the western cattle industry. For more than sixty years, cattlemen and the politicians who serviced their needs and parroted their philosophies had resisted any sort of federal control over what could and could not be done on the public grazing lands of the West, save for those laws—such as the Desert Land Act of 1877, the Enlarged Homestead Act of 1909, or the Stock Raising Homestead Act of 1914—that either had been designed to service the cattle industry's desires or could be manipulated easily to its advantage. Armed with protestations of "frontier independence," however conditional, and driven by the fear, however exaggerated, that government action might seriously weaken its hegemony, the cattle industry at first refused to have anything to do with the New Deal's relief programs. But not for long.

More than half the public grasslands of the West were overgrazed and eroded by the end of 1933. Further, it was becoming clear that the rapidly spreading drought was going to inflict terrible losses on cattle from the Mississippi River Valley to the Rocky Mountains, and as far west as Arizona. There was, the Department of Agriculture estimated, a surplus of anywhere from 8 to 10 million head of cattle alone, and by December 1933 the industry was ready to adjust its position with regard to government help.[10] "Traditionally independent though he be, whether he likes it or not," warned F. E. Mollin, secretary of the American National Livestock Association, "the cattleman today is very much in the 'new deal,' entirely unable to cope single-handedly."[11]

What the cattlemen wanted, and what the government ultimately gave them in the Jones-Connally Farm Relief Act of 1934, including an appropriation of $200 million in drought relief money, was a simple purchase program that would reduce the number of existing animals without interfering with any cattleman's desire to keep his cows popping out more calves.

The program began on a fairly modest scale in the last week of May 1934 with the planned purchase of 50,000 animals per week in 121 counties in Minnesota, North Dakota, and South Dakota. The payment schedule was a complicated, two-part system that gave the seller anywhere from one to fourteen dollars a head as an outright purchase price, in addition to anywhere from three to six dollars a head as a relief "benefit." The two figures, each dependent on the age and condition of the animal being sold, could not add up to more than twenty dollars. Those cattle too diseased or emaciated to be useful as food were shot at the point of purchase and buried in pits; the rest were shipped off to the Federal Surplus Relief Corporation for slaughter and distribution as relief food.

A maximum of twenty dollars a head was not as much as cattlemen might have liked, but it was a good deal better than nothing, a conclusion many found inescapable as the drought burned its way across the plains with increasing ferocity. By the end of the summer, the Drought Relief Service had

received so many demands for help that the government expanded its relief territory all the way to Texas and began to purchase cattle with extravagant fervor, particularly after Congress obliged by appropriating another $500 million for general drought relief. Soon, nearly 600,000 cattle were being shot or shipped every week. Goats and sheep were added to the list, and before long thousands of those animals also became government issue.

Too many animals, some Department of Agriculture officials began to worry, and at the end of August it was announced that the program would cease when the total number of purchased cattle reached 7.3 million, after which relief efforts would concentrate on helping to feed and care for remaining herds. This plan did not sit well with the industry. "It would be doubtful economy," the *American Cattle Producer* editorialized, "to do a good job of buying seven million cattle and then quit just when the program could be properly rounded out with a little more time and money."[12] In November, after the industry had turned the full force of its political artillery on the Roosevelt administration, the government rethought the situation and agreed to buy another 1.5 million cattle before shutting the program down entirely at the end of January 1935, when even the cattle industry had to admit that it was no longer needed. So many cattle had been killed that market prices were on the rise, soon surpassing the twenty-dollar maximum per head provided by the law.

While arguably necessary to keep the cattle industry from being annihilated by the drought, the cattle purchase program—even though it reduced the number of grazing animals significantly for a while—would have done nothing to help the land itself if the livestock industry had been left to return to the unchained habits of the past, as many conservation-minded folk feared it would do. Among those people was Interior Secretary Harold L. Ickes, who in October 1934 declared that "an evil that is the twin of the destruction of our forests is the destruction of the public range through over-grazing. Herds of sheep and cattle, totaling more heads than the range can reasonably support, literally stand about hungrily waiting for a venturesome blade of grass to stick its head through the soil."[13]

By the time the secretary had come up with this cunning image, he had, he hoped, gone a long way toward helping to ensure that such overgrazing could finally be brought under control. The instrument of this hope was the Taylor Grazing Act, legislation largely contrived by Senator Edward S. Taylor of Colorado, a man who had spent most of his public life rejecting the notion that the livestock industry or any other aspect of western life should be subject to any sort of federal interference. But the drought years had changed his thinking, he said. Saving the land was a job "too big and interwoven for even the states to handle with satisfactory co-ordination. On the western slope of Colorado and in nearby states I saw waste, competition, over-use, and abuse of valuable range lands and watersheds eating into the very heart of western economy. . . . Erosion, yes, even human erosion, had taken root. The livestock industry . . .was headed for self-strangulation."[14]

Taylor introduced the first version of his legislation early in 1933, but since one of its provisions would have given the states veto power over federal regulations, Ickes testified against it, and the bill died in committee. The second version, introduced in 1934, did not include the offensive stipulation, and Ickes supported it vigorously. The bill called for 173 million acres of vacant and otherwise unappropriated public domain lands outside Alaska to be withdrawn from entry by any existing land laws. Eighty million acres of these lands were to be divided up into grazing districts that also would include national forest grazing lands. Those individuals "within or near a district who are landowners engaged in the livestock business, bona fide occupants or settlers, or owners of water or water rights" would be allowed to graze animals on these district lands under a ten-year permit system that stipulated the number of animals permitted on the land (subject to revision in any given year at the discretion of the secretary of the interior). For the first time, stockmen would be required to pay for this privilege—five cents per cow per month and one cent per sheep per month. Half the money would go back to the states for redistribution in the affected counties, while one-fourth would be reserved for range improvements. The program would be managed by a Division of Grazing in the Interior Department, and on the local level by district managers selected from the local population, with the aid and counsel of local advisory boards consisting of seven cattlemen and seven sheepmen appointed by the holders of grazing permits in the region, with a Division of Grazing employee appointed by the Interior Department.

As the bill came up for hearings, many livestock organizations waded in against it. A. A. Jones, head of the Arizona Wool Growers Association, went so far as to testify that "there isn't any such thing in the Southwest as overgrazing." Opposition among livestock interests was not unanimous, however, with stockman Farrington Carpenter of northwest Colorado speaking for many others when he said the bill was the industry's "only chance against being completely wiped out of existence."[15] Indeed, some of the most significant objections to Taylor's bill came not from the livestock industry but from Chief Forester Ferdinand A. Silcox and Agriculture Secretary Henry Wallace, both of whom complained that the bill gave far too much power to local stockmen and issued a specific warning against one particular amendment offered successfully by Nevada senator Pat McCarran. This stipulated that no grazing permit could be denied if doing so would "impair the value of the livestock unit of the permittee, if such unit is pledged as security for any bona fide loan."[16]

The amendment had been revised to provide that no "permittee complying with the rules and regulations laid down by the Secretary of the Interior" would be denied any permit. Still, Silcox worried that the stipulation remained loose enough to be open to interpretation. As in the earlier legislation that Ickes had opposed, the chief feared that the clause would give these permits the status of private property "which the Secretary could neither diminish, restrict, nor impair."[17] In passing along Silcox's comments that the bill—as presented to the president for his signature—was a bad idea, Wallace added: "An empire of 173 million acres should not be disposed of in language of doubtful meaning. It should be conserved by a law expressed in direct, specific, and unequivocal terms. This is not a measure of that kind."[18]

Nevertheless, it was the only measure likely to get through a Congress whose public land committees were occupied almost entirely by western congressmen and senators, most of whom were supported by the livestock industry. It also was a measure that Roosevelt wanted, however imperfect it might have been. On June 28, 1934, he signed it. On November 24, he issued an executive order withdrawing the first 80 million acres (an additional withdrawal of 62 million acres would be made in December 1935), while Ickes appointed bill proponent and stockman Farrington Carpenter head of the Division of Grazing. "In more ways than one," Ickes told an assembly of stockmen in Denver on February 12, 1935, "the Taylor Grazing Law is not merely a regulatory measure to upbuild and maintain the public range and to control its use

in the interest of the stockmen of the nation. It is a Magna Charta upon which the prosperity, well-being, and happiness of large sections of this great western country of ours will in the future depend."[19]

Ickes believed it, and so it might have been in a perfect world. But in the long run, the worries of Silcox and Wallace proved closer to the mark. The Division of Grazing (later renamed the U.S. Grazing Service) did erect some useful guidelines for the future management of the public grazing lands that especially foresightful ranchers were happy to follow, and this redounded to the benefit of some land in some areas. But Silcox was right about the permit system. The grazing permits were regarded as property, and any attempts to revoke or reduce them were viewed as assaults on private property rights. Not that the managers of most grazing districts were inclined to do anything of the sort. They were local people who had to answer to their neighbors for their actions, and even if they disagreed with the advice offered by the industry-weighted boards looking over their shoulders, few were brave enough to contradict it. On most of the grazing lands of the West, short-term profit (or the dream of it) continued to come at the expense of long-range protection, a condition that would prevail even when the Grazing Service was wedded to the General Land Office to create the Bureau of Land Management (BLM) in 1946. Nearly sixty years after passage of the Taylor Grazing Act, a study undertaken by the Natural Resources Defense Council and the National Wildlife Federation could estimate that no less than 100 million acres of BLM grazing land were still in "unsatisfactory" condition.[20]

Nor have the welfare traditions of the cattle industry been seriously tampered with in the decades since the New Deal (as have Aid to Dependent Children and other New Deal programs for the deserving poor, most of which are now left to the tender mercies of the individual states). Today, it costs only $1.35 per AUM (animal unit month—the amount of forage required by a cow-calf pair for a month) to graze cows on federal land; in some areas, this is as little as one-tenth the cost of doing the same on private lands. Grazing private cows on public land—the industry piously insists against all logic—is a right, not a privilege, and any restriction on that freewheeling tradition strikes at the heart of the West. And its politicians still rise up to defend that curious position with the passion of acolytes in some obscure religion, exerting so much muscle that the Department of the Interior fairly trembles when men like Senator Conrad Burns of Montana come to call.

It probably would be too much to say that Bernard DeVoto's famous outburst in "The West Against Itself" in *Harper's* magazine—"Cattlemen and sheepmen, I repeat, want to shovel most of the West into its rivers"—holds true today with the same force it did when he wrote it in 1947, but there is still validation for his conclusion. The fulfillment of "the great dream of the West, mature economic development and local ownership and control," he wrote,

envisions the establishment of an economy on the natural resources of the West, developed and integrated to produce a steady, sustained, permanent yield. While the West moves to build that kind of an economy, a part of the West is simultaneously moving to destroy the natural resources forever. That paradox is absolutely true to the Western mind and spirit. But the future of the West hinges on whether it can defend itself against itself.[21]

Severely eroded land, Coronado National Forest, Arizona.

PILLAGED PRESERVES
Livestock in National Parks and Wilderness Areas

Andy Kerr and Mark Salvo

Livestock grazing is currently permitted in thirty-two units of the national park system and, under the
provisions of the 1964 Wilderness Act, is not restricted in any area on the basis of its status as designated
wilderness. This notable exception to the preservation goals of national parks and wilderness areas is due
to a combination of factors: western history; the political power of stock growers; the public's general
unfamiliarity with pristine native grasslands and deserts; and a conservation movement that
has traditionally disregarded public lands livestock grazing in favor of other issues.

Andy Kerr *of the Larch Company writes, consults, and agitates on
environmental issues. He spent twenty years with the Oregon Natural
Resources Council, the group that helped make the northern spotted owl
a household name. Disguised as a hiking book, his* Oregon Desert Guide:
70 Hikes *is a plea and argument to preserve more than 7 million acres
of public lands.*

Mark Salvo *serves as grasslands and deserts advocate for the American
Lands Alliance in Portland, Oregon, where he coordinates American Lands'
campaign to protect the sage grouse, the "spotted owl of the desert."*

Domestic livestock grazing in the National Wilderness Preservation System
is—in all cases—inimical to the wilderness concept. Nevertheless, it is
allowed. Livestock grazing in the national park system is—in almost all
cases—inimical to the purpose of national parks. Nevertheless, it is allowed.
Livestock grazing is currently permitted in thirty-two units of the park
system.[1] Six are Civil War monuments (grazing occurred at the time of designation, indeed at the time of the war) or units surrounded by sprawling urban
landscapes and are not considered further here.

The political history of public land livestock grazing, wilderness, and national
parks is closely intertwined. To understand the sequence of decisions to permit
grazing in the wilderness and park systems is to appreciate the dilemma conservationists face in removing domestic livestock from these otherwise protected landscapes. In addition to lobbying, legislating, and agitating to remove
livestock from present and future parks and wilderness, a less confrontational
approach in the form of grazing permit retirement may be necessary to coax
public lands livestock grazers from these lands.

Cowboy Power

Although livestock grazing on the public lands is ecologically destructive,
economically irrational, and contrary to the wishes of the vast majority
of the American people, it still occurs—even in the most sacred of national parks and wilderness areas. We believe there are four major reasons
for the status quo.

1. *History.* Livestock (acting on behalf of cattle and sheep barons) were
   (ab)using the public lands for 50 to 150 years before any such lands
   were designated as parks or wilderness areas. Our political system
   usually grants great advantage to prior appropriation, and grazing
   is no exception.
2. *Political power.* Historically, cattle and sheep barons were extremely
   powerful politically and held public office in vast disproportion to their
   numbers. Our political system grants great advantage to the formerly

Livestock are allowed to graze in one of America's most beloved
national parks—Grand Teton, in Wyoming—and are given
priority over the needs of wildlife and the interests of visitors.
Grizzly bears have been killed, and wolves have been captured
and relocated, to protect these privately owned cattle.

powerful because the democratic system of checks and balances tends to resist change.

3. *Unknowing public.* Because cattle have been so pervasive throughout the American West for so long, few examples of ungrazed arid ecosystems are readily visible to the public. People are accustomed to seeing "cow bombed" landscapes. In contrast, examples of standing virgin forest are numerous, though not as numerous as clearcuts, and the public can easily appreciate the difference. Given the nature of arid lands, cow-damaged landscapes are often perceived as aesthetically pleasing, even though ecologically wounded.

4. *Unknowing conservation movement, apathy, and other priorities.* Most of the conservation movement knows little more than the general public about the ecological costs of livestock grazing. Historically, and to the present day, conservationists have chosen to ignore damage caused by livestock grazing to address instead what are perceived to be more acute threats to biodiversity. Efforts against logging, road building, mining, and development are higher priorities to most conservationists than livestock grazing.

Grazing in the National Park System

Prior to their designation as national parks or monuments, most National Park Service units were used for livestock grazing. In 1916, Congress passed the National Park Service Organic Act, creating the National Park Service and providing direction for managing the national parks.

> The service thus established shall promote and regulate the use of the Federal areas known as national parks, monuments, and reservations hereinafter specified, except such as are under the jurisdiction of the Secretary of the Army, as provided by law, by such means and measures as conform to the fundamental purpose of the said parks, monuments, and reservations, which purpose is to conserve the scenery and the natural and historic objects and the wild life therein and to provide for the enjoyment of the same in such manner and by such means as will leave them unimpaired for the enjoyment of future generations.[2]

The same law concedes grazing in parks:

> Provided, however, that the Secretary of the Interior may, under such rules and regulations and on such terms as he may prescribe, grant the privilege to graze livestock within any national park, monument, or reservation herein referred to when in his judgment such use is not detrimental to the primary purpose for which such park, monument, or reservation was created, except that this provision shall not apply to the Yellowstone National Park.[3]

Before the creation of the National Park Service, the U.S. Army managed the parks with a definitive dislike for domestic livestock. The army excluded cattle from Yellowstone National Park after its establishment in 1872. The army also defended Sequoia National Park against livestock.

In the winter of 1917–1918, after the passage of the Organic Act, then Interior Secretary Franklin K. Lane sent a letter to Park Service Director Stephen Mather implementing a new grazing policy. The Lane letter authorized cattle grazing in parks in "isolated regions not frequented by visitors" and where "natural features" would not be harmed.[4] It forbade sheep in the parks, however.

The Organic Act and the Lane letter codified grazing in the National Park System.[5] Given the era, one can understand the allowance of limited cattle grazing, especially considering wartime pressures for beef production and the newness of the National Park Service. The agency had yet to establish itself as a sustainable bureaucracy capable of demanding adequate funds from Congress, commanding public support, and setting its own course.

The grazing provision in the Organic Act remains on the books today, although, mercifully, it has been mitigated by administrative regulation that disfavors livestock grazing:

> (a) The running-at-large, herding, driving across, allowing on, pasturing, or grazing of livestock of any kind in a park area or the use of a park area for agricultural purposes is prohibited, except:
> (1) As specifically authorized by Federal statutory law; or
> (2) As required under a reservation of use rights arising from acquisition of a tract of land; or
> (3) As designated, when conducted as a necessary and integral part of a recreational activity or required in order to maintain a historic scene.[6]

"Historic scene" generally refers to park system units associated with colonial times or the Civil War. A hostile administration could overturn this regulation.

Grazing in the National Wilderness Preservation System

When Aldo Leopold, the nation's greatest ecological thinker and cofounder of the Wilderness Society, wrote his management proposal to establish the nation's first formally protected wilderness area in the Gila country of New Mexico, he grandfathered in livestock grazing. Forest Service historian Dennis M. Roth noted:

> In May 1922, Leopold, now assistant district forester in Albuquerque, made an inspection trip into the headwaters of the Gila River. When he returned, he wrote a Wilderness plan for the area that excluded roads and additional use permits, except for grazing. Only trails and telephone lines, to be used in case of forest fires, were to be permitted.[7]

Regarding the Gila, Leopold's biographer Curt Meine added: "Some cattle grazed there, but Leopold considered this an asset in that frontier grazing operations were themselves of recreational interest. The cattlemen, too, would benefit by the exclusion of new settlers and hordes of motorcars."[8]

Meine also observed that Leopold was seeking ranchers as allies in his efforts to regulate hunting as part of an overall game management regime, which included predator control at the time. This was before Leopold killed his last wolf and watched the "fierce, green fire" die in its eyes.[9] However, as with wolves, Leopold's thinking on livestock grazing evolved. Meine noted that "in his later years, he would place increasing emphasis on Wilderness as a 'land laboratory,' a place to understand how biotic communities are able to function in a state of health." After visiting de facto wilderness in northern Chihuahua in 1936–1937, Leopold wrote:

> I sometimes wonder whether semi-arid mountains can be grazed at all without ultimate deterioration. I know of no arid region which has ever survived grazing through long periods of time, although I have seen individual ranches which seemed to hold out for shorter periods. The trouble is that where water is unevenly distributed and feed varies in quality, grazing usually means overgrazing.[10]

Leopold's change of heart could not save the wilderness system from hungry livestock. Once the precedent favoring grazing was established, it became impossible to change later in more formalized Forest Service wilderness rules. As Roth noted:

> Grazing is the oldest and best-established use of national forest areas. Until the 1920s, grazing fees were the largest source of income from all national forest system lands. Stockmen were a potent political force in the West and exerted their power whenever the Forest Service threatened to raise grazing fees or cut back on overgrazing. Under these circumstances the Forest Service had allowed controlled grazing in Wilderness areas under the L-20 and U Regulations.[11]

The first draft of what became the Wilderness Act, written by Wilderness Society Executive Secretary Howard Zahniser, characterized livestock grazing in wilderness as a "nonconforming" use that should be terminated "equitably."[12] In subsequent versions of the bill, Congress stated that "grazing of domestic livestock . . . *may* be permitted to continue subject to such restrictions *as the Secretary of Agriculture deems desirable*" (emphasis added).[13] However, the final language in the Wilderness Act of 1964 states "the grazing of livestock, where established prior to the effective date of this Act, *shall* be permitted to continue subject to such reasonable regulations *as are deemed necessary by the Secretary of Agriculture*" (emphasis added).[14]

At the time the Wilderness Act passed in 1964, conservationists were more concerned about ongoing Forest Service attempts to declassify existing administrative wilderness areas to allow new road building and logging than they were about the continued grazing of livestock. Robert Wolf, who served on the staff of Senator Clinton Anderson (D-NM), then chair of the Senate Interior and Insular Affairs Committee, says Anderson went along with the compromise to ensure passage of the wilderness bill. Anderson, a former secretary of agriculture, knew that grazing was subject to reduction for purposes of conserving range condition. Anderson also felt that grazing was increasingly uneconomic and would decline in the future.

In 1980, Congress again took up the matter of wilderness grazing in the Colorado Wilderness Act, stating that:

> The Congress hereby declares that, without amending the Wilderness Act of 1964 . . . with respect to livestock grazing in National Forest wilderness areas, the provision of the Wilderness Act . . . relating to grazing shall be interpreted and administered in accordance with the guidelines contained under the heading "Grazing in National Forest Wilderness" in the House Committee Report . . . accompanying this act.[15]

This is a very unusual provision of law. It states that Congress is not amending the Wilderness Act, but it effectively does. It also incorporates, by reference, language in a committee report. As with all obtuse, confounding, and unclear congressional language, there are reasons for this.

In 1980, the conservation community was fighting dreaded "hard release" legislation. Such legislation would have prevented the Forest Service from ever again considering wilderness designation for roadless areas. If enacted, the agency's final environmental impact statement on its second Roadless Area Review and Evaluation (RARE II) would stand for wilderness areas for all time. A compromise was struck in which Congress enacted "soft release" language, which prohibited further wilderness consideration for a specified time. Part of the compromise was what became known as the "Colorado grazing language."

The Colorado grazing language entrenches livestock interests in our national wilderness preservation system.[16] It ratifies, in stronger terms, the grandfathering of livestock grazing in wilderness areas. It expands the Wilderness Act grazing provision to include wilderness areas managed by any federal agency. It allows the use of motorized equipment to service livestock and the construction of new fences and water developments in wilderness. It allows for increased numbers of livestock, and any authority previously conferred on the secretary of agriculture to require reasonable regulation of grazing to protect wilderness values is weakened. Finally (and incredibly), the Colorado language states that there is no restriction on domestic livestock grazing—no matter how reasonable—in any wilderness area as a result of its designation as such.[17]

Current Congressional Trends

Every relevant wilderness bill enacted by Congress, except one, has included language to provide for livestock grazing.[18] In 2000, the Steens Mountain Cooperative Management and Protection Act set a new direction—it created, in southeastern Oregon, the nation's first wilderness area that explicitly excludes domestic livestock. As the Oregon congressional delegation debated wilderness designation for the Steens, Oregon conservationists were adamant that livestock be removed from the fragile mountain meadows and federally designated wild and scenic rivers that descend from three sides of the mountain.[19]

For the national park system, congressional grazing policy has slowly improved. In 1994, Congress enacted the California Desert Protection Act. While grazing in the new Death Valley National Park and Mojave National Preserve was permanently grandfathered (at no more than current levels and subject to Park Service regulations), authority was granted to the National Park Service to acquire base properties (those private lands to which federal grazing permits have traditionally been attached) in order to end grazing on adjacent park lands.[20]

With fits and starts, Congress has also begun setting a time-certain end to grazing in some new parks. In 1999, Congress established the Black Canyon of the Gunnison National Park, grandfathering livestock grazing in the park (1) for the lifetime of the individual permit holder in the case of an individual permittee; or (2) for the lifetime of the individual permit holder, or dissolution of the partnership or corporation, in the case of a commercial permit holder.[21]

Until recently, rangelands have received scant attention from conservationists. Conservation efforts have focused on forests, and issues such as habitat fragmentation have been described in terms of forest islands in a sea of agriculture. Yet rangelands, defined here as wildland landscapes in which the dominant plants are not trees, make up roughly 70 percent of the terrestrial surface of the earth. . . . Without rangelands, the biodiversity of the world and of North America would be quite incomplete.

— Reed F. Noss and Allen Y. Cooperrider,
*Saving Nature's Legacy: Protecting and
Restoring Biodiversity*, 1994

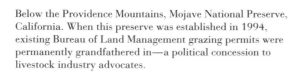

Below the Providence Mountains, Mojave National Preserve, California. When this preserve was established in 1994, existing Bureau of Land Management grazing permits were permanently grandfathered in—a political concession to livestock industry advocates.

52–53: Cattle "moonscape," Grand Staircase–Escalante National Monument, Utah.

FREE SPEECH
The Cowboy and His Cow

Edward Abbey

In April 1985, "Cactus Ed" recounted his foolish youth to a Missoula, Montana, audience. He told
of idolizing cowboys and hanging out with a gun-happy bronc rider named Mac during his college days.
Eventually, however, the cowboy myth wore thin, and Abbey concluded that the livestock industry, which
had long abused the public lands while taking the taxpayer for a ride, was overdue for being
turned out to pasture. It's clear not everyone in his audience agreed with him.

Edward Abbey *(1927–1989) was a writer, iconoclast, and passionate defender of the American West and wildness. Among his most famous works are* Desert Solitaire *(1968) and* The Monkey Wrench Gang *(1975), which have helped to inspire legions of environmental activists and desert rats.*

54–55: Cattle raze desert vegetation, Sonoran Desert National Monument, Arizona.

Opposite: Ranch hand chasing cows on a four-wheeler, Warner Valley, eastern Oregon.

When I first came West in 1948, a student at the University of New Mexico, I was only twenty years old and just out of the Army. I thought, like most simple-minded Easterners, that a cowboy was a kind of mythic hero. I idolized those scrawny little red-nosed hired hands in their tight jeans, funny boots, and comical hats.

Like other new arrivals in the West, I could imagine nothing more romantic than becoming a cowboy. Nothing more glorious than owning my own little genuine working cattle outfit. About the only thing better, I thought, was to be a big-league baseball player. I never dreamed that I'd eventually sink to writing books for a living. Unluckily for me—coming from an Appalachian hillbilly background and with a poor choice of parents—I didn't have much money. My father was a small-time logger. He ran a one-man sawmill and a submarginal side-hill farm. There wasn't any money in our family, no inheritance you could run ten thousand cattle on. I had no trust fund to back me up. No Hollywood movie deals to finance a land acquisition program. I lived on what in those days was called the GI Bill, which paid about $150 a month while I went to school. I made that last as long as I could—five or six years. I couldn't afford a horse. The best I could do in 1947 and '48 was buy a third-hand Chevy sedan and roam the West, mostly the Southwest, on holidays and weekends.

I had a roommate at the University of New Mexico. I'll call him Mac. He came from a little town in southwest New Mexico where his father ran a feed store. Mackie was a fair bronc rider, eager to get into the cattle-growing business. And he had some money, enough to buy a little cinderblock house and about forty acres in the Sandia Mountains east of Albuquerque, near a town we called Landfill. Mackie fenced those forty acres, built a corral and kept a few horses there, including an occasional genuine bronco for fun and practice.

I don't remember exactly how Mackie and I became friends in the first place. I was majoring in classical philosophy. He was majoring in screw-worm management. But we got to know each other through the mutual pursuit of a pair of nearly inseparable Kappa Kappa Gamma girls. I lived with him in his little cinderblock house. Helped him meet the mortgage payments. Helped him meet the girls. We were both crude, shy, ugly, obnoxious—like most college boys.

[Interjection: "Like you!"]

My friend Mac also owned a 1947 black Lincoln convertible, the kind with the big grille in the front, like a cowcatcher on a locomotive, chrome-plated. We used to race to classes in the morning, driving the twenty miles from his house to the campus in never more than fifteen minutes. Usually Mac was too hung over to drive, so I'd operate the car, clutching the wheel while Mac sat beside me waving his big .44, taking potshots at jackrabbits and road signs and billboards and beer bottles. Trying to wake up in time for his ten o'clock class in brand inspection.

I'm sorry to say that my friend Mac was a little bit gun-happy. Most of his forty acres was in tumbleweed. He fenced in about half an acre with chicken wire and stocked that little pasture with white rabbits. He used it as a target range. Not what you'd call sporting, I suppose, but we did eat the rabbits. Sometimes we even went deer hunting with handguns. Mackie with his revolver, and me with a chrome-plated Colt .45 automatic I had liberated from the US Army over in Italy. Surplus government property.

On one of our deer-hunting expeditions, I was sitting on a log in a big clearing in the woods, thinking about Plato and Aristotle and the Kappa Kappa Gamma girls. I didn't really care whether we got a deer that day or not. It was a couple of days before opening, anyway. The whole procedure was probably illegal as hell. Mac was out in the woods somewhere looking for deer around the clearing. I was sitting on the log, thinking, when I saw a chip of bark fly away from the log all by itself, about a foot from my left hand. Then I heard the blast of Mac's revolver—that big old .44 he'd probably liberated from his father. Then I heard him laugh.

"That's not very funny, Mackie," I said.

"Now don't whine and complain, Ed," he said. "You want to be a real hunter like me, you gotta learn to stay awake."

We never did get a deer with the handguns. But that's when I had my first little doubts about Mackie, and about the cowboy type in general. But I still loved him. Worshipped him, in fact. I was caught in the grip of the Western myth. Anybody said a word to me against cowboys, I'd jump down his throat with my spurs on. Especially if Mac was standing near by.

Sometimes I'd try to ride those broncs that he brought in, trying to prove that I could be a cowboy too. Trying to prove it more to myself than to him. I'd be on this crazy, crackpot horse going up, down, left, right, and inside out. Hanging on to the saddle horn with both hands. While Mac sat on the corral fence throwing beer bottles at us and laughing. Every time I got thrown off, Mac would say, "Now get right back on there, Ed. Quick, quick. Don't spoil 'im."

It took me a long time to realize I didn't have to do that kind of work. And it took me another thirty years to realize that there's something wrong at the heart of our most popular American myth—the cowboy and his cow.

[Jeers]

You may have guessed by now that I'm thinking of criticizing the livestock industry. And you are correct. I've been thinking about cows and sheep for many years. Getting more and more disgusted with the whole business. Western cattlemen are nothing more than welfare parasites. They've been getting a free ride on the public lands for over a century, and I think it's time we phased it out. I'm in favor of putting the public-lands livestock grazers out of business.

First of all, we don't need the public-lands beef industry. Even beef lovers don't need it. According to most government reports (Bureau of Land Management, Forest Service), only about 2 percent of our beef, our red meat, comes from the public lands of the eleven Western states. By those eleven I mean Montana, Nevada, Utah, Colorado, New Mexico, Arizona, Idaho, Wyoming, Oregon, Washington, and California. Most of our beef, aside from imports, comes from the Midwest and the East, especially the Southeast—Georgia, Alabama, Florida—and from other private lands across the nation. More beef cattle are raised in the state of Georgia than in the sagebrush empire of Nevada. And for a very good reason: back East, you can support a cow on maybe half an acre. Out here, it takes anywhere from twenty-five to fifty acres. In the red-rock country of Utah, the rule of thumb is one section—a square mile—per cow.

[Shouts from rear of hall]

Since such a small percentage of cows are produced on public lands in the West, eliminating that part of the industry should not raise supermarket beef prices very much. Furthermore, we'd save money in the taxes we now pay for various subsidies to these public-lands cattlemen. Subsidies for things like "range improvement"—tree chaining, sagebrush clearing, mesquite poisoning, disease control, predator trapping, fencing, wells, stockponds, roads. Then there are the salaries of those who work for government agencies like the BLM and the Forest Service. You could probably also count in a big part of the overpaid professors engaged in range-management research at the Western land-grant colleges.

Moreover, the cattle have done, and are doing, intolerable damage to our public lands—our national forests, state lands, BLM-administered lands, wildlife preserves, even some of our national parks and monuments. In Utah's Capitol Reef National Park, for example, grazing is still allowed. In fact, it's recently been extended for another ten years, and Utah politicians are trying to make the arrangement permanent. They probably won't get away with it. But there we have at least one case where cattle are still tramping about in a national park, transforming soil and grass into dust and weeds.

[Disturbance]

Overgrazing is much too weak a term. Most of the public lands in the West, and especially in the Southwest, are what you might call "cowburnt." Almost anywhere and everywhere you go in the American West you find hordes of these ugly, clumsy, stupid, bawling, stinking, fly-covered, shit-smeared, disease-spreading brutes. They are a pest and a plague. They pollute our springs and streams and rivers. They infest our canyons, valleys, meadows, and forests. They graze off the native bluestem and grama and bunch grasses, leaving behind jungles of prickly pear. They trample down the native forbs and shrubs and cacti. They spread the exotic cheatgrass, the Russian thistle, and the crested wheat grass. *Weeds*.

Even when the cattle are not physically present, you'll see the dung and the flies and the mud and the dust and the general destruction. If you don't see it, you'll smell it. The whole American West stinks of cattle. Along every flowing stream, around every seep and spring and water hole and well, you'll find acres and acres of what range-management specialists call "sacrifice areas"—another understatement. These are places denuded of forage, except for some cactus or a little tumbleweed or maybe a few mutilated trees like mesquite, juniper, or hackberry.

I'm not going to bombard you with graphs and statistics, which don't make much of an impression on intelligent people anyway. Anyone who goes beyond the city limits of almost any Western town can see for himself that the land is overgrazed. There are too many cows and horses and sheep out there. Of course, cattlemen would never publicly confess to overgrazing, any more than Dracula would publicly confess to a fondness for blood. Cattlemen are interested parties. Many of them will not give reliable testimony. Some have too much at stake: their Cadillacs and their airplanes, their ranch resale profits and their capital gains. (I'm talking about the corporation ranchers, the land-and-cattle companies, the investment syndicates.) Others, those ranchers who have only a small base property, flood the public lands with their cows. About 8 percent of federal land permittees have cattle that consume approximately 45 percent of the forage on the government rangelands.

Beef ranchers like to claim that their cows do not compete with deer. Deer are browsers, cows are grazers. That's true. But when a range is overgrazed, when the grass is gone (as it often is for seasons at a time), then cattle become browsers too, out of necessity. In the Southwest, cattle commonly feed on mesquite, cliff rose, cactus, acacia, or any other shrub or tree they find biodegradable. To that extent, they compete with deer. And they tend to drive out other and better wildlife. Like elk, or bighorn sheep, or pronghorn antelope.

[Sneers, jeers, laughter]

How much damage have cattle done to the Western range lands? Large-scale beef ranching has been going on since the 1870s. There's plenty of documentation of the effects of this massive cattle grazing on the erosion of the land, the character of the land, the character of the vegetation. Streams and rivers that used to flow on the surface all year round are now intermittent, or underground, because of overgrazing and rapid runoff.

Our public lands have been overgrazed for a century. The BLM knows it; the Forest Service knows it. The General Accounting Office knows it. And overgrazing means eventual ruin, just like stripmining or clear-cutting or the damming of rivers. Much of the Southwest already looks like Mexico or southern Italy or North Africa: a cowburnt wasteland. As we destroy our land, we destroy our agricultural economy and the basis of modern society. If we keep it up, we'll gradually degrade American life to the status of life in places like Mexico or southern Italy or Libya or Egypt.

In 1984 the Bureau of Land Management, which was required by Congress to report on its stewardship of our rangelands—the property of all Americans,

remember—confessed that 31 percent of the land it administered was in "good condition," and 60 percent was in "poor condition." And it reported that only 18 percent of the rangelands were improving, while 68 percent were "stable" and 14 percent were getting worse. If the BLM said that, we can safely assume that range conditions are actually much worse.

[Shouts of "Bullshit!"]

What can we do about this situation? This is the fun part—this is the part I like. It's not easy to argue that we should do away with cattle ranching. The cowboy myth gets in the way. But I do have some solutions to overgrazing.

[A yell: "Cowboys do it better!" Answered by another: "Ask any cow!"
Coarse laughter]

I'd begin by reducing the number of cattle on public lands. Not that range managers would go along with it, of course. In their eyes, and in the eyes of the livestock associations they work for, cutting down on the number of cattle is the worst possible solution—an impossible solution. So they propose all kinds of gimmicks. Portable fencing and perpetual movement of cattle. More cross-fencing. More wells and ponds so that more land can be exploited. These proposals are basically a maneuver by the Forest Service and the BLM to appease their critics without offending their real bosses in the beef industry. But a drastic reduction in cattle numbers is the only true and honest solution.

I also suggest that we open a hunting season on range cattle. I realize that beef cattle will not make sporting prey at first. Like all domesticated animals (including most humans), beef cattle are slow, stupid, and awkward. But the breed will improve if hunted regularly. And as the number of cattle is reduced, other and far more useful, beautiful, and interesting animals will return to the range lands and will increase.

Suppose, by some miracle of Hollywood or inheritance or good luck, I should acquire a respectable-sized working cattle outfit. What would I do with it? First, I'd get rid of the stinking, filthy cattle. Every single animal. Shoot them all, and stock the place with real animals, real game, real protein: elk, buffalo, pronghorn antelope, bighorn sheep, moose. And some purely decorative animals, like eagles. We need more eagles. And wolves. We need more wolves. Mountain lions and bears. Especially, of course, grizzly bears. Down in the desert, I would stock every water tank, every water hole, every stockpond, with alligators.

You may note that I have said little about coyotes or deer. Coyotes seem to be doing all right on their own. They're smarter than their enemies. I've never heard of a coyote as dumb as a sheepman. As for deer, especially mule deer, they, too, are surviving—maybe even thriving, as some game and fish departments claim, though nobody claims there are as many deer now as there were before the cattle industry was introduced in the West. In any case, compared to elk the deer is a second-rate game animal, nothing but a giant rodent—a rat with antlers.

[Portions of the audience begin to leave]

I've suggested that the beef industry's abuse of our Western lands is based on the old mythology of the cowboy as natural nobleman. I'd like to conclude this diatribe with a few remarks about this most cherished and fanciful of American fairy tales. In truth, the cowboy is only a hired hand. A farm boy in leather britches and a comical hat. A herdsman who gets on a horse to do part of his work. Some ranchers are also cowboys, but most are not. There is a difference. There are many ranchers out there who are big-time farmers of the public lands—our property. As such, they do not merit any special consideration or special privileges. There are only about 31,000 ranchers in the whole American West who use the public lands. That's less than the population of Missoula, Montana.

The rancher (with a few honorable exceptions) is a man who strings barbed wire all over the range; drills wells and bulldozes stockponds; drives off elk and antelope and bighorn sheep; poisons coyotes and prairie dogs; shoots eagles, bears, and cougars on sight; supplants the native grasses with tumbleweed, snakeweed, povertyweed, cowshit, anthills, mud, dust, and flies. And then leans back and grins at the TV cameras and talks about how much he loves the American West. Cowboys also are greatly overrated. Consider the nature of their work. Suppose you had to spend most of your working hours sitting on a horse, contemplating the hind end of a cow. How would that affect your imagination? Think what it does to the relatively simple mind of the average peasant boy, raised amid the bawling of calves and cows in the splatter of mud and the stink of shit.

[Shouting, laughter, disturbance]

Do cowboys work hard? Sometimes. But most ranchers don't work very hard. They have a lot of leisure time for politics and bellyaching (which is why most state legislatures in the West are occupied and dominated by cattlemen). Any time you go into a small Western town you'll find them at the nearest drugstore, sitting around all morning drinking coffee, talking about their tax breaks.

Is a cowboy's work socially useful? No. As I've already pointed out, subsidized Western range beef is a trivial item in the national beef economy. If all of our 31,000 Western public-land ranchers quit tomorrow, we'd never even notice. Any public school teacher does harder work, more difficult work, more dangerous work, and far more valuable work than the cowboy or the rancher. The same applies to the registered nurses and nurses' aides, garbage collectors, and traffic cops. Harder work, tougher work, more necessary work. We need those people in our complicated society. We do not need cowboys or ranchers. We've carried them on our backs long enough.

[Disturbance in rear of hall]

"This Abbey," the cowboys and their lovers will say, "this Abbey is a wimp. A chicken-hearted sentimentalist with no feel for the hard realities of practical life." Especially critical of my attitude will be the Easterners and Midwesterners newly arrived here from their Upper West Side apartments, their rustic lodges in upper Michigan. Our nouveau Westerners with their toy ranches, their pickup trucks with the gun racks, their pointy-toed boots with the undershot heels, their gigantic hats. And of course, their pet horses. The *instant rednecks.*

[Shouts]

To those who might accuse me of wimpery and sentimentality, I'd like to say this in reply. I respect real men. I admire true manliness. But I despise arrogance and brutality and bullies. So let me close with some nice remarks about cowboys and cattle ranchers. They are a mixed lot, like the rest of us. As individuals, they range from the bad to the ordinary to the good. A rancher, after all, is only a farmer, cropping the public rangelands with his four-legged lawnmowers, stashing our grass into his bank account. A cowboy is a hired hand trying to make an honest living. Nothing special.

I have no quarrel with these people as fellow humans. All I want to do is get their cows off our property. Let those cowboys and ranchers find some harder way to make a living, like the rest of us have to do. There's no good reason why we should subsidize them forever. They've had their free ride. It's time they learned to support themselves.

In the meantime, I'm going to say good-bye to all you cowboys and cowgirls. I love the legend too—but keep your sacred cows and your dead horses out of my elk pastures.

[Sitting ovation. Gunfire in parking lot]

A cowboy is a hired hand on the middle of a horse contemplating the hind end of a cow.
— Edward Abbey

Bovine backside. Not just malodorous, but also a major source of bacterial contamination and water pollution throughout the West.

LOOKING
THE ARID

ACROSS
WEST

What's Wrong with This Picture?

PART III

One of the most problematic obstacles for those advocating an end to public lands livestock grazing is the subtle nature of livestock abuse. Unlike the clearly visible damage to the land in a clearcut forest, the effects of livestock production on rangelands are far less obvious to the untrained eye. While someone with no ecological background can be moved to tears by the destruction of centuries-old trees and the loss of a forest ecosystem, the equivalent devastation of a grassland or shrub ecosystem engenders no remorse, no sad commentary, no outrage. "Overgrazing" to most people may conjure up images of a Saharan wasteland. Yet only in the very worst situations does livestock grazing create a barren landscape, devoid of all vegetation. Rather, most changes wrought by livestock are gradual, with the effect on plants being the replacement, over time, of more desirable species (for wildlife habitat and food as well as, often, for livestock consumption) with less desirable plant species.

But the alteration of plant communities is only the beginning of what livestock grazing does to the land. Other, even more subtle effects include compaction of soils, leading to lower water infiltration and greater runoff; loss of hiding cover for small mammals and birds; and removal of flowers, seeds, and leafy vegetation that are food for such species as butterflies, birds, and herbivorous mammals. Other problems caused by livestock production are fencing that hinders wildlife movement; disturbance of plant communities that favors weed invasion; dewatering of streams that reduces the width of riparian areas; draining of wetlands to create hay fields; trampling of stream banks and degradation of fish habitat; development of springs and removal of water on which frogs, birds, and other native species depend; and other effects that are not apparent to the uneducated observer.

Yet for someone trained to "read" the landscape, the ecological wounds caused by livestock production are clear and abundant. George Wuerthner first presents a critical analysis of range management techniques, especially traditional rangeland health evaluation methods. His essay helps to explain why so many range managers tend not to perceive fully the damage caused by livestock. The photographs and text of the following "How to Look . . . and See" section amply illustrate, and will begin to train your eyes to see, what is happening on the West's arid lands. Once you can start to read the "unnatural history" of the West—a tragic tale of greed, ignorance, and malfeasance—you will see it is sadly ubiquitous. The story is written in the eroding gullies, the fishless streams, the river valleys converted to hay pastures, the dried-up springs, the crumbling riverbanks, the silent and abandoned prairie dog colonies, and the countless Grizzly Creeks, Buffalo Meadows, and Wolf Mountains, the names of which are the only reminders of the vibrant life that once graced our lands.

62–63: Land grazed by livestock, New Mexico.

Opposite: Beaverhead County, Montana. Looking at fencelines is an easy way to begin seeing differences between land grazed by livestock and land that either has not been grazed or has been given time to recover (as on the left side of this photo).

One of the penalties of an ecological education is that one lives alone in a world of wounds. Much of the damage inflicted on the land is quite invisible to laymen.

—Aldo Leopold, *A Sand County Almanac*, 1949

UNDERSTANDING RANGE MANAGEMENT

George Wuerthner

Range managers assess land health by classifying range into four condition categories: excellent, good, fair, and poor, or some equivalent terminology. The system is based on an understanding of what plants should naturally be present on a given site. The scheme contains a number of biases against a full evaluation of ecosystem well-being. Nevertheless, the majority of public rangelands are rated in "fair" or "poor" condition, which means that most lands have lost half or more of the plant species expected.

One of the fundamental concepts in range management is the idea of "range condition." This is a generalized measure of the ecological health of the landscape. Although different government agencies have their own preferred methods for determining range condition, all are predicated on the notion of a vegetative and soil climax state that naturally exists in the absence of disturbance by livestock. This idealized condition is a benchmark against which any changes or declines in health are compared. Given a certain set of physical parameters, such as precipitation, soil type, slope and aspect, range managers expect to see particular groups of plants growing on a site.

Ideally, range condition is evaluated by measuring the percentage of key plant species in the total vegetation within test plots. These plots should be located along random transects. The plant species composition on any individual site is compared with the theoretical climax plant community for that specific location, and the amount of deviation from the ideal is "range condition." The usual means of producing credible results involves clipping all plants within a test plot, then weighing each species separately. The percentages of the desirable or climax plants for the site indicate how the site should be rated.

Range condition is indicated by four condition classes, which the Bureau of Land Management (BLM) terms excellent, good, fair, and poor. The Forest Service gives the categories different names, but the condition indicated is the same: potential natural community, late seral, midseral, and early seral. Any grazing allotment condition other than "excellent" or "potential natural community"—barring other factors such as wildfire or drought—is usually considered to be a result of overgrazing.

There are criticisms of this system. Some ecologists believe that the notion of a climax community is an overly restrictive benchmark. There may be far more natural variability on a healthy site than is recognized by the current assessment methodology. Also, plant communities do not necessarily move through recognizable and predictable stages. On some sites, overgrazing may have precipitated such severe erosion that plants may never progress to "higher" successional stages.

Some range professionals argue that their management goals may specifically aim for "lower" successional stages. For instance, changing a grassland site to one dominated by sagebrush may be desirable if managers wish to promote mule deer numbers. In some places, native wildlife, such as prairie dogs, may prefer heavily grazed sites. Of course, where livestock are overgrazing the range, it is convenient for managers to state they are managing for lower ecological stages to "benefit" wildlife. All in all, however, the system of measuring range condition can work reasonably well to indicate how good a job range managers are doing—if the measuring is done objectively.

Range condition is very site-specific. Anyone trying to assess range condition has to understand how all physical and biological factors come together in a particular area to produce the vegetation observed. For example, a lot of bare soil or cactus on a site—conditions often indicative of overgrazing—may lead people to assume there's a problem. However, in some areas, on certain types of sites, cactus and bare dirt may be natural and healthy.

Range managers also observe the trend of range condition. By repeatedly visiting a site over a number of years, managers can usually determine whether a particular site is improving or declining in productivity. Among the factors considered are the amount of bare soil; the occurrence of weedy, invasive species; and the presence of young seedlings of climax species. An upward trend means greater productivity; a downward trend means less productivity. Trend may also be stable—in other words, conditions are not changing. It is important to realize that stable does not necessarily mean good. On BLM lands, 64 percent of the rangelands are considered to be in stable condition, but much of this land is also in poor condition. Too often, the range simply cannot be in any worse shape because it is already at "rock bottom."

Though the above-described method of evaluating range condition is reasonably reliable, the ideal procedure (random placement of test plots, clipping and weighing of vegetation, and so forth) typically is not used, because of time

Cattle manure and trampled spring, Bureau of Land Management lands, Lemhi Valley, Idaho.

and funding constraints. On most public lands, the "ocular" method is often the only way in which range condition and trend is evaluated. To use the ocular method, a range manager merely walks—or in far too many cases, drives by—an allotment, eyeballing the range. Although experienced range managers can estimate range condition with a fair degree of accuracy, evaluation is still subject to many biases of the observer. It is entirely possible that an allotment could be rated in "fair" condition by one person, and in "good" or even "excellent" condition by another. Referring to such ocular methods, many range professionals say that range management is as much art as science. In response, critics have said that "range management is heavy on the art and lean on the science, and like art everywhere, beauty is in the eye of the beholder."

Certainly the ocular method is open to much abuse. A range manager who wants to show how his or her management schemes have improved the range has powerful incentive to see improvement when doing an ocular evaluation. Reporting can be purposefully deceptive, or the bias may not even be recognized by the observer. In any case, since most allotment management plans only call for "improvement," no matter how small or how long it takes, there is plenty of wiggle room to allow officials to do their jobs without really bettering the conditions on the ground.

The definitions of range condition categories, in and of themselves, allow for rather deceptive evaluations. For instance, a site can have as little as 51 percent of the expected plants and still be classified as in "good" condition. Most people equate the term *good* with a situation that is quite desirable, yet can a site that has lost nearly half of its desirable plant species really be termed *good*? Even the word *fair* is a euphemism since a site can have lost as much as 74 percent of its preferred species—by any reasonable measure, an ecological disaster—yet still be called "fair-condition" range.

The lag time between the onset of degradation and visible downward trend in range condition also makes traditional range evaluation procedures problematic. Accelerated soil erosion and soil compaction usually precede loss in plant vigor and changes in plant composition. By the time a change in range condition is detected, critical thresholds may have been exceeded.

Another problem in range management is reliance on plant productivity and composition as the sole measures of range health. Other values are not con-

sidered. Even range rated as being in "good" or "excellent" condition can still be deficient in terms of its animal community and what it provides for wildlife habitat. For example, cattle might be brought onto a site with ten-inch-tall grass of the preferred climax species. The cattle quickly chomp away, reducing the overall height of the grass to one inch. Then the cows are immediately withdrawn before they can damage the plants further. Since the desired plants are still on the site, albeit shortened considerably in height, the site would rate as "excellent." Yet, for a sharp-tailed grouse that requires more than eight inches of grass cover to hide successfully from predators, the site is certainly not in excellent shape. Or, the range may be below par in its ability to hold soil moisture. The lack of standing grass stems may allow snow to blow away instead of being trapped on the site. Thus, the soil gradually dries out—which almost certainly signals an eventual change in the plant community, among other things, even if there is no immediate alteration of the species present.

In most cases, range condition refers to the entire allotment. In the West, many parts of an allotment scarcely ever receive livestock grazing. They are too steep, too far from water, too high, too rugged. Yet these unused lands are averaged in with the overall range condition for the allotment. Heavily impacted areas, such as wet meadows and riparian areas, are considered with parts that receive little or no grazing pressure. The result is that many allotments are rated as "good" or "fair" while areas within the allotment actually used by livestock are severely overgrazed. These same beat-up portions of an allotment also tend to be the most ecologically important, the most critical to supporting a wide array of native species.

Even operating under the biases described above, the majority of public rangelands in the West have been rated in "fair" or "poor" condition by range managers. Less than 3 percent are rated "excellent." This is an enormous indictment of the livestock industry.

Finally, in addition to all the problems already described, a significant number of allotments are monitored only infrequently, or not at all. According to a General Accounting Office (GAO) study, two-thirds of BLM allotments and one-fourth of Forest Service allotments do not have management plans or data. Another GAO study states that as many as one-third of BLM allotments on some districts have never been visited by range managers at all!

Debating about the proper number of cows to have on an allotment is like arguing over which seat to take on the deck of the Titanic.
—Andy Kerr, livestock grazing activist, 2001

This excellent-condition rough fescue grassland is in Glacier National Park, Montana, which is off-limits to livestock. Rough fescue is a grass species characteristic of areas with higher summer moisture. Under natural conditions, it grows commonly and abundantly within its range. However, it tends to disappear under even moderate livestock grazing pressure, and the species has declined significantly throughout the West.

HOW TO LOOK...

Cattle evolved in moist Eurasian woodlands. In the arid West, they are attracted to the sites most resembling their ancestral habitats—the areas along streams and around springs and seeps, otherwise known as riparian habitat. By concentrating in these places so heavily, cattle have caused an immense amount of damage. This photo shows a barren, rocky landscape that many people think is normal for such an arid place as Nevada. However, the wide, shallow stream *(1)*, the dearth of streamside vegetation *(2)*, and the soils compacted by cattle that rarely wander more than a few hundred yards from the meager water source *(3)* are all indications of a naturally dry land that has been made more desertlike by abusive livestock production.

Cattle are very large animals, supported by relatively small-diameter hooves. When nearly their entire weight is applied to the ground via their footsteps, stream banks can crack and eventually fall apart. Over time, the result is a flattening of the stream profile and a wider, more shallow channel in which water warms and evaporates more rapidly. Cattle strip the vegetation next to and close by the stream, which results in greater soil erosion and faster runoff. Cattle hooves also act on the land around the stream, compacting soils and leading to diminished water infiltration. Plants that are more drought-resistant, that stand up to trampling better, and that are less palatable to livestock end up being the dominant species. In extreme cases of overgrazing, and on naturally very arid sites, nearly all vegetation may be eliminated (as in this photograph).

Bureau of Land Management lands, Nevada.

.AND SEE

HOW TO LOOK...

The photo at the upper left shows what looks like a nice trout stream. However, despite the green vegetation growing along the banks, the stream has been seriously degraded by a hundred years of livestock use. Grazing of the riparian area has created a stream channel that is approximately six to eight feet wide and about a foot deep *(1)*. The stream bank slumps and falls into the stream because of livestock trampling and the removal of vegetation that held the bank in place *(2)*.

The photo at the upper right shows the same stream entering a fenced exclosure, where livestock have been kept out for almost a century. Note how the stream channel suddenly narrows inside the exclosure *(3)*. It also becomes much deeper. This is far better habitat for trout and other aquatic organisms.

The photo at the lower left was taken approximately 50 feet inside the exclosure. Note the abundance of shrubby cinquefoil *(4)*, which is absent outside the fence. Such vegetation helps to slow water velocity during floods and reduces the scouring effects of high flows. The stream is very narrow *(5)*.

The final photo, at the lower right, shows the same stream 300 feet from the fence, inside the exclosure. The stream channel is now three feet deep and only a foot wide *(6)*. The dominant plants along the stream are sedges, with fibrous root systems that are better at holding stream banks together. The stream is so narrow and well vegetated that it is nearly invisible (the photographer accidentally fell in at one point because of this fact). Also, note the willows and other shrubs *(7)* that serve, among other things, as nesting habitat for songbirds.

Pole Creek, Sawtooth National Recreation Area, Idaho.

.AND SEE

HOW TO LOOK...

"Big Sky" country? Yes. But look closer, at the ground and across the water, where native wildlife must compete with livestock for habitat and resources. This landscape may be spacious, but it has been made less productive and less diverse by livestock grazing.

Note the stream bank that is slumping into the river *(1)*. Cattle moving in and out of the water, trampling and eating riparian vegetation, have weakened the stream banks. Over time, what were once well-vegetated, overhanging banks—excellent hiding cover for trout—have become exposed and laid back.

Vegetation is cropped low *(2)*. Willows and other riparian shrubs have disappeared. The ground is exposed to greater solar radiation, which in turn means faster evaporation of moisture from the soil. Cattle also compact the soil with their hooves. Water infiltration from rain and snow is diminished; soils are drier for this reason also. Shallow-rooted Kentucky bluegrass, an exotic, dominates the site, instead of deep-rooted native sedges. The bluegrass does a much poorer job of holding the soil together than do the sedges.

Because of the livestock impacts just mentioned, along with other effects of livestock grazing, the stream channel has become shallow and wide *(3)*. The water heats more rapidly in the summer, creating less favorable conditions for coldwater fish, such as trout.

Cattle grazing Bureau of Land Management lands, Red Rock River, Montana.

.AND SEE

3

HOW TO LOOK...

These two pairs of photos show the dramatic changes that can occur when livestock are introduced to an area, or when livestock are permanently removed and recovery is allowed to occur. The top photo on this page was taken a few weeks before livestock were brought to a site in the Wasatch Range in Utah. Note the thick, verdant vegetation *(1)*. The bottom photo shows the same place, right after livestock had finished grazing the area. The plant cover is almost entirely gone, and the soil is mostly bare *(2)*.

The top photo on the facing page was taken on the San Pedro River in Arizona in 1984, when cattle were allowed to graze the Bureau of Land Management lands along the river. Note the wide, sandy river channel *(3)* and the lack of young cottonwood trees or other riparian vegetation. The bottom photo was taken from the same spot in 1997, after livestock had been excluded from the site for eleven years. The river channel has narrowed *(4)*, and cottonwoods now dominate the banks. Within five years after livestock had been removed, research showed a spectacular increase in numbers of western wood peewees, yellow warblers, yellowthroats, yellow-breasted chats, summer tanagers, and song sparrows. There were healthy increases in yellow-billed cuckoos and Bell's vireos, as well. All these bird species are dependent on riparian habitat.

Wasatch-Cache National Forest, Utah.

Opposite: San Pedro River, Arizona.

AND SEE

HOW TO LOOK...

A small exclosure on a spring *(1)* contrasts with a Bureau of Land Management grazing allotment *(2)*. The allotment permit is held by the Nature Conservancy, which purchased the 214-acre "base property." Unfortunately, when the Nature Conservancy took over the permit, it did not put an end to livestock abuse of the associated 68,000-acre public lands allotment.

Inside the exclosure, Nevada angelica, a rare plant, is abundant and grows two feet tall. Outside, Nevada angelica and all other herbaceous riparian plants are grazed to ground level, and an extensive spring/seep complex is being irreparably damaged and dried up *(3)*.

The Nature Conservancy installed a solar-powered pump to move spring water above the canyon rim-rock, theoretically to try to lessen cattle damage to the spring by providing an artificial water source to cattle in the uplands. However, negative impacts of the water development include reducing water flow from the spring and extending livestock damage in the Owyhee River California Bighorn Sheep Area of Critical Environmental Concern and the uplands of the Owyhee River Canyon Wilderness Study Area. Plus, as the photo shows, the riparian area is still being pummeled by livestock.

45 Ranch Allotment, Bureau of Land Management land, Spring Creek Basin, southern Idaho.

IDAHO

Mountain Springs Allotment
Sage Creek

Although the livestock industry tries to portray public lands grazing as mostly small Ma-and-Pa-type operations, most of the grazed acreage on public lands, by far, is utilized by large corporations and wealthy individuals. Thus, the top five hundred, or 2.3 percent, of all Bureau of Land Management (BLM) permittees control an astounding 46.5 percent of all BLM lands leased for livestock grazing. Among the larger permittees for many years were computer innovators William Hewlett and David Packard, who controlled the sprawling Mountain Springs/San Felipe Allotment. Today, Hewlett's daughter, Mary Hewlett Jaffe, owns the base ranch and holds the grazing permit.

The allotment has been the center of controversy for decades. During that time there has been no significant change in management. In a 1979 report, the BLM wrote that the allotment was "nonfunctioning" as a natural area and called for the AUMs (animal unit months—one AUM being the amount of forage required by a cow-calf pair for a month) to be halved to halt range damage. In 1991, the BLM stated in another report that "severe utilization and loss of riparian vegetation is evident on the upper springs sources." The agency again cited the importance of reducing the number of cattle. Finally, in the late 1990s, modest changes in stocking rates were implemented, but livestock impacts were still very much in evidence when this photo was taken in August 2000.

COW-DAMAGED

TEXAS

Big Bend National Park

Big Bend exemplifies the potential for ecological restoration—given enough time—even after decades of devastating overgrazing. Scattered livestock operations existed in the Big Bend area for centuries, but following the arrival of the railroad in the 1880s, large-scale livestock grazing began. By 1886, more than 60,000 head of cattle ranged over the region. Cattle and domestic sheep transformed grasslands into sparse, rocky landscapes dominated by creosote bush. They accelerated soil erosion and arroyo cutting and made tree-lined creeks into sandy desert washes. By the 1930s, following decades of severe overgrazing, ranching in Big Bend country was marginal at best.

The park was established in 1944, and since that time Big Bend has been off-limits to livestock (except for areas along the Rio Grande River, where trespass livestock from Mexico have long been a serious problem). Although the effects of livestock grazing are not erased, the land has recovered some measure of its former glory. Certain wildlife species, such as black bear and javelina—made rare by the incursion of livestock—have recovered from population lows. Others, such as bighorn sheep, are still largely absent from the ecosystem.

Whether Big Bend can be made whole following its severe wounding by abusive ranching remains a question for both science and society.

ARIZONA

Tonto National Forest

Livestock proponents will often admit that during the early years of ranching in the West there was overstocking and overgrazing. But today, they assert, range management is much more careful and scientific, and range condition is greatly improved.

Forest Ranger Fred Croxen described how Tonto Creek, in central Arizona, was once "timbered with the local creek bottom type of timber from bluff to bluff, the water seeped rather than flowed down through a series of sloughs, and fish over a foot in length could be caught with little trouble." But at the time of his writing in 1926, "this same creek bottom is little more than a gravel bar from bluff to bluff. The old trees are gone. Some were cut for fuel, while many others were cut for cattle during droughts and for winter feed, and many were washed away during the floods that rushed down the stream nearly every year since the range started to deplete. The same condition applies to practically every stream in the Tonto."

This photo of a creek in the Tonto Forest was taken in 1997, seventy-one years after Croxen wrote about stream conditions there. You may judge whether things have changed for the better.

COW-DAMAGED

NORTH DAKOTA

**Theodore Roosevelt National Park
Little Missouri River**

Healthy riparian areas are critical to the survival of many native wildlife species in the West. A majority use these streamside zones during some part of their life cycle; some live exclusively in these narrow, but biologically rich strips of habitat. This cottonwood gallery forest, composed of large, old trees and a lush carpet of understory vegetation, provides homes for cavity-nesting birds and mammals and during dry periods offers shade, moisture, and lush, green forage when surrounding areas may be hot and brown.

The abundant vegetation is good not just for wildlife, but also for retaining moisture in the soil and slowing erosion. During floods, the roots of the trees and other plants help to hold the soil in place. The vegetation also slows rushing water when the river tops its banks. With water velocity reduced, the erosive power of the current is also lessened. Another benefit of the slowed current is that sediments then drop out, and the land's fertility is renewed. Plants grow even more vigorously, and the natural community is maintained.

Unlike native grazers, such as the bison, elk, and deer that are all found in Theodore Roosevelt Park, cattle tend to "camp out" in riparian areas, repeatedly eating the same plants and trampling the same soils and stretches of stream bank. A downward cycle begins in which plants are weakened and die out, erosion accelerates, soils become drier, and floods become more intense.

MONTANA

Big Hole River

Flyfishing guides, tackle shop proprietors, lodge and restaurant owners, and the people who work in them all depend for their livelihood on the healthy fisheries of the Big Hole River. Of course, the fish themselves and a host of other aquatic organisms depend for their very lives on the Big Hole's continued flow and high water quality. Yet, *one* rancher was allowed to bulldoze a diversion dam across nearly the entire river to put the water into his irrigation ditch. Unfortunately, this type of situation is very common in the West, where ranchers have historically appropriated large quantities of water and few have questioned their ability to do so.

A system of "water rights" prevails in the West. However, a water right does not constitute ownership of water; it merely determines who can divert water and how much. The water itself is legally the property of the public. But this public ownership and interest is rarely asserted. Consumptive water users, farmers and ranchers in particular, are permitted to degrade and even completely remove water from streams, while water quality, fisheries, recreation, and wildlife habitat all suffer.

Gradually, the public has awakened to the desirability of maintaining water in rivers, but most corrective measures are still based on the erroneous idea that water rights mean water ownership. Thus, the government or a conservation organization need not, and should not, lease water back from a water rights-holder to preserve an in-stream flow. State and federal agencies have a vested authority, and duty, to uphold the public's interest in clean water, wildlife habitat, and recreation.

Every western state's constitution or water code provides that water is a public resource—it belongs to everyone.
—Marc Reisner and Sarah Bates,
Overtapped Oasis, 1990

COW-DAMAGED

ARIZONA–
NEW MEXICO

Coronado National Forest

The area pictured here is located in the Peloncillo Mountains in the Malpai Borderlands region along the Arizona–New Mexico border. This photograph of the grazed area outside of the Blair exclosure *(upper left)* was taken after a wet winter in 2001, yet cattle had still grazed the vegetation down to dirt. Forage for native herbivores is greatly reduced. Hiding cover for small mammals and birds is significantly diminished outside the exclosure, whereas inside the exclosure, foot-tall grass grows. The reduced vegetation outside the exclosure leads to higher soil surface temperatures and evaporation rates, with a subsequent reduction in soil moisture. Soils compacted by many cattle hooves absorb less rainwater; more water is lost to the site in the form of runoff. Greater runoff increases soil erosion rates. In multiple ways, an already dry landscape is made even more arid.

In the case of livestock grazing, prudence is especially warranted but has long been in short supply. There have been warnings down through the years—from seers as well as the land itself—about the impacts of grazing in the arid West. None has been heeded.

—Debra L. Donahue, *The Western Range Revisited,* 1999

COW-DAMAGED

OREGON

**Trout Creek Mountains
Bureau of Land Management Lands
Upper Little Whitehorse Creek**

Little Whitehorse Creek is home to the Lahontan cutthroat trout, a threatened species. In the 1980s, studies revealed the fish were declining because of habitat degradation by livestock. But instead of closing the grazing allotments, the Bureau of Land Management (BLM) designated the area a "cooperative demonstration site" with continued livestock grazing. The prospect of enforcement of the Endangered Species Act to protect the trout compelled the ranchers in the area to reduce the severity of their cattle grazing in some areas. A temporary reversal in fish population declines followed these changes, but in recent years fish numbers have recommenced their downward trend.

The BLM has spent more than $1 million, and thousands of hours of BLM employee time, on range developments designed to reduce livestock impacts. These developments include numerous water developments that drain springs to provide cows with water, a major new water pipeline, new fencing in a proposed wilderness area, and other alterations to the natural landscape. Despite the expenditure, cows continue to wreak havoc on the land. Although touted as a "success story" by ranching advocates, this project has actually cost taxpayers more money and provided less protection for the trout and for the ecosystem than if the BLM had simply closed the allotments. Incidentally, the cattle on the main allotment are owned by a millionaire from southern California.

Little Whitehorse Creek exemplifies the usual agency response to habitat destruction caused by livestock: conduct additional studies and spend more money on more range developments.

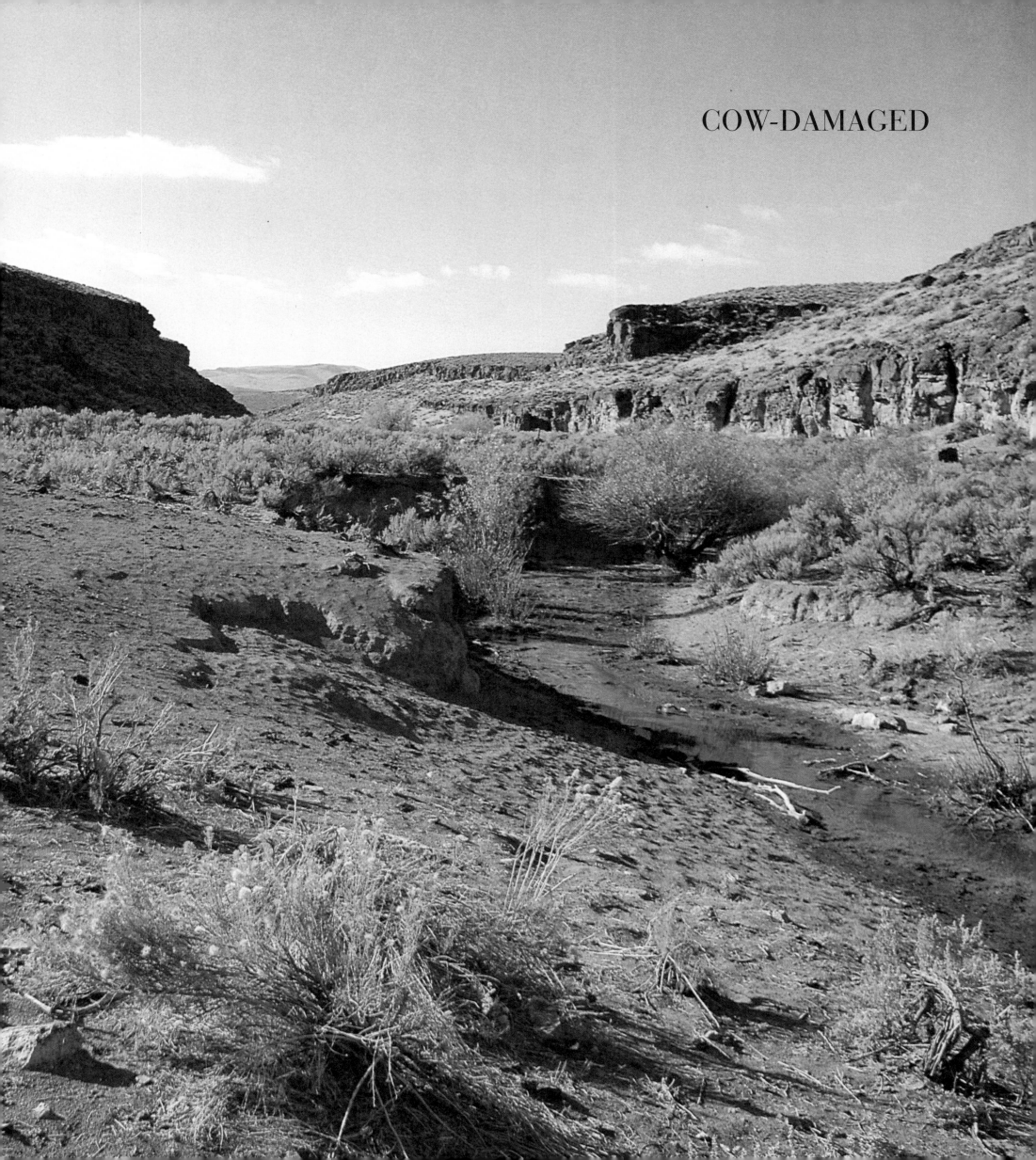

ARIZONA

**State Lands Southwest of Tucson
Altar Wash**

Even if livestock grazing were to end today, some areas of the West are so severely damaged that the time required for recovery may be on the order of centuries, or longer. Particularly within the most arid regions, such as the Southwest, critical ecological thresholds have been crossed already through excessive soil erosion, heavy invasion by exotic plants, and extirpation of keystone species, such as beaver. Nonetheless, that some places are beyond hope of complete and perfect recovery does not justify continued abuse. Given enough time to rest from the impacts of livestock, and perhaps some measured, carefully considered intervention (such as reintroduction of native species), many areas could still become viable wildlife habitat.

In addition, restoration efforts must occur on a watershed level. It is not enough to focus on a ravaged downstream area while upstream, waterways continue to be trashed and soil erosion continues unabated. The need for focus on the watershed as a whole is particularly great in the Southwest, where summer thunderstorms cause sudden, intense runoff. A powerful downpour could instantly wipe out years of recovery and active restoration efforts.

The damage that began in the 1800s, and which continues to this day, has so changed the land that it should not be called grazing, but mining. Over vast areas of the West, the soil is gone. Viewed in the human scale by which we measure civilizations, such soil has become a non-renewable resource. It has been mined.
—Steve Johnson, *Defenders*, 1985

COW-DAMAGED

IDAHO

Boise National Forest

Your average, run-of-the-mill cow-trashed stream in the West needs to be recognized for the ecological disaster that it is. This murky, shallow little creek and the close-cropped meadow alongside it have all the earmarks of severely degraded riparian habitat. The stream banks are trampled and denuded of vegetation in many places; riparian vegetation that would be expected here (such as willow) is scarce; and drought-tolerant shrubs grow close to the stream margin (as seen on the left side of the photo), indicating a dropped water table. What should be a haven of greenery and biological diversity is instead a pulverized pasture, dominated by one exotic animal.

This is federal land, owned by all Americans. Federal employees oversee its management, and are duty-bound to manage for the good of the land and for the good of all citizens. Yet when public land looks like this, it doesn't seem possible that anybody could be benefiting, except those who own the cattle and pay very little money for the privilege of stripping the land of its vegetation, and its ecological integrity.

The partial information that is available shows that there are tens of thousands of miles of riparian areas in the West, with only a small portion of them in good condition. Poorly managed livestock grazing is the major cause of degraded riparian habitat on federal rangelands.
—U.S. General Accounting Office, 1988

COW-DAMAGED

COLORADO

Dolores River

What does the public get in return for allowing ranchers to graze livestock on the public lands? For every cow-calf pair grazing on an allotment for a month (otherwise known as an animal unit month, or AUM) there is a fee. The federal rate is notoriously underpriced, often eight to ten times lower than fees charged on comparable private grazing land. For many years, the federal grazing fee has been set at $1.35 per AUM—or as others have noted, less than it costs to feed a gerbil for a month. At various times, there have been grazing reform efforts focused on raising the grazing fee, which ranchers have vociferously, and successfully, resisted.

But the bargain-basement price of grazing on public lands is only one aspect of the grazing fee subsidy. Typically, 50 to 63 percent of the fees are returned to local agency districts, to be placed in the (so-called) Range Betterment Fund. Money from this fund goes to such projects as fencing, spring development, herbicide spraying, and other activities designed to make public lands better serve the needs of privately owned livestock. Occasionally, these projects help mitigate for the damaging effects of livestock grazing, though even this is debatable (for example, water developments that encourage livestock to graze in upland areas and away from streams may only be transferring impacts from one part of the watershed to another). The bigger issue is that without livestock grazing on the public land, such mitigation would not be necessary.

COW-DAMAGED

NEVADA

Sheldon National Wildlife Refuge

Until the mid-1990s, Sheldon Refuge was grazed by livestock. The riparian areas suffered greatly. According to historical accounts, the stream shown in these two photos was once bordered by a wet meadow. Old beaver-cut aspen boles can still be found amid the sagebrush by the stream—further testimony to an earlier time when livestock had not yet altered the hydrology of the stream and the plant communities found alongside it. In the early 1990s, when the photo on the left was taken, no beaver, aspen, or other riparian vegetation could be found. There was very little water flowing in the creek. The predominance of sagebrush and the severely incised channel were a result of many decades of livestock grazing and trampling of the stream area.

Since the closure of all grazing allotments on the Sheldon, the condition of many areas has improved substantially. The photo on the right was taken in the summer of 2001, and streamside vegetation is now so thick that the stream isn't even visible. Because the water table has risen, sagebrush on the adjacent stream banks is dying out, and a wet meadow is starting to reclaim the margins of the stream. Today the 571,000-acre Sheldon National Wildlife Refuge is the largest livestock-free area in the entire Great Basin.

COW-DAMAGED

LIVESTOCK-FREE

UTAH

**Wasatch-Cache National Forest
Blacksmith Fork**

In this upper portion of the watershed, riparian vegetation ought to be thick and lush. Instead, nearly all the riparian plants have been eliminated. Note the relict willow shrubs, lonely reminders of what the entire streamside zone should look like. Livestock grazing has gradually caused this former wet meadow to be invaded by sagebrush; removal of overhanging vegetation along the watercourse causes the water temperature to rise more rapidly and also accelerates evaporation.

The Blacksmith Fork is one of the few rivers within the range of the Bonneville cutthroat trout still to contain a small population of the fish. The refugia for the trout lies downstream from this photo in a stretch of river that threads a narrow canyon, and is not subject to livestock grazing and dewatering. Yet despite the Bonneville cutthroat's imperiled status, the watershed for this fish is still subjected to abuse and degradation from livestock production.

Elsewhere, the Bonneville cutthroat trout has been eliminated through hybridization and competition with exotics, along with habitat loss, degradation, and fragmentation. Many of the exotics that have replaced the cutthroat are less sensitive to livestock-induced habitat changes than the native fish. Cutthroat trout require higher water quality than some other trout species—one reason the fish is today largely relegated to tiny headwater streams.

The official destruction committed in the sacred precincts of this massive range would be called vandalism if others had done it. The damage is so vast, so incredible, so awful that it has a permanent effect.
— Former Supreme Court justice William O. Douglas, *My Wilderness: East to Katahdin*, 1961

COW-DAMAGED

NEVADA

**Humboldt National Forest
Snake Range**

A fence protecting a spring has been demolished by cattle, while on the hillside, sagebrush, rather than native grassland, dominates. This is a landscape that has little tolerance for heavy grazing pressure by large herbivores, such as cows. In most of the Intermountain West—for example, the Great Basin, the Palouse grasslands of eastern Washington, the Southwest—and the grasslands of California, there were never large herds of bison, nor were other ungulates, such as elk and deer, typically found in high numbers in these places.

The introduction of cattle transformed many plant communities dramatically. In the Great Basin, for instance, the proportion of sagebrush surged, while the area covered by native grasses shrank to a pittance.

Up there on Gold Creek, Walt Martin used to cut his winter hay by mowing the ridges. The native grass was thick enough that all he had to do was find a smooth place. Prior to the turn of the century, I lived on the Evans Ranch with my uncle. It was wonderful feed country. It was primarily a grass range. All the smoother ridges were covered with bunch grass. There wasn't any sagebrush to speak of, just grass. The creek bottoms from the present highway (Highway 225) to the mountain were continuous narrow meadows. There was no sagebrush in them. The vegetation was mostly redtop and white clover.
    —Sid Tremewan, first supervisor of the Humboldt National Forest, describing years later the Jarbidge Mountains of 1910

COW-DAMAGED

ARIZONA

COW-DAMAGED

**Bureau of Land Management Lands and
Buenos Aires National Wildlife Refuge**

These two images were taken on the same day on the same slope, approximately 50 feet apart. The photo on the left is Bureau of Land Management land grazed by livestock, whereas the area on the right is within the boundaries of the livestock-free Buenos Aires National Wildlife Refuge. Such a juxtaposition offers a rare opportunity to witness the differences between a livestock-dominated landscape and a natural landscape, where livestock influence is minimal or absent.

The grass-covered slope in the refuge has more hiding cover for small mammals and birds, as well as less soil erosion, than the land grazed by livestock. In addition, studies done in Arizona have demonstrated that a vegetative cover reflects heat and acts as a mulch, contributing to higher soil moisture. Loss of vegetation as a consequence of livestock grazing, as seen in the photo on the left, results in higher average soil surface temperatures, which leads to greater water evaporation from the soil and ultimately creates an even harsher, drier, more desertlike regime.

Places such as the Buenos Aires National Wildlife Refuge are still unusual in the livestock-dominated West. The dearth of livestock-free lands, by which we could better measure the full consequences of livestock production, offers one more reason to remove cattle and sheep from the public lands. Such action would not lead to the precise re-creation of the natural communities that existed prior to the advent of Europeans and ranching activity, but it would almost certainly help scientists, and society as a whole, to develop better insight into how native ecosystems function. This understanding has its own inherent value; it also has the potential to show us how we can improve the ways we use and inhabit the magnificent landscapes of the American West.

OREGON

Hart Mountain National Antelope Refuge
Poker Jim Ridge

These bunchgrasses have never been grazed by livestock because the surrounding cliffs and steep terrain prevent access by cows. Some bunchgrasses will live as long as fifty to one hundred years, and where this occurs, can be thought of as "old-growth" grass. The beauty and complexity people associate with old-growth forests exists equally in these ancient grasslands. It is as much a travesty to "clear-cut" these grasses as it is to saw down a stand of thousand-year-old trees.

Livestock were removed from the 250,000-acre Hart Mountain Refuge in the 1990s. It is now the second largest livestock-free zone in the Great Basin, after Sheldon National Wildlife Refuge in Nevada.

An intact, natural grassland is a wonderland of life and beauty. A healthy bunchgrass community may hold anywhere from a few to over twenty species of bunchgrass, a great variety of herbaceous, flowering plants, many brush and cactus species, trees along drainages and perhaps scattered around the landscape, yuccas, a carpet of soil lichen and mosses between the larger plants, even mushrooms—in all, hundreds of species all growing together, along with an amazing variety of animals, as a complex yet harmonic intermingling of life forms.
—Lynn Jacobs, Waste of the West, 1991

LIVESTOCK-FREE

COLORADO

Rocky Mountain National Park

Significant parcels of open, undeveloped land, off-limits to cattle and sheep, are a rarity in the West. Livestock dominate nearly all places that have the potential to be grazed (in other words, that offer at least some minimal amount of forage and are not so steep or rocky as to be inaccessible). With so little land that is livestock-free and still available to native wildlife, the effects of livestock on many animal species are unknown. Few scientists have taken up this subject.

One recent set of studies on water voles in Montana and Wyoming failed to show any viable vole populations where livestock grazing occurred. What would research show about the impacts of livestock on butterflies, hummingbirds, marmots, or a host of other species that are potentially affected by them? The more "obscure" creatures of the West are still awaiting investigation of their basic biology and ecological requirements.

Only sanctuaries such as national parks offer the possibility of discovering how species and ecosystems operate free of the influence of significant habitat-altering activities, such as livestock grazing. They may also offer the only possibility, for some species, that they will survive at all.

LIVESTOCK-FREE

ARIZONA

State Lands, Near Carefree

Solitary cow, left behind after a roundup, searches for grass.

The Eleventh Commandment

Thou shalt inherit the Holy Earth as a faithful steward, conserving its resources and productivity from generation to generation. Thou shalt safeguard thy fields from soil erosion, thy living waters from drying up, thy forests from desolation, and protect thy hills from overgrazing by thy herds, that thy descendants may have abundance forever. If any shall fail in this stewardship of the land, thy fruitful fields shall become sterile stony ground and wasting gullies, and thy descendants shall decrease and live in poverty or perish from off the face of the earth.

—W. C. Lowdermilk, Conquest of the Land for 7000 Years, 1957

COW-DAMAGED

NEW MEXICO

Below Manzano Mountains

Domestic livestock have decimated the natural plant and animal communities of the West, but they should not be blamed for it. Though living beings, they have been highly manipulated and cruelly treated to obtain profits for the livestock industry. They have been bred to be slow, helpless, and lacking in intelligence. Because they are poorly adapted to the dry, rugged conditions of the West, they do damage. But also because of this, they suffer, too. They are essentially used as tools for extracting forage from the land and converting it into a product that ranchers sell.

Some consumers of meat who are interested in a more wholesome, "politically correct" food think that "range-fed" beef is the answer. They believe that cattle out on the open range lead happier, healthier lives than factory farm animals, and are better to eat, too. This might be marginally true. However, what cattle on the open range must endure could hardly be said to be paradise. Subject to injury, predation, poisonous plants, thirst, and starvation, and usually unattended and unsupervised for months at a time, range livestock lead difficult and short lives. In another context, such treatment would constitute animal abuse.

If a rural resident withholds food from his caged pet geese for weeks at a time, he may be subject to county legal proceedings. But if a rancher puts a hundred cattle onto an overgrazed range and half of them starve to death, then the government gives him emergency assistance.

—Lynn Jacobs, *Waste of the West*, 1991

114

COW-DAMAGED

COLORADO

**White River National Forest
and Rocky Mountain National Park**

Aspen groves can be magical oases of shade and relative dampness in a land that is often sharply bright and searingly dry. The leaves quiver in the lightest breeze, and songbirds dart through the foliage. In regions dominated by open rangeland or coniferous forest, aspen offer an alternative, and often crucial, type of habitat for certain wildlife species. The grove on the right, in Rocky Mountain National Park, is off-limits to livestock grazing.

In contrast, the grove on the left, in White River National Forest, has been stripped of its rich understory vegetation by grazing cows. Young aspen, which sprout up from the roots of mature trees, are eaten. Over many years of heavy livestock grazing and the repeated elimination of new trees, the regeneration of the grove may be jeopardized, as individual aspen do not tend to live very long. Many of the habitat advantages that aspen groves offer to wildlife, not to mention their serene and verdant beauty, disappear with livestock grazing.

Under optimal soil and moisture conditions, aspen are often tall and robust, and the understory is a lush tangle of shrubs and herbaceous vegetation. . . . In an aspen community with a long history of use as livestock range, the understory is represented by fewer species and by the competitive dominance of the least palatable species.
—Audrey DeLella Benedict, *A Sierra Club Naturalist's Guide: The Southern Rockies*, 1991

COW-DAMAGED

MONTANA

Mount Haggin Wildlife Management Area

Many state wildlife areas are purchased with fishing and hunting license fees, and are ostensibly dedicated to the management and conservation of hunted animals such as elk. Yet cows often get the best, while native wildlife gets the rest. Typically, 50 to 60 percent of all forage is allotted to livestock on public lands, and often more is taken. Yet every blade of grass going into the belly of a cow is that much less to support a grasshopper that might feed a trout, or an elk that might feed a grizzly or wolf.

Even where cattle do not consume the bulk of the forage, the mere presence of cows can negatively affect wildlife. Studies done on this state game area have demonstrated that elk avoid areas actively grazed by livestock; the elk are socially displaced. The same has been demonstrated for other wildlife species as well. Often, native species are relegated to lands with less forage, less hiding cover, and less water and, overall, have to cope with lower-quality habitat. Native animals can therefore be under greater physical stress and are consequently more vulnerable to predators, inclement weather, or disease. When hunters blame predators or ignorant wildlife biologists for declining populations of "game animals," perhaps they should focus their attention on livestock—roaming lands presumably managed for conservation of wildlife.

Cattle and domestic sheep are getting all the gravy, while deer, pronghorn, bighorns, and other wildlife are left to lick the bowl. . . . It is amazing to me that the American people, including the bulk of this nation's livestock industry, allow relatively few grazing permittees to defile public property and destroy fish and wildlife to such a degree.
—Lonnie Williamson, *Outdoor Life*, 1983

COW-DAMAGED

WYOMING

Highway South of Lander

The highway right-of-way is on the left. Bureau of Land Management lands grazed by livestock are to the right of the fence. Although a fenced roadside leaves some things to be desired as a "control" that one can compare with livestock-grazed land, it nonetheless provides some idea of the impact of cattle and sheep on vegetation. Sadly, in most regions of the West, these are the only places one can see grasses and other plants growing out of the reach of hungry livestock.

Next time you are driving some lonely stretch of highway with nothing better to do, check out the fencelines and the vegetation on either side as you go by. And if there isn't a fence, watch out. In most western states, it is your responsibility to avoid colliding with cattle or other livestock. It is not the rancher's responsibility to fence his animals in. If you hit a free-roaming cow with your car, you—or your estate—have to pay for it.

One friend reports having a flash of understanding when he stood by a fence that separated grazed and ungrazed portions of the same creekbed. One side was lust and verdant. The other side looked like the face of the moon. Moo.
—Donald M. Peters, *Arizona Republic,* 1990

LIVESTOCK-FREE

COW-DAMAGED

IDAHO

Near Howe

Because of climatic conditions, most livestock operations require additional feed produced by irrigated croplands. Yet the growing of hay, alfalfa, and other crops is one of the biggest and least-appreciated impacts associated with livestock production. Most people look at the green alfalfa field shown here and fail to see the ecological disaster it represents. The native vegetation has been plowed up and replaced with a monoculture. The plant grown is an exotic. With the loss of the native plant community most of the native wildlife is also lost, from butterflies to bison.

It takes supplemental water to grow hay and most other forage crops in the arid West, making irrigation for livestock feed production the number one consumptive use of water in every western state. To store this water, thousands of dams have been constructed, fragmenting river ecosystems and changing flow regimes—with disastrous ecological consequences. Ranchers take water not only from larger rivers, but also from creeks, springs, seeps, and underground aquifers. Wetlands and waterways are diminished, or even dried up. Water-dependent species, from fish to frogs, must cope with less habitat, of poorer quality.

Most of the water applied to fields evaporates, or is transpired from the crops into the atmosphere. Thus, most water is lost from the local watershed. What little water eventually returns to rivers and streams is often polluted with fertilizers, cow manure, and excessive sediments and is usually far warmer than it was when drawn out of the stream. Ecologically speaking, growing hay and alfalfa is perhaps the most environmentally damaging aspect of western livestock production. It is also the one thing western livestock producers —if they are to remain economically viable— cannot do without.

COW-DAMAGED

MONTANA

Makoshika State Park

Some livestock advocates suggest that cows merely replace the teeming masses of bison and other wildlife that once lived on the Great Plains. There are, however, vast differences in the way native wildlife and exotic livestock use the landscape, including the greater mobility of native animals versus the sedentary behavior of cattle and, in particular, the way in which cows spend nearly all their time in riparian areas.

Beyond these differences, however, is the issue of whether plant communities on the Great Plains "need" to be grazed to stay healthy—a common belief among livestock proponents. Because Great Plains plants evolved with large herds of grazing herbivores and can tolerate grazing, some people erroneously conclude that these plants require grazing. Most Great Plains plants are able to recover more quickly from an injury—a grazing event—than other plants can, but this ability should not be interpreted as a requirement for continued injury!

When a plant leaf is removed, either by a cow or by some other animal, the plant replaces the photosynthetic "equipment" it has lost by redirecting existing resources to leaf growth. This is usually at the expense of other parts of the plant, such as the roots or reproductive parts (that is, seed production).

Our present camp is a beautiful one. A rich and open plain of luxuriant grass, dotted with buffalo in all directions, a high picturesque hill in front, and a lovely stream of cold mountain water flowing at our feet. On the borders of this stream, as usual, is a dense belt of willows and under the shade of these we sit and work.
—John Kirk Townsend, *Journey Across the Rocky Mountains to the Columbia,* 1835

124

LIVESTOCK-FREE

ARIZONA

Arivaca Creek

Riparian areas are critical to the native wildlife of the West. In more arid regions, they provide habitat for 70 to 80 percent of vertebrate species, while occupying less than 1 percent of the landscape.

These two photos were taken from the same spot at the same time. One view *(right)* looks upstream past a fence to an area currently grazed by livestock; the other *(left)* looks downstream to a stretch of creek that had been protected from livestock for seven years at the time the photo was taken. In the upstream shot, note the absence of young cottonwood and willow; the wide, shallow stream; and the expansive, flood-washed stream bed. In the downstream photo, young cottonwood and willow are abundant, and the stream channel is noticeably narrower. There is much less exposed, rocky ground.

Livestock-induced alterations in stream hydrology have also resulted in diminished cottonwood regeneration. The big, old trees seen along many western streams are relicts. Once they die, the seed source for new generations of trees will be gone, and cottonwood will be permanently eliminated from those watersheds. In fact, this has already happened in some drainages. Of course, it is possible to replant such places, but replanting will be successful over the long term only if livestock grazing is eliminated. Ideally, cottonwood forests will be allowed to replenish themselves, as is occurring in the photo on the left, before it is too late.

COW-DAMAGED

UTAH

**Bureau of Land Management Lands
Adjoining the Nature Conservancy's
Dugout Ranch**

Some 250,000 acres of public land are grazed by cattle belonging to the Dugout Ranch in southern Utah. The 5,167-acre base ranch was purchased by the Nature Conservancy in 1997. The Nature Conservancy believed the private land was vulnerable to subdivision and thus engineered a deal that now keeps the ranch off-limits to development. However, livestock grazing continues on both the private and public land. The Nature Conservancy claims a victory for environmental protection but minimizes the well-documented impacts of livestock production on arid ecosystems.

In "saving" 5,000 acres of land from the possibility of subdivision, the Nature Conservancy perpetuates the sacrifice of fifty times that acreage to the impacts of livestock grazing. The conflicts with ecosystem protection are myriad. For example, note in the photo the nearly complete removal of vegetation. Cattle have removed food and habitat for numerous native species. They have trampled biological soil crusts, which are especially important in the Southwest for nutrient cycling, retention of soil moisture, and protection of soils from excessive erosion. Cattle trampling also significantly increases soil compaction, thereby reducing water infiltration. Water runoff is greater, flooding is more intense, and less water is retained on-site to meet native plant and animal needs.

On other parts of the public land grazed by Dugout Ranch cattle, desert seeps and riparian areas are trampled, and stream channels are broken down. Finally, to maintain livestock in this arid environment, irrigation water is used to produce pasture and hay. The loss of this water through evaporation and transpiration results in significantly less water for downstream public lands.

COW-DAMAGED

ARIZONA

**Gila Box National Riparian Conservation Area
Bonita Creek**

In the 1820s, mountain man James O. Pattie traveled down the Gila Box, which he described as lined with cottonwood. On this stretch of the Gila River, he caught dozens of beaver and saw dozens of bighorn sheep. He even reported grizzly bears living along the river. Today there are no beaver, bighorn, or grizzlies, and the few large cottonwood left are historic relicts that are not being replaced because of livestock grazing.

At the time this photo was taken, the Bureau of Land Management (BLM) was attempting to re-establish cottonwood by planting young trees that they irrigated and individually fenced to protect from cattle. Although livestock grazing was the cause of shrinking cottonwood stands, the BLM was paying for the planting project with money earmarked for wildlife habitat projects, not with range management dollars. In effect, mitigation for damage caused by privately owned livestock was being paid for with taxpayer money—money that could have been going to projects making a positive change for wildlife, not just making up for ongoing mismanagement.

Unfortunately, this is a common occurrence with government agencies overseeing the public lands, and one more little-known way that taxpayers subsidize the livestock industry.

Even though wildlife riparian surveys are far from complete, the Arizona Game and Fish Department now identifies at least 137 species of fish, amphibians, reptiles, birds, and mammals that may face extinction if current habitat trends continue. . . . Grazing abuse is a leading cause of riparian decline.
—Steve Johnson, testifying for Defenders of Wildlife before the U.S. House of Representatives Subcommittee on National Parks and Public Lands, 1987

COW-DAMAGED

MONTANA

Red Rock Lakes National Wildlife Refuge

Cows are pervasive on public lands in the West. Even many wildlife refuges are grazed by domestic livestock, much to the detriment of the wildlife they are supposed to be protecting. Of 109 refuges in the U.S. Fish and Wildlife's Region Six—which includes Montana, Wyoming, Colorado, North Dakota, South Dakota, and Nebraska—103 have some degree of livestock production on them.

A recent General Accounting Office report on national wildlife refuges found that a majority of refuge managers considered livestock grazing to be in serious conflict with their mandate to protect wildlife. Conflictive activities related to accommodating livestock on refuges include—in addition to grazing—predator control, production and harvest of hay, and fire suppression.

Certain wildlife refuges have managed to reduce or even eliminate livestock. Ecological benefits are usually apparent fairly quickly, though some indicators of recovery can take years to develop. At Red Rock Lakes National Wildlife Refuge, a 50 percent reduction in livestock numbers since the mid-1970s has allowed aspen and willow to regenerate in places where they had not sprouted for many decades. Retention of higher, thicker grasses and other vegetation means that winter snows do not blow away as readily; soil moisture is higher, and plant productivity is consequently greater. Populations of various wildlife species have gone up, including ground-nesting birds, hawks, and falcons.

One wonders what additional recovery, and taxpayer savings, would accrue if livestock were completely removed from the refuge.

CALIFORNIA

Giant Sequoia National Monument

In 2000, then-President Bill Clinton established a national monument to assure the long-term survival of some of the last stands of unprotected sequoia trees in the world. Unfortunately, some logging will be allowed to continue within the monument boundaries. Also disappointing, but not as widely publicized, is the fact that livestock grazing also will be allowed to continue, unabated. Battered, muddied streams have no more place in a national monument than sawed tree stumps.

Interestingly, one of the agency justifications for logging in the monument is reduction of fire hazard. In the past, sequoias typically grew in parklike stands, with few understory trees. Fire suppression is commonly blamed for the infiltration of dense-growing, shade-tolerant trees into these previously open forests. Greater fuel density, and the ability of fires now to reach into the sequoia canopy via the understory "ladder" trees, sets the stage for intense blazes in the future that could kill the giant trees. (Sequoias easily survived the frequent, light fires that characterized the natural fire regime in the southern Sierra.)

However, livestock grazing may also be a significant factor in the alteration of forest stands and fire regimes. Scientists have documented that livestock grazing allows the seedlings of coniferous trees to gain a toehold in areas where healthy grasses would otherwise exclude them. The consumption of grasses by livestock also reduces the amount of fine fuels available to carry a fire. Fires become less frequent as a result, even in the absence of effective fire suppression. When fire does occur, it is more likely to be very hot and very big.

COW-DAMAGED

NEVADA

**Stillwater National Wildlife Refuge
and Private Land, Near Fallon**

These two photographs were taken on the same day, at immediately adjacent sites. The photo on the left is of a dry lake bed on the Stillwater Refuge. The photo on the right is of a flood-irrigated alfalfa field. The Carson River, flowing off the east slope of the Sierra Nevada, used to sustain the Stillwater Marsh, which, along with nearby lakes and other wetlands, was a major nesting area and stopping point for water birds on the Pacific Flyway.

In the early twentieth century, construction of the ambitious Newlands Reclamation Project commenced. It was intended to support irrigated agriculture in the desert—primarily crops for livestock fodder. This project not only caused drastic shrinkage of the Stillwater Marsh but is responsible for the disappearance of nearby Winnemucca Lake, which was delisted as a national wildlife refuge in 1962, and for the drying up of wetlands at Fallon National Wildlife Refuge.

What water does reach the Stillwater Refuge is mostly drainage from adjacent farm fields. It is laden with minerals leached out of the irrigated soils that can be poisonous to plants and animals. In part because of the leaching, the dream of turning the desert into lush farmland is itself collapsing. Even with massive subsidies, farmers cannot make their operations profitable.

Legislation passed in 1990 allows the Stillwater Refuge to purchase water rights from willing sellers—a program that has brought some relief to the marsh and its wildlife but which is still far from restoring the wetlands to their former glory. The irony of the Stillwater is that without the ill-conceived federal reclamation project, most farmers and ranchers would never have been able to obtain the water that they are now selling back to the government.

COW-DAMAGED

OREGON

Whitehorse Butte Allotment
Bureau of Land Management Lands
Trout Creek Mountains

Public land leased for livestock grazing in the Trout Creek Mountains area of eastern Oregon has been the setting for environmental conflict for years. Environmentalists have long contended that livestock are destroying riparian areas and aquatic habitat and are stripping the land of vegetation critical for the survival of many native species. A cooperative project between the Bureau of Land Management and grazing permittees in the Trout Creek area was finally initiated in the late 1980s and early 1990s in response to environmentalist pressure. This project included the removal of cows from some areas for several years, and the construction of more than a million dollars' worth of water pipeline, fencing, and other so-called range "improvements." Taxpayers footed the bill.

These cattle belong to the Whitehorse Ranch. In 1999, the Whitehorse ranch manager, quoted below, rebutted accusations that livestock are still overgrazing. In 2001, a decade after the Trout Creek cooperative project was started, and two years after the ranch manager claimed stock growers have gotten "smarter," this photograph was taken. You be the judge.

But a lot of this environmental stuff is bullshit. Every time you turn around, you hear ranchers are ruining everything. I certainly don't think we are ruining anything. The hardest part is having environmental people, with no clue what they are talking about, come out here and tell us everything we're doing is wrong. Granted, there were mistakes made in the past. Certainly, there was overgrazing from time to time. But like any business, you either get smarter or you don't get anywhere.
—Britt Lay, manager of the Whitehorse Ranch, quoted in *High Country News*, 1999

COW-DAMAGED

NEVADA

Great Basin National Park
Lehman Creek

The luxuriant vegetation along this stream creates a scene more reminiscent of the Cascade Mountains in the Pacific Northwest than of a desert range in Nevada. Yet this site—livestock-free for several decades—shows what could be expected, even in an extremely dry region such as Nevada, if livestock were removed.

In most of the rest of Great Basin, however, like other, newer national parks and monuments, livestock grazing was allowed to continue after the park's designation in 1986. For a period of time, cattle were free to foul park campgrounds and trample meadows and streams, whereas a park visitor could be fined for dropping a gum wrapper or picking a few wildflowers. Eventually, the federal government, with the help of the Conservation Fund, bought out most of the permittees, enabling officials to keep management more consistent with the mission of the National Park Service.

One permittee does remain: a California corporation that runs domestic sheep. Their presence threatens the survival of bighorn sheep in the park, as wild sheep are notoriously susceptible to diseases of domestic sheep.

CALIFORNIA

**Inyo National Forest
South Fork Kern Wild and Scenic River
Golden Trout Wilderness**

More than 150 years of livestock grazing along the South Fork of the Kern River has resulted in broken-down stream banks and the creation of a shallow, wide channel that provides little habitat for coldwater fish such as the Volcano Creek golden trout. The trout is a candidate for listing under the Endangered Species Act and is the state fish of California.

Until a ten-year rest period was ordered by the Forest Service in 2001, Anheuser-Busch Corporation (a well-known beer purveyor) grazed over 2,500 cattle on the Templeton and Whitney Allotments within the Golden Trout Wilderness. Years of citizen activism and the threatened listing of the golden trout spurred the government's decision.

According to the Forest Service's own reporting, livestock grazing had damaged over 80 percent of the Kern Plateau with gully erosion and had reduced habitat for the golden trout, the mountain yellow-legged frog (proposed for federal protection), and the willow flycatcher (a state-listed endangered species). Livestock grazing and associated structures, such as fences, were in conflict with the ecological, recreational, and scenic values of the wilderness area.

Though resting the land from livestock for a decade is a step in the right direction, given the ecological and public values at stake, the Forest Service should close the allotment permanently. In ten years, it can be hoped the agency will have the political will to do so.

COW-DAMAGED

CALIFORNIA

Yosemite National Park
Tuolumne Meadows

In his day, John Muir—founder of the Sierra Club and visionary conservationist—saw many mountain meadows ravaged by livestock. He also witnessed predators poisoned and shot, and hillsides being turned to dust by hordes of sheep and cattle. Muir considered livestock grazing the greatest threat to natural landscapes. He became an ardent national park proponent, in part because park establishment ensured the removal of livestock. (Ironically, despite greater scientific understanding of the harm done by livestock, today many newer national parks and monuments permit livestock grazing to continue.)

It is interesting to compare this meadow in Yosemite, which has not been grazed by livestock for a century, with the preceding photo of the South Fork Kern River. Although the two meadows are at approximately the same elevation, both in the Sierra Nevada Range, and both influenced by the same basic geology, the stream in Yosemite is narrow and deep, and bordered by thick grasses and other greenery. Along the South Fork of the Kern, bare gravel dominates a wide, washed-out river channel.

The world, we are told, was made for man. A presumption that is totally unsupported by the facts. There is a very numerous class of men who are cast into painful fits of astonishment whenever they find anything, living or dead, in all God's universe, which they cannot eat or render in some way what they call useful to themselves.
—John Muir, "Man's Place in the Universe," 1867

144

LIVESTOCK-FREE

NEVADA

**Humboldt National Forest
Jarbidge Wilderness**

While logging, mining, road building, permanent structures, motorized vehicles, and bicycles are all banned in federally designated wilderness, livestock grazing is allowed. Livestock grazing is arguably as destructive and unnatural as logging and mining and certainly has greater overall impact on the local ecology than off-road vehicles —as unappealing as ORVs are to many wilderness devotees. Fences and stock tanks, too, are regular features of many wilderness areas.

The 1964 Wilderness Act states that livestock grazing may continue in designated wilderness areas. This loophole was inserted into the bill at the insistence of a few powerful western congressional members—who also happened to be ranchers. The compromise has haunted wilderness lands and wilderness activists ever since. In all subsequent wilderness legislation, livestock have been permitted to stay. The Wilderness Act does state that livestock grazing can be banned if it compromises wilderness values, but this provision is not very specific and is rarely invoked, either by conservationists or by land management agencies.

"Wilderness is a resource that can shrink but not grow," Aldo Leopold once remarked, and went on to observe that it takes intellectual humility to understand the cultural value of nature unaltered and unimproved. Nobody ever accused a government agency of intellect or humility (or, for that matter, the capacity to manage land), but we have reached a point in our historical development when stale jokes about the "Forest Circus" and the "Bureau of Livestock and Mining" and the principles of "multiple abuse" and "sustained greed" no longer serve to mask bemusement with amusement.
—Page Stegner, *Outposts of Eden*, 1989

COW-DAMAGED

NEW MEXICO

**Gila National Forest
Near Mule Creek**

The gully stops right at the fenceline, where livestock grazing also stops. A coincidence?

Livestock production is a major cause of accelerated soil erosion in the West. And among all the consequences of growing cattle and sheep in an arid land, soil loss is likely the most devastating, long-lasting impact. Topsoil cannot be effectively replaced in anything less than a geological time frame: the development of one inch of soil takes a thousand years or more. Yet under conditions of severe overgrazing, an inch of soil can erode away in a few years.

Soil is the basis for plant life, and thus the foundation for the entire biotic community. Not only has the livestock industry mistreated the land for over a century, it has squandered the fertility, productivity, and biological diversity of the western lands for many, many centuries into the future.

This nation and civilization is founded upon nine inches of topsoil and when that is gone there will no longer be any nation or any civilization.
—Dr. Hugh Bennett,
U.S. Soil Conservation Service

148

COW-DAMAGED

NEW MEXICO

North of Grants

A fenced highway right-of-way (left of fence) contrasts with Bureau of Land Management lands that are grazed by livestock (right of fence). Some livestock proponents claim that "over-rest" leads to grassland deterioration and eventually greater and greater areas of bare soil.

This claim can easily be refuted by observing the thousands of miles of thick grass growing along fenced rural roads and highways of the West. Many of these roadsides are unmowed, unwatered, and untended, yet still sustain beautiful, healthy native grasses and other vegetation.

On the other side of the fences, where livestock are doing their job of "improving" the vigor of the plant life, bare soil is abundant.

Resolved, that none of us know, or care to know, anything about grasses, native or otherwise, outside the fact that for the present there are lots of them, the best on record, and we are getting the most out of them while they last.
— Unanimous declaration, passed at a West Texas cattlemen's meeting, 1895

LIVESTOCK-FREE

150

COW-DAMAGED

WYOMING

**Yellowstone National Park
Lamar Valley**

Yellowstone's northern range is grazed, but not by domestic livestock. Native ungulates—tens of thousands of elk, thousands of bison, and hundreds of pronghorn, deer, moose, and bighorn sheep—along with other herbivores such as grasshoppers, ground squirrels, and marmots, all feed on the lush lower-elevation meadows.

But wildlife use the landscape differently from livestock. Most ungulates, particularly elk and bison, graze Yellowstone's lower, northern range predominantly in the winter, when plants are dormant, riverbanks are hidden under snow, and soil is frozen. Plants are deprived of the dead leaf material of the previous summer but are not stripped of actively growing parts. Opportunities to produce flowers or seeds, or to put valuable energy reserves into their roots, are not lost.

Impacts to riparian areas are minimized, too, because they are protected by layers of snow and ice, and frozen soils are not easily displaced. In the summertime, when both plants and stream channels are more susceptible to animal impacts, grazing by large herbivores is minimal in the valleys and foothills. The hooved mammals of Yellowstone tend to migrate to higher elevations as they follow the spring and early summer green-up. They disperse over a much wider area than they occupy in winter and thereby reduce their collective effect on any given site.

LIVESTOCK-FREE

NEVADA

Bureau of Land Management Lands

This cow is defecating into a stream. Cattle excrete between 29 and 70 pounds of feces per day, and between 30 and 49 pounds of urine. Much of this ends up in surface waters. Cattle tend to congregate around streams and other water bodies. Studies have shown that when cattle are brought into an area, fecal coliform concentrations go up tremendously: livestock use can increase stream bacteria counts as much as 1,000 percent. Among the organisms found in cattle waste that potentially can cause human disease are *Cryptosporidium*, *Giardia*, and *Listeria*.

Livestock production is the number one source of nonpoint water pollution in the West, not only contributing bodily wastes but also accelerating erosion and stirring up sediments on stream and lake beds. Although land management agencies are legally obligated to protect and restore water quality, few do even occasional monitoring of rangelands to determine current and changing levels of pollution and contamination. In addition to being a potential threat to human health, of course, degraded water supplies place wildlife in jeopardy. Exposure to disease-causing organisms and higher turbidity are just two water quality problems posed by the livestock industry to native species.

I can imagine the splendor of a not-so-distant past when Westerners could drink from streams without fear of Giardia *and other water-borne illnesses. But now, whether it be drinking from an alpine lake, a Rocky Mountain waterfall, or a canyon creek, that opportunity has been lost due to indiscriminate cattle grazing.*
—Ken Rait, *High Country News*, 1990

154

COW-DAMAGED

OREGON

Sycan Marsh Preserve

On the dry east side of the Cascade Range, Sycan Marsh is an ecologically significant wetland that provides habitat to endangered bull trout, redband trout, and unique populations of freshwater mollusks—including a species of land snail new to science. The Nature Conservancy has acquired more than 30,000 acres in and around Sycan Marsh—its largest preserve in Oregon. Although the Nature Conservancy states that it is focusing on ecological restoration of the marsh and surrounding uplands, and has restored small parcels of the marsh, it has allowed ranching operations to continue. (The land is leased to the ZX Ranch, owned by J. R. Simplot—considered the richest person in Idaho and holder of grazing permits to more than 2 million acres of public land.)

Although some parts of Sycan Marsh may be in better shape than nearby lands, what this photo shows is that the preserve is maintained at far below its biological potential because of the presence of livestock. Cattle have mowed the vegetation down to stubble only a few inches in height. Habitat for small mammals, snakes, and ground-nesting birds has been radically transformed. Cattle have certain unavoidable impacts on soils and streams, whether they graze on commercial ranches, public land, or private "preserves." Their heavy hooves compact soils and break apart stream banks. Their excrement pollutes the water. They remove forage that could otherwise be feeding native species.

Scientific justifications for livestock grazing simply do not hold up under close scrutiny. The Nature Conservancy may have political or economic reasons for maintaining ranching on its preserves. However, the organization does a great disservice to citizens and scientists working for the protection and integrity of western ecosystems when it portrays livestock grazing as essential to land health, or beneficial.

COW-DAMAGED

ARIZONA

Tonto National Forest
Dutchwoman Butte

This 100-acre mesa top in central Arizona is surrounded by steep cliffs and has never been grazed by livestock. Recent studies show that the soils on the butte are very fertile, with abundant organic matter, stable soil aggregates, and low bulk density. The soils also have high water infiltration rates, making them excellent for root growth.

In a 2000 paper published in *Rangelands*, Forest Service researchers wrote that the most "striking aspect of Dutchwoman Butte is the diversity, density, and vigor of the grasses." Twelve species of grass were found on the butte, compared with only two on a typical grazed area of a nearby livestock allotment. On the butte, grasses covered 35 to 40 percent of the ground, with most of them reaching knee to waist high; but on the area grazed by livestock, grass cover was only 16 percent.

Curly mesquite, a plant species viewed as "undesirable" for forage, and therefore indicative of overgrazing when abundant, was rare on the butte (5 percent cover). However, it took up 90 percent of the vegetative canopy in nearby areas grazed by livestock. Another "undesirable" forage species, snakeweed, was common on livestock-grazed lands below the butte but was sparse on top.

Although the study occurred during a severe drought, the authors found that "Dutchwoman Butte was producing about four times as much forage as similar areas with a long-term history of [livestock] grazing."

LIVESTOCK-FREE

CALIFORNIA

Carrizo Plain National Monument

In 1886, a visitor to Carrizo Plain wrote: "In the spring, native bunch grasses, reaching as tall as the side of a horse, grew thick on the undulating land, turning to naturally cured hay in the summer. Wild horses, elk, deer, and antelope were abundant in the plain and large flocks of sandhill cranes spent each winter at Soda Lake."

Today, the Carrizo Plain is dominated by exotic grasses and weeds, as are most of California's grasslands. Livestock grazing and farming are the leading causes of the disappearance of the state's native grasses. Although Carrizo Plain's relative geographic isolation and aridity spared it the intensive agricultural development characteristic of the Central Valley just to the east, it was agriculture nonetheless—primarily ranching—that put the plain's diverse and once-abundant flora and fauna at risk.

Most of the 250,000 acres of the recently declared Carrizo Plain National Monument are still grazed by livestock. And though Carrizo Plain is sometimes called "California's Serengeti," parts of it are sadly more reminiscent of sub-Saharan Africa.

COW-DAMAGED

A CENTURY OF
PUBLIC LANDS

TRASHING

Ecological Impacts of Livestock Production in Arid Western Landscapes

PART IV

At the heart of this book lies the heart of our concerns: the land, the rivers, and the wild things that inhabit them. Here are the species for which public lands grazing policy may truly be a matter of life or death. Here are the rivers drying up, the water tables dropping, the soils washing away. Three hundred million acres of public land are at stake in the public lands livestock grazing debate. In this section, you will see and read in detail about what is to be found on these 300 million acres, and what has vanished from them.

The following essays attempt to redress the ways in which the harm done by livestock in the West have been obscured. Livestock-induced change tends to be visually subtle (at least, in comparison with other types of environmental alteration, such as clearcut logging or urban sprawl) and incremental. Throughout the arid western United States, livestock production is nearly ubiquitous (with the exception of a few protected areas and places extremely inhospitable) and has been going on for a century or more. Even areas ungrazed by livestock are affected because they have become islands in a sea of livestock production as well as other types of human use. They are subject to the problems typical of fragmented habitats, such as edge effect and genetic isolation. Altogether, the prevalence of livestock production means that most people accept the present look of the land as the norm and regard the diminished populations of many of the West's native species as "the way things have always been."

This set of essays on the ecological impacts of livestock in the West is led by a discussion on the nature of science itself—what science can tell us, and what it cannot. We feel it is important to acknowledge the limitations of science, as well as to harness its strengths, because ultimately policy change and restoration of the public lands will depend on much more than science alone. But before action must come understanding, and before understanding, a witnessing of the facts. And so, read on. The facts are here.

162–163: Cattle-battered Bureau of Land Management lands, eastern Oregon.

Opposite: This satellite photo shows Arizona's Buenos Aires National Wildlife Refuge *(top),* which is fenced off from livestock, and adjacent Bureau of Land Management lands grazed by livestock *(bottom).* (The grass-dominated refuge is light-colored because of the abundance of dry stems of the previous season's growth. The BLM land is dominated by shrubs, and therefore is dark in color.) Although the contrast between areas grazed and ungrazed by livestock is seen on a landscape level here, such opportunities are rare throughout the West, thanks to livestock's near ubiquity. When studying the effects of livestock grazing, most researchers utilize very small exclosures (an acre or less) as their "controls." This practice is far from ideal for scientific purposes, as the small exclosures are subject to many external influences from the surrounding livestock-grazed land. Nonetheless, research shows, among other things, that where abundant grass cover is retained—as it is on the Buenos Aires Refuge— soils are less subject to erosion and provide better growing conditions.

THE SCIENCE OF NATURE AND THE NATURE OF SCIENCE
Questions of Fact and Value

Mollie Matteson

Scientific research on physical and biological systems offers credible information about how
the natural world works. However, researcher assumptions and study parameters and controls must
be evaluated carefully, as these will affect research results. In using science to aid policymaking,
as in the case of public lands livestock grazing, we must keep in mind that even good science
is not free of certain biases, and that, ultimately, human values—what people desire
and care about—provide the guidelines for decision making.

Mollie Matteson *is a writer, editor, and environmental activist in Eugene, Oregon. She obtained a master's of science degree in wildlife biology from the University of Montana, studying a recovering population of gray wolves. Her work has included advocacy for wolf recovery and wildlands protection and restoration, and against predator control.*

Sawtooth National Recreation Area, Idaho. Small, fenced exclosures are occasionally used by range managers and scientists to examine the effects of livestock grazing on plant communities. Unfortunately, many exclosures are not maintained over long periods of time, which negates much of their value as no-livestock controls.

To ask, What are the ecological impacts of livestock grazing in the arid West? is to plunge, metaphorically, into a thicket of diverse, sometimes contradictory and confusing, and often recondite research findings in disciplines as various as plant ecology, mammalogy, ornithology, soil science, geomorphology, hydrology, and even invertebrate biology and microbiology. The scientists who study grazed ecosystems are far from pronouncing any consensus on the above question, in part because few have actually posed the question (at least in a public forum), and in any event, few forums exist to bring together researchers laboring in such seemingly unrelated fields. Also, such a broad query demands a willingness to synthesize and even generalize— something most scientists are loath to do. The nature of scientific investigation is usually narrow, limited in geographic purview and time, as well as in most other parameters. The majority of researchers will claim expertise only in their given specialty, reasonably so, and rarely attempt to extend their studies, let alone their conclusions, to areas beyond those with which they are very familiar.

Yet, although the credibility of scientific knowledge clearly depends on the assiduous avoidance of overgeneralization, just as important is the fact that the usefulness of scientific knowledge depends on the extent to which wider inferences can be made from specific results.

The essays that follow attempt to summarize the best scientific knowledge to date, in a variety of fields, on what livestock and the livestock industry do to the land, water, wild animals, and plants of the American West. The scientists who wrote these essays are experts in their chosen disciplines, and not necessarily prepared, or willing, to speak to the overall effect of livestock production on arid land ecosystems. Yet, it is exactly this broad problem that all those who care about the ecological health of these native systems must confront. Not for grasshopper sparrows, gray wolves, or bluebunch wheatgrass alone should the fate of public lands livestock grazing be debated. Not on the basis of a certain degree of stream sinuosity, level of late-season water flow, or bacterial threshold in a river, but on the basis of all these things together (and much more) must the question of what livestock do, and what should be done about it, be pondered.

The scientific evidence presented here is solid, gathered together by thoughtful, committed investigators who know the power of the scientific method to uncover truths about the world and the way it works. However, it should be stated honestly and up front that science, as a human endeavor, is not value-free. Complete objectivity simply is not possible. Even in the act of asking a particular question, the scientist states by implication that the subject is worthy of contemplation (and that others are perhaps less so). Thus, the scientists who contributed to this book come to their topics with certain perspectives, and one that is ubiquitous among them is that livestock grazing is a variable, not a given.

Western ecosystems in the not-so-distant past were without domestic livestock, and a worthy project is the comparison of lands where livestock grazing is ongoing versus those places where it never occurred, or where cows and sheep are now absent. This viewpoint guides the following discussions, whereas, in contrast, much of the range science and range management literature examines ways to mitigate for livestock grazing, compares varying levels of grazing intensity, or looks for ways to improve the productivity of the land for the purpose of raising livestock. In general, it can be said that range scientists, and the discipline within which they work, assume the presence of livestock. Although helpful ecological data still come out of their research, it is highly problematic to make conclusions about the larger implications of livestock grazing for native species and natural processes when "no grazing" controls are absent.

Unfortunately, many studies of the impacts of livestock do not include adequate comparisons between areas ungrazed and grazed by livestock. Even when comparisons are made, ungrazed areas are often small, isolated exclosures, or they may have been protected from livestock grazing for only a short time. Such controls are far from ideal, as a multiplicity of ecological processes may still be absent or out of whack, and thus the "controls" are not truly representative of complete ecosystems uninfluenced by the effects of livestock production.

Nonetheless, in the first of the following essays, researcher Allison Jones manages to find studies of adequate quality and quantity, comparing ungrazed and grazed sites, to begin assessing livestock impacts across a diversity of western landscapes and in a wide range of categories. Her meta-analysis tallies data on differences between ungrazed and grazed sites with regard to, for example, rodent diversity, soil bulk density, and tree seedling survival. This attempt at synthesis has its limitations but also offers the opportunity to gain a greater perspective on how livestock affect larger ecological systems in terms of not just one or two parameters.

The remaining essays are windows into a variety of scientific disciplines and what researchers have to say about the impact of livestock on the particular species, taxonomic group, or ecological or physical system they study. Typically, debates over livestock center on whether there is overgrazing (how plants are affected) and whether damage is being done to stream morphology and thus stream-dependent species. In addition to these topics, you will find essays on the effects of livestock on sage grouse, prairie dogs, terrestrial snails, grizzly bears, and cryptogamic crusts, as well as on exotic weeds, endangered fish, native herbivores, birds, and water quality.

Despite the wealth of evidence we present here—supporting our contention that livestock grazing is highly destructive to ecosystems of the arid West, is not amenable to mitigation, and therefore is not appropriate to continue on America's *public* lands, at the very least—some will dispute our science. Some opponents will find data that appear to contradict the conclusions put forth in this work. They may take exception to a particular assertion, for a particular case, and they may see such exceptions as invalidating the totality of our argument.

We have three responses. First, we ask skeptics to go back and evaluate their own science in light of the above critique. Any good science should have reasonable controls and should be conducted at a scale (time, area, distance, and so forth) that is likely to give reliable results. Where research assumes the presence of livestock, rather than seeing it as a variable, we believe the parameters of the investigation are inadequate.

Second, we assert that on the public lands, public values—ecological integrity, biological diversity, environmental services, recreational and spiritual pursuits—should take precedence over narrow, commercial agendas. Given that scientific knowledge is never complete, and that there will always be a degree of uncertainty regarding the impacts of human use, we strongly urge the precautionary principle as the first and greatest maxim for public land use policy. Land managers should always err on the side of protection, perhaps even "too much" protection, of public values, rather than risking too little protection, as is usually the case. (Imagine how refreshing it would be to have industry representatives complaining that government officials were keeping our public waters "too clean," wildlife populations "too abundant" and "too healthy," and our deserts and mountains "too beautiful.")

Finally, science cannot give us all the answers. In fact, the most difficult questions, the most persistent problems, are not ultimately matters of science, but of values. What humans hold in their hearts, what they dream of and defend most passionately—this is the raw, vital, and often troublesome material out of which society must fashion its rules, priorities, and direction for the future. Knowledge can lead to caring, so the science of nature does provide ways to deepen our understanding of and regard for more and more elements of the natural world. We have been pleased in this book to offer essays not just on the charismatic fauna—bears, elk, birds, and the like—but also on the small, the obscure, and even the largely invisible: native frogs and mollusks, water chemistry and bacterial counts, stream channel change over time, microbes on the surface of soil.

We hope that in learning and understanding, you will come to care, to see that all these things are part of something larger. The native ecological systems in the West, in their intricacy, diversity, and complexity, are beautiful and awe inspiring. We believe that many elements of these systems, and even entire ecosystems themselves, are in danger of disappearing or disintegrating, perhaps permanently. Science can tell us this is happening. It cannot tell us to care. It cannot make us do anything about it. Rather, we must go back to nature, something in our own, human nature, for the proper guidance. Within—call it conscience, call it faith, or love, or sense of duty—that is where the answers lie.

It is that simple, and that daunting.

A botanist keys out plants on Steens Mountain, eastern Oregon.

SURVEYING THE WEST
A Summary of Research on Livestock Impacts

Allison Jones

The results of a quantitative analysis of the scientific literature strongly suggest that
livestock grazing has detrimental effects on arid western ecosystems. In particular, the analysis
shows that soil infiltration rates, rodent species diversity, amount of vegetative cover,
and vegetation biomass were negatively affected on sites grazed by livestock,
as compared with sites not grazed by livestock.

Allison Jones *received a master's of science degree in conservation biology
from the University of Nevada, Reno, where she studied the effects of cattle
grazing on small mammals in the Great Basin. She is currently staff
conservation biologist for the Wild Utah Project in Salt Lake City and is
analyzing grazing impacts in southern Utah.*

In the debate regarding the effects of livestock on the arid lands of the West, both critics and advocates of grazing often turn to the scientific literature to bolster their arguments. Yet most studies are done on a very site-specific basis, and few are designed to give insight into broader questions, the kind crucial to discussions of land management policy and the public interest: What are the general, ecological impacts of livestock grazing on arid western lands? Are trends negative, positive, or neutral overall?

To begin to address these questions, I undertook a meta-analysis of the existing scientific literature. This synthetic approach, as I describe below, produced new and stronger support for the assertion that cattle grazing in the West consistently leads to significant adverse ecological effects.

I preferred to make a quantitative, rather than qualitative, examination of the grazing literature to bolster the objectivity of the findings. Qualitative assessments are subject to accusations of bias, both in the selection of studies to include in the survey and in the interpretations of results. My analysis, however, drew on the pool of all research articles I could locate on the effects of livestock grazing on arid lands in the American West between 1945 and 1996.

I established a number of screening criteria for studies to include in my analysis. First, I rejected studies that included grazers other than cattle. In addition, I included only studies that simultaneously compared grazed areas with nearby ungrazed controls. These criteria eliminated all investigations that compared only different intensities or levels of grazing (that is, heavy versus light grazing) and studies that made comparisons of the same site before and after livestock grazing. I took this approach because studies that only compare different levels of grazing are less likely to detect any grazing effects.

Additionally, I included only work conducted in arid environments of the western United States and with site descriptions that included xeric vegetation types (that is, plants adapted to dry conditions). Most studies used were conducted west of the Rocky Mountains, but a few occurred in arid shrub/grasslands of the western Great Plains or the Southwest. Study sites ranged from alpine to desert ecosystems; vegetation types ranged from forest ecosystems to grasslands.

Coyote pups.

To achieve pooled sample sizes large enough to analyze, I needed to lump data from seemingly disparate study areas. Similarly, to achieve pooled categories of sufficient size, I needed to lump studies that used different systems of grazing (for example, a deferred-rotation system, or a spring, summer, or winter grazing regime). Although pooling data from such disparate study sites may seem to gloss over important ecosystem differences, the diversity of data sources actually strengthens this analysis. That is, differences between grazed and ungrazed sites that prove consistent over many kinds of habitats and landscapes stand as powerful evidence of the impact of livestock grazing across the arid West.

In reviewing the research, I analyzed sixteen grazing effect categories separately: rodent species diversity (total numbers of species); rodent species richness (relative abundances of species); vegetation diversity (shrubs, grasses, and forbs); total vegetation cover; shrub cover; grass cover; forb cover; total vegetation biomass; seedling survival of trees; seedling survival of nontrees (shrubs, grasses, and forbs); cryptogamic crust cover; litter cover; litter biomass; soil bulk density; infiltration rate of water into soil; and soil erosion.

Initially, I found 112 studies on the effects of grazing on fauna, flora, and soil properties, but after applying the screening criteria described above, I used only 54 of these studies. Several papers included data that were useful for more than a single analysis, such as articles documenting grazing effects on both vegetation and animals. Some papers also contributed two or more observations to a given analysis, such as studies with multiple data points for vegetation cover. I therefore ended up analyzing 196 different sets of grazed/ungrazed comparisons, grouped into the sixteen different "grazing effect categories" listed above.

In all analyses, I tested the hypothesis that grazing has no effect on the measured variables with a t-test—a form of statistical analysis that tests for obvious differences between two sets of data (see Table 1). To do so, I tested for significant positive or negative deviations from zero in ungrazed/grazed paired comparisons. For example, grazing would generally be considered detrimental if it significantly reduces plant or animal diversity; if it reduces cover or biomass of plant litter or cryptogamic crusts; if it reduces seedling survival; or if it reduces infiltration rates of water into soil. However, for two of the variables analyzed—soil bulk density and erosion—grazing-induced *increases* would instead be considered detrimental effects.

Table 1 shows, for each of the sixteen response variables analyzed, the percentage of total observations showing detrimental effects with livestock grazing, the identity of each paper included in the analysis, and the number of data points utilized per paper. A finding is described as statistically significant when it can be demonstrated that the probability of obtaining such a difference by chance alone is very low. The "P value" is the statistical tool used to gauge the likely significance of the test results. In the case of this analysis, P values less than 0.05 indicate 95 percent confidence that the t-tests indicate real effects of grazing for that response variable. Eleven of the sixteen analyses (69 percent) revealed significant detrimental effects of livestock grazing on arid rangelands (see bolded percentage figures in column 2 of the table). In addition, many of the other five analyses were quite close to being significant (that is, the largest P value for any category was 0.111).

On its surface, this meta-analysis of data gleaned from the literature firmly suggests that livestock grazing has detrimental effects on North American arid

Table 1. Results of Tests for Detrimental Effects of Livestock Grazing on Arid Ecosystems

Category	Percentage of Total Observations Showing Detrimental Effects with Grazing[a]	Literature Sources[b]
Rodent species diversity	**87%**	3, 18 *(7)*, 19, 24, 30, 31, 34 *(2)*, 46
Rodent species richness	**59%**	3, 16 *(3)*, 18 *(7)*, 24, 30, 31, 34 *(2)*, 46
Vegetation diversity (shrubs, grasses, forbs)	47%	3, 4, 11, 15, 20, 22, 30, 31, 39 *(3)*, 41, 46, 50, 51
Shrub cover (%)	**56%**	3, 4, 7, 11, 15, 20, 21, 22, 30, 31, 39 *(2)*, 41, 46, 47 *(2)*, 50, 51
Grass cover (%)	**71%**	3, 4, 7, 11, 15, 20, 22, 30, 31, 32, 39 *(3)*, 41, 46, 50, 51
Forb cover (%)	53%	3, 4, 7, 11, 12, 20, 22, 30, 31, 32, 39 *(3)*, 41, 46, 50, 51
Total vegetation cover (%)	50%	1, 4, 6, 7, 11, 20, 21, 23, 27, 39 *(3)*, 41, 44, 45, 50
Total vegetation biomass (kg/ha)	**91%**	8, 14, 16 *(2)*, 25, 32 *(2)*, 38, 39 *(3)*
Seedling survival, trees (%)	75%	10, 17, 26, 28 *(4)*, 49
Seedling survival, nontrees (%)	**100%**	33 *(2)*, 36, 38
Cryptogamic crust cover (%)	**83%**	1, 2, 7, 21, 23, 41
Litter cover (%)	50%	2, 7, 12, 14, 23, 35, 39 *(3)*, 41, 47 *(2)*
Litter biomass (kg/ha)	**86%**	5, 12, 16 *(2)*, 27, 42, 43
Soil bulk density (g/cm³)	**78%**	8, 27, 29 *(2)*, 32 *(2)*, 35, 43, 52
Soil water infiltration rate (cm/hr)	**80%**	5 *(2)*, 8, 9, 13, 14, 27, 32 *(2)*, 37 *(2)*, 40, 43, 52, 53
Soil erosion (kg/ha)	**100%**	13, 14, 32 *(2)*, 37 *(2)*, 40, 52, 54

Note: Statistical tests employed were paired-comparison t-tests on actual data for all categories, except for seedling survival (nontrees), litter cover, litter biomass, and soil erosion, for which paired-comparison t-tests were done on data standardized by ungrazed means, and soil/water infiltration rate, for which a Wilcoxon matched-pairs signed ranks test was conducted.

[a] Percentages in column 2 illustrate the percentage of observations (usually individual studies) in the particular category showing a detrimental effect in the dependent variable due to grazing. A detrimental effect for all variables but soil bulk density and soil erosion is defined as a *decrease* in values; for soil bulk density and soil erosion, a detrimental effect is defined as an *increase* in values. The boldfaced numbers represent statistically significant results (P <0.05).

[b] Numbers within parentheses indicate numbers of observations or data points utilized per literature source; no parentheses are present for sources from which one only observation was utilized. Otherwise, numbers refer to literature sources as listed in the endnotes for this essay.

ecosystems. Because the data are drawn from various studies conducted at different times and in different environments, these effects are applicable to North American xeric systems in general, rather than being peculiar to specific locations and/or study periods. On closer inspection of the results, it appears that various features of arid land soils are highly sensitive to cattle grazing, with statistical evidence for an effect of grazing on soil, erosion, infiltration, and cryptogamic crusts.

The analysis also revealed that rodents react negatively to grazing. Effects of cattle on rodents are probably manifested indirectly through effects on soils and/or vegetation. For example, some desert rodent species are specialized at foraging for seeds in certain soils and thus prefer particular soil properties. Grazing-induced changes in physical properties of soils could thus lead to the loss of such specialized species or their replacement by species more suited to the new soil conditions. Similarly, reduction in organic litter due to grazing may explain the loss of some species; western harvest mice, for example, exhibit a strong affinity for grass litter. Moreover, the analyses indicated that grazing in these arid ecosystems reduces total vegetation biomass as well as shrub and grass cover. Reductions in grass, shrub, and/or total vegetation cover over time can have profound effects on desert rodent densities and species composition. Thus, it is quite possible that reduced vegetation cover in grazed areas drives the responses of the local rodent community.

The responses of vegetation itself to grazing were more variable, with four of the eight vegetation response variables testing significantly. Because many studies included in the analyses provided data only for vegetation categories, such as "shrubs," "forbs," or "grasses," the analyses were necessarily limited to broad designations. Although forb cover and vegetation diversity were statistically similar between grazed and ungrazed areas, much of this apparent lack of response to grazing may simply be an artifact of lumping plant species into broad vegetation categories. For example, no grazing effect on forbs may appear if a depletion of palatable forb species is compensated by an increase in unpalatable forb species or grazing-tolerant exotic weeds. Similarly, the vegetation diversity category would have been more useful in this analysis if it had been possible to include all of the grazing studies that reported vegetation diversity in terms of numbers of native and nonnative actual, separate species.

The conclusion that North American arid systems are highly sensitive to livestock grazing is perhaps unsurprising. Whereas large herbivores that might be considered ecological counterparts to domestic livestock are native to many other arid regions of the world, there is a paucity of large native grazers in contemporary North American dry environments. American bison occurred very rarely in the arid West. A worldwide review of the effects of grazing by large herbivores concluded that an evolutionary history involving grazing animals and the local environment was the most important factor in determining negative impacts of grazing on productivity, or the amount of photosynthesis that occurs in a plant community.[1] North American arid lands lack such an evolutionary history. Until Europeans introduced cattle and other grazers to our arid West, it was relatively free of large grazing mammals for 10,000 years.

Certainly, teasing apart effects of herbivory by native species versus livestock is a concern to scientists. However, it is notable that native grazers, such as jackrabbits, and native browsers, such as mule deer and pronghorn antelope, are usually, by design, able to access the interior of grazing exclosures (such as those used in the studies compiled in this review). Hence, the absence of grazing and browsing by native herbivores is usually not a confounding factor in assessments of cattle grazing effects.

When biologists are faced with an abundance of very disparate studies or individual studies that yield no significant effects (as is often found in the grazing literature), meta-analysis allows for the detection of broad patterns by determining whether there are trends consistent among studies. I used meta-analysis to glean more objective information from the grazing literature than had been ferreted out in the past. It seems that certain soil-related and vegetative cover variables are most sensitive to grazing in arid systems. These findings may prove useful to managers, who traditionally have only used one or two metrics to assess rangeland health. Perhaps investigation of a whole suite of connected variables, such as cryptogamic crust cover, soil infiltration rates, and litter cover, will give managers a more complete picture of ecosystem integrity in grazed landscapes.

It is imperative that conservation biologists work closely with range scientists and land managers. Livestock grazing is the most widespread land management practice in western North America. Seventy percent of the western United States is grazed, including wildlife refuges, wilderness areas, and parts of our national park system. There remains much to study and understand about the influence of livestock grazing on arid ecosystems. Quantitative reviews can help provide more insight as to the ecological effects of grazing on various regions or ecosystems. Additionally, conservation biologists could contribute much toward understanding impacts of livestock grazing on biodiversity and ecosystem function by executing more sophisticated grazing studies than those conducted in the past.

LIFEBLOOD OF THE WEST
Riparian Zones, Biodiversity, and Degradation by Livestock

J. Boone Kauffman, Ph.D.

Riparian, or streamside, areas are critical habitat for many plants and animals in the arid West.
Livestock grazing is the leading cause of riparian degradation. Impacts to vegetation, stream hydrology,
and geomorphology can separately or synergistically affect stream functioning and many
wildlife species. Thus, riparian restoration, including the removal of livestock,
must be a high priority for the conservation of biodiversity.

J. Boone Kauffman *is professor of ecosystem sciences in the Department of Fisheries and Wildlife, Oregon State University, Corvallis. He studies the ecology and restoration of riparian zones in the arid West, as well as tropical forest ecology in Central and South America, and has written more than a hundred scientific papers on natural ecosystems. He grew up on the West Texas plains and holds a Ph.D. degree in forest ecology from the University of California at Berkeley, as well a B.S. degree in range management and an M.S. degree in range ecology.*

Riparian zones are a unique wetland environment adjacent to rivers or streams. People have long recognized that riparian zones and rivers are the lifeblood of the western landscape, being more productive and home to more plants and animals than any other type of habitat. Scientists refer to riparian zones as hotspots of biodiversity, a characterization that is particularly apparent in arid and semiarid environments, where such zones may be the only tree-dominated ecosystems in the landscape. The presence of water, increased productivity, favorable microclimate, and periodic flood events combine to create a disproportionately higher biological diversity than that of the surrounding uplands.

In the Intermountain West and Great Basin, about 85 percent of native animal species are dependent on riparian zones for all or part of their life cycles. In these same riparian zones, more than 100 plant species commonly can be found on a single gravel bar of about 150 feet in length. In Oregon and Washington, about 71 percent of the native animal species utilize riparian zones.[1] Given that riparian areas make up only 0.5 to 2 percent of the landscape, their value in terms of biological diversity is incomparable.

Healthy riparian zones are of special importance to native fish, aquatic insects, and other stream-dwelling organisms. Particularly in headwater areas, most of the nutrients and energy used by aquatic organisms come from riparian zones—either plant materials that fall into the stream or nutrients dissolved in groundwater flows. Riparian zones are also the source of large pieces of wood that provide important in-stream habitat. Riparian vegetation gives shade over creeks, strongly influencing water temperature and thereby the distribution of coldwater species such as bull trout and other salmonids. Roots bind soil together and create resistance to stream erosion, resulting in complex habitat features such as overhanging banks, deep pools, and clean gravels. Finally, riparian zones are important in influencing water quality through nutrient uptake, chemical transformation (such as the conversion of nitrogen compounds into forms more useful for a variety of plant and animal species), and the mechanical filtering of sediments when flood waters flow high over stream banks.

Cattle graze in a wet meadow along Blacktail Creek, Beaverhead National Forest, Montana. Streamside vegetation is cropped low, parts of the stream bank are bare, and other parts are sloughing into the creek.

Many of the same attributes of riparian zones that result in high productivity and high biodiversity are of great economic value to human society. Unfortunately, many current uses of riparian corridors and wetlands by society do not correspond with preservation of these places as wildlife habitats or as providers of important natural services, such as the reduction of flood velocity and intensity. Broad floodplains formed along streams through the millennia have been productive not only because of their complex wildlife habitats and linkages to the aquatic biota but also because of their nutrient-rich soils. In fact, although the best lands for tilled agriculture, livestock forage, and timber growth are riparian zones and wetlands, these same activities, along with a variety of others, have been conducted in such an abusive fashion as to diminish greatly the value of riparian and stream ecosystems, from both utilitarian and ecological viewpoints.

A multitude of land uses show a lack of respect for, or ignorance of, the value of healthy and vigorous riparian zones and have resulted in the deterioration of not only riparian areas, but the entire landscape. In general, land abuses that have degraded riparian zones include logging, water diversion for irrigation or municipal uses, mining, roads, channelization, urbanization, industry, and agriculture. In the western United States, it is likely that livestock grazing has been the most widespread cause of ecological degradation of riparian/stream ecosystems.[2] More riparian areas and stream miles are affected by livestock grazing than by any other type of land use.

Livestock, especially cattle, prefer riparian zones for many of the same reasons that so many species of wildlife use them: high plant productivity, proximity to water, favorable microclimate, and level ground. As much as 81 percent of the forage removed by livestock within a grazing allotment can come from the 2 percent of land area occupied by the riparian zone.[3] Without controls on animal numbers, timing, and duration of use, cattle can rapidly and severely degrade riparian areas through forage removal, soil compaction, stream bank trampling, and the introduction of exotics. These factors have been defined as the direct effects of domestic livestock grazing on ecosystems.[4]

Through time, the direct effects of livestock can have many additive or even synergistic impacts that dramatically change the structure, function, and composition of the riparian zone. Of particular importance are the effects of livestock on streamside forests of cottonwood, aspen, and willow. The highest densities of breeding songbirds in the West are found in these habitats. Long-term overgrazing can eliminate these stands, which are of inestimable value as centers of biological diversity. In the short term, herbivory can depress both plant growth and reproductive output. Depressing the vigor of native plant species, along with increased soil disturbance due to livestock trampling, facilitates the spread of exotic weeds. Herbivory also causes a corresponding decline in the root biomass of riparian vegetation.

At stream edges, the combination of root loss and trampling weakens and collapses banks. Bank loss and the resulting sediment loads contribute to down-

cutting, channel widening, and degradation of water quality and fish habitats. As the channel downcuts, overbank flows cease, and subsurface water exchange between stream and floodplain is lost. Floodplain forests evolved to grow and develop in the environment created by large floods. By altering or eliminating the natural flood regime, channel downcutting impedes or halts the development of multi-layered, multi-aged—or "gallery"—forests, such as those composed of cottonwood trees and willows, along with other riparian plants. Loss of the riparian forests negatively affects not only the terrestrial wildlife, but the aquatic biota as well. Loss of shade and organic inputs from riparian vegetation results in increased stream temperature, altered water quality, and a change in composition and abundance of the aquatic biota.[5]

Although occupying a small portion of the landscape, riparian zones are keystone ecosystems because of their high level of biodiversity and provision of other ecosystem services. The restoration of riparian zones would yield many positive benefits, including the return of flood events to something resembling their natural patterns. Because riparian plants have adapted to survive frequent floods and other natural disturbances, they often show great resilience after the cessation of human activities that are causing degradation. Such removal of harmful activities is termed passive restoration, and in the arid West, the most significant act of passive restoration would be the removal of grazing livestock.[6] Logically, passive restoration should be implemented first, and its effectiveness assessed, before the initiation of more active measures, such as structural modifications and reintroductions of species.

Among the greatest barriers to effective riparian recovery are political and social factors. Land and river managers have often been limited to, or limited themselves to, band-aid approaches that do not address the real causes of degradation. For example, salmon have continued to decline in the Columbia Basin of the Pacific Northwest, despite the input of billions of dollars for restoration projects and mitigating measures, because, among other reasons, livestock continue to degrade riparian zones. A prominent and popular project on public lands in the Columbia Basin has been the installation of artificial structures in smaller streams, in an attempt to re-create aquatic habitat that has been lost to decades of poor resource management. However, artificial stream structures can be expensive and often are sited and constructed poorly.

In many cases, the most effective, cheapest, and simplest approach to restoring these river courses would be to halt grazing damage and allow the streams to recover their own natural vegetative and morphological characteristics over time. But it can be extremely difficult politically for managers to make such decisions. And restoration results can take a long time to appear, whereas political demands arise much more rapidly. Yet, given the inestimable natural values that arise from healthy riparian zones, a long-term commitment to riparian restoration, preservation, and sustainable management should receive high priority. The reduction or removal of livestock from vital riparian and wetland habitats throughout the West needs to be given serious consideration by all those concerned about ecosystem health.

This stream in northern New Mexico has become "entrenched." Over time, grazing and trampling of the soils and banks by livestock have caused the stream to widen and cut downward. Typical results of this stream degradation process include lowered water table, drier soil in the zone adjoining the waterway, and riparian-type plants (such as willows) gradually replaced by more drought-resistant plants (such as sagebrush).

WHAT THE RIVER ONCE WAS
Livestock Destruction of Waters and Wetlands

Joy Belsky, Ph.D., Andrea Matzke, and Shauna Uselman

Livestock grazing is responsible for damage to a majority of streams and riparian areas in
the arid West. New grazing practices do little, if anything, to improve streams damaged decades ago
by overstocking; in fact, government reports state that many riparian areas are in their worst condition
ever. Livestock grazing does not appear to be compatible in any way
with healthy streams and wildlife habitat.

Joy Belsky *was a well-respected grassland ecologist and outspoken critic
of traditional range management methods. At the time of her death in 2001,
she was staff ecologist at the Oregon Natural Desert Association, where
she reviewed federal resource management plans and worked to develop
scientific bases for ecosystem protection. She held a Ph.D. degree in plant
ecology from the University of Washington and published over forty-five peer-
reviewed scientific papers on North American as well as African rangelands.*

Andrea Matzke *holds a master's of public health degree in environmental
health sciences from the University of Michigan. She has worked for the
U.S. Forest Service as a hydrologic technician and is currently basin
monitoring coordinator for the Oregon Department of Environmental
Quality in Medford.*

Shauna Uselman *holds a master's degree in environmental management
from Duke University and is currently a doctoral candidate in ecology,
evolution, and conservation biology at the University of Nevada, Reno.
Her research interests include biogeochemistry, forest and wetland
ecology, and global climate change.*

Bureau of Land Management lands along a tributary of the
Humboldt River in northern Nevada. Livestock are a leading
cause of stream alteration throughout the West. Damage is
so severe in some places that removing livestock, although a
key first step, will have to be accompanied by other, active
restoration efforts.

Grazing by livestock has damaged 80 percent of the streams and riparian
ecosystems in arid regions of the western United States. Because riparian
and stream ecosystems represent only 0.5 to 1 percent of the surface area of
arid lands, they were historically ignored by land managers. In fact, until
the late 1960s, western land managers viewed riparian habitats as "sacrifice" areas, being dedicated primarily to providing food and water for domestic livestock. This attitude began to change as environmental advocates
focused national attention on the degraded state of western streams and
related losses of biodiversity.

To counter environmentalists working to protect these stream and riparian
ecosystems, spokespersons for the livestock industry assert that most of the
damage to streams occurred a hundred years ago and is no longer occurring,
and that new grazing techniques, such as rest-rotation or seasonal grazing,
actually help streams and riparian zones to recover.

Scientific research and reports by dozens of rangeland experts refute these
arguments.

Western Streams in Worst Condition Ever

Although evidence is undeniable that early grazing practices were highly
destructive, current grazing practices remain a key factor in the continued
degradation of riparian habitats. As recently as 1990, a U.S. Environmental
Protection Agency report found that "extensive field observations suggest that
riparian areas throughout much of the West are in their worst condition in history."[1] In addition, a joint Bureau of Land Management (BLM) and U.S. Forest
Service report concluded that "riparian areas have continued to decline" since
grazing reforms in the 1930s.[2]

No-Grazing Better Than "New" Grazing

In our survey of the scientific literature,[3] we reported on the effects of livestock
grazing on western streams and riparian zones. In reviewing over two hundred
peer-reviewed scientific papers and reports, we found none reporting a positive impact (or benefit) of cattle on riparian areas when those areas were compared with ungrazed controls. Most claims of "benefits" resulted from studies

comparing reduced stocking rates and/or newer grazing systems with older, more destructive techniques. None of these "new" methods was better than no grazing at all.

Cattle are more damaging to riparian zones than their often low densities on arid public lands would suggest. Cattle evolved in cool, wet meadows of northern Europe and Asia; as a result, they are a true riparian species, avoiding dry, hot environments and congregating in wet, cool areas where forage is more succulent and shade more available than in uplands. One study found that a riparian zone in eastern Oregon made up only 2 percent of the grazing allotment by area but produced 21 percent of the total forage in the allotment. And of all the forage consumed by cattle in the allotment, 81 percent came from the riparian zone.[4]

As summarized in Table 1, livestock harm streams, their associated riparian zones, and the fish and other wildlife that inhabit them in multiple ways. Cattle contribute to water degradation by adding fecal matter, increasing stream bank erosion, and increasing water temperatures by altering steam width and depth.

Hydrological Processes Harmed

Among the most damaging effects of livestock grazing is their alteration of stream depth and flow. Normally, intact plants on undisturbed uplands and streamsides slow the downhill flow of rainwater, promoting its infiltration into soils. This water gradually moves through the subsoil and seeps into stream channels throughout the year, helping feed year-round flows.

But when livestock consume or trample upland and riparian vegetation and compact hillside soils with their hooves, less rainwater can enter the soil. During storms, more water flows overland, and more quickly enters streams, creating high peak flows.

These peak flows are highly erosive, causing channel downcutting (see Figure 1). As the channel deepens, water drains from the surrounding floodplain into the channel, lowering the associated water table and leaving the roots of riparian plants suspended above the water table in dry soils. Eventually, these plants and their associated wildlife species die out and are replaced by drought-tolerant upland species, such as sagebrush and juniper. In addition, less water is available in the soil to contribute to late-season stream flows. Consequently, high-intensity spring floods are followed by a reduction or complete loss of water flow in late summer, in sharp contrast to ungrazed stream systems.

Biodiversity Endangered

The changes in water quality, quantity, and seasonal flow caused by livestock grazing have enormous impacts on biodiversity. Those species that can tolerate dry soils and seasonally dry stream beds increase in number. Those that depend on year-round water flows; cool, clean water; deep, shaded pools; and moist soils decline in abundance or disappear. These latter species include riparian plants, such as willow and sedges, that rely on wet soils; salmon and trout that require cool, clean water to spawn and feed; and the 80 percent of native animal species in southeastern Oregon that depend on riparian ecosystems.[5] A 1994 U.S. Forest Service report found livestock grazing to be the fourth major cause of species endangerment in the United States and the second major cause of endangerment of plant species.[6]

Recovery Fastest with Livestock Exclusion

Nearly all scientific studies refute the claim that livestock grazing benefits streams. Previously denuded stream banks may revegetate, and erosion may decline with improved livestock management, but recovery will take longer than if grazing were terminated completely. New studies suggest that new grazing systems may slow the rate of degradation but do not reverse it.

BLM riparian expert Wayne Elmore and Oregon State University riparian specialist Boone Kauffman best summed up the available evidence by stating that "livestock exclusion has consistently resulted in the most dramatic and rapid rates of ecosystem recovery."[7] Other riparian specialists point out that two to fifteen years of total livestock exclusion may be required just to initiate the recovery process.

Although some streams may eventually recover under lighter grazing management, Professor Robert Ohmart of the University of Arizona is concerned about this approach. He questions whether degraded riparian communities throughout the West can "hang on to their thread of existence for another 30–50 years" while waiting for grazed systems to recover slowly.[8]

We undertook our survey of the scientific literature because we believed that the public was being misinformed by federal agencies and the ranching industry about the current impacts of cattle grazing in the arid West. We sought to answer the question, Is livestock grazing ever compatible with healthy streams, abundant wildlife, and riparian zones? The answer, backed by scores of scientific studies, is an unequivocal no.

Figure 1. Stream Degradation Due to Livestock Grazing

A, Healthy stream morphology; **B**, grazing-caused degradation begins to lower water table; **C**, further degradation, lost wet meadow, formation of gully; **D**, wider banks from trampling, lost streamside meadow.

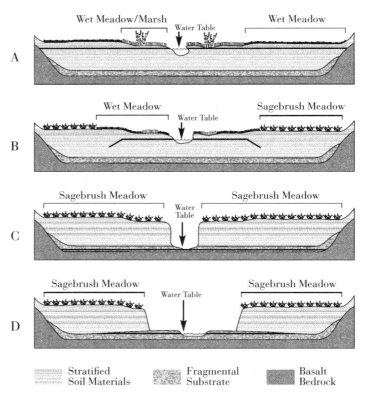

Source: U. S. Department of the Interior, *Riparian Area Management, Process for Assessing Proper Functioning Condition*, TR 1737-9 1993 (Denver: BLM, 1993).

Table 1. Effects of Livestock Grazing in Riparian Zones

Effects of Livestock Grazing on Stream Channel Morphology

INFLUENCE	CAUSES	IMPACTS
Stream bank undercuts: reduced quality and quantity	Stream bank breakdown by livestock and loss of stabilizing vegetation	Fewer hiding spaces and pools for fish
Channel form: fewer meanders and unvegetated gravel bars	Increased water velocity; removal of stabilizing vegetation; erosion of stream banks	Increased erosion; fewer pools for fish; decreased stream bank roughness

Effects of Livestock Grazing on Hydrology (Stream Flow Patterns)

INFLUENCE	CAUSES	IMPACTS
Overland flows (runoff): increased	Reduced water infiltration into soil due to compaction and loss of streamside vegetation	Increase in sheet and rill erosion; increased flooding; reduced groundwater recharge; lowered water table
Peak flows: increased	Larger volume of runoff flowing directly into channel	Increased stream energy for channel erosion; downcutting of channel bed; gully formation

Effects of Livestock Grazing on Riparian Zone Soil

INFLUENCE	CAUSES	IMPACTS
Bare ground: increased	Vegetation consumed and trampled by livestock	Drier soil surfaces; higher erosion and sediment delivery to streams and aquatic habitats
Erosion (water, ice, and wind): increased	Soil compaction; removal of vegetation; trampling disturbance	Increased sediment load to receiving stream; loss of fertile topsoil; suffocation of fish eggs; loss of pools
Compaction: increased	Trampling by livestock on wet, heavy soil; reduced litter and soil organic matter	Decreased infiltration rates and more runoff; reduced plant productivity and vegetation cover

Effects of Livestock Grazing on Stream Bank Vegetation

INFLUENCE	CAUSES	IMPACTS
Herbaceous cover, biomass, productivity, and native diversity: declined	Grazing and trampling by livestock; selective grazing on palatable species; loss of vulnerable species; lowered water table; drier, warmer, more exposed environment	Less detritus (food inputs) for stream and aquatic organisms; higher water temperatures in summer and cooler temperatures in winter; degraded habitat for fish and wildlife; reduced biodiversity; loss of moisture- and shade-dependent species
Tree and shrub biomass and cover: declined	Browsing by livestock on shrubs and tree saplings when they are most vulnerable	Decline in stream bank stability; increased erosion; reduced stream shade and higher water temperatures
Species composition: altered	Lowered water table; warmer, drier environment; livestock selection of palatable species; compacted soil	Replacement of riparian species by upland species and exotic weeds; reduction in riparian area

Source: Adapted from A. J. Belsky, A. Matzke, and S. Uselman, "Survey of Livestock Influences on Stream and Riparian Ecosystems in the Western United States," *Journal of Soil and Water Conservation* 54 (1999): 419–431. Source notes for the table can be found in the *Journal* article.

Table 1. Effects of Livestock Grazing in Riparian Zones (continued)

Effects of Livestock Grazing on In-stream Vegetation

INFLUENCE	CAUSES	IMPACTS
Algae: increased	More sunlight; higher temperatures; higher concentrations of dissolved nutrients	Low levels of dissolved oxygen after algal bloom die-off
Higher plants (submerged and emergent): often decline in abundance	Often trampled; buried in deposited sediments; uprooted by strong flows	Reduced trapping of sediments; less food for aquatic organisms; higher water velocity and erosive force

Effects of Livestock Grazing on Water Quality

INFLUENCE	CAUSES	IMPACTS
Bacteria/protozoa: increased	Direct fecal deposition into water; fecal material in runoff; sediments containing buried microorganisms churned up by hoof action	Higher human and wildlife disease-producing potential from pathogens; human health endangered by swimming and other contact
Water temperature: increased	Increased solar exposure due to reduced shade from streamside vegetation and increased channel width-to-depth ratio	Poor-to-lethal environment for salmonids and other temperature-sensitive, coldwater species

Effects of Livestock Grazing on Aquatic and Riparian Wildlife

INFLUENCE	CAUSES	IMPACTS
Fish—diversity, abundance, and productivity: decreased	Higher water temperatures increase salmonid mortality and negatively affect fish spawning, rearing, and passage; greater water turbidity, increased siltation, and bacterial counts; lower summer flows; less protective plant cover; fewer insects and other food	Loss of salmonids and other coldwater species; loss of avian and mammalian predators; replacement of coldwater species with warmwater species
Birds—diversity, abundance, and species composition: altered	Reduction in food, water quality, and water quantity; loss of perches, nesting sites, and protective plant cover; loss of complex vegetational structure	Reduction in biodiversity; replacement of riparian specialists by upland species and generalists; loss of some neotropical migrants
Mammals (large and small)— diversity and species composition: altered (sometimes but not always)	Loss of riparian habitat and food sources; warmer, drier, more exposed environment; behavioral characteristics such as avoidance of livestock	Habitat use shifts by wildlife; suboptimal nutrition for females and offspring; lower beaver activity in creating wetlands; riparian species replaced by upland species and generalists

Middle Fork Gila River, Gila National Forest, New Mexico. The lush vegetation along this stream, where no livestock have ever grazed, belies a common belief that the Southwest is naturally a thinly vegetated, unproductive desert.

CATTLE AND STREAMS
Piecing Together a Story of Change

Suzanne Fouty

Cattle influence streams and watersheds in ways that are complex and interrelated. Degraded
stream systems differ from healthy systems in terms of channel morphology and amount, distribution,
and type of riparian vegetation. Cattle trample banks, compact soils, and consume vegetation, which can
lead to more erosion, greater runoff, faster stream downcutting, larger floods, and lower water tables.
Cattle removal may be only the first step to recovery in some highly degraded watersheds;
other actions, such as beaver reintroduction, may be needed.

Suzanne Fouty *is a Ph.D. candidate in geography at the University of
Oregon. Her research examines how changes in cattle and elk grazing
pressure and in beaver activity influence stream channel morphology.
Her focus is on determining the cultural and physical constraints to
stream and ecosystem restoration and identifying ways to accelerate
the recovery process. She has also worked for the U.S. Forest Service as
a hydrologist/hydrologic technician examining grazing allotments.*

Centennial Valley, Montana. Scientific research has shown that the
biological diversity of the western United States is highly dependent on
stream and riparian areas, and that livestock are very destructive of these
habitats. Despite a better appreciation today than in the past for the
ecological importance of riparian zones, management practices have
changed little, and livestock are still degrading streams.

Cattle graze on about 300 million acres of public land in the United States
and have been identified as a key cause of stream channel change.[1] Yet there
is minimal long-term research assessing rates, magnitudes, and the nature of
stream channel change. I suggest that the reasons for this dearth have to do
with the speed at which landscapes change, the historical timing of those
changes, and our distorted image of what healthy streams and riparian areas
should look like.

The Influence of Cattle on Streams

To understand how cattle influence stream-riparian processes and channel
pattern and shape (morphology), we must first identify the physical characteristics of healthy stream systems versus degraded systems (see Table 1). The
following overview applies only to streams that flow through wide, flat valleys
and that do not have beaver. The important influence of beaver on channel
morphology, stream processes, and the restoration of cattle-degraded watersheds will be seen later on.

Among the main differences between degraded and healthy streams are the
type, density, and distribution of riparian vegetation along the stream banks
and on the valley floor. Vegetation is a key element because there is a feedback
relationship between vegetation and bank stability, and between vegetation
and the trapping and stabilization of sediment. Anything that degrades or
hinders the establishment and growth of densely rooted and abundant vegetation does so to the detriment of the stream, water quality, fish and other
wildlife habitat, and flood control.

The impact of cattle on stream morphology and watershed hydrology is
complex. Some of the cattle impacts are direct, such as trampling of banks,
compaction of soils, and the consumption and removal of riparian vegetation.
Other cattle impacts are indirect and the result of processes that cattle set in
motion as they alter upland and stream bank vegetation and compact soils.
Examples of indirect impacts include increased runoff from storms and
decreased stream bank resistance to erosion. The first is a result of soil compaction and removal of upland vegetation, and the second is due to removal of
riparian vegetation.

Reversing these physical processes requires, as a starting point, major changes in cattle use patterns, often accompanied by a reduction in the number of livestock, or their complete removal. However, in many places, damage to and alteration of the systems is so severe that simply removing cattle will not be enough and more active restorative efforts will be necessary.

Stream Processes

Streams are dynamic systems. They continuously adjust to changes in runoff and sediment inputs by modifying their channel morphology. In turn, channel morphology controls how frequently a valley floor will flood if all other variables are held constant. For example, a narrow channel will overflow more frequently than a wide channel of the same depth. When a stream overflows its banks and spreads across the valley, the water sinks into the valley sediments as it would a huge sponge. Some of this water is used by the valley vegetation, some is stored, and some reenters the stream as subsurface flow. This subsurface flow contributes water to the channel during the dry season. If the valley floods every year or two, the valley water table remains high. The result is lush valley floor vegetation, deeply rooted and abundant vegetation along the stream banks (especially willows, sedges, and cottonwoods), and a perennial stream. Banks remain resistant to stream erosion, channels are narrow, water temperatures are cooler, and the stream and valley floor are hydrologically connected.

Therefore, channel morphology and the relationship of a stream to its valley floor control the following factors:

- The condition and extent of the riparian zone
- The quality and diversity of fish and other wildlife habitat
- Water table levels

- The magnitude and duration of floods
- Dry-season, in-stream flow volumes.

As a stream channel deepens or widens because of livestock grazing or other human disturbances, it begins to disconnect hydrologically from its valley floor. Greater amounts of water must now fill the channel before the stream tops its banks. Because large-flow events (for example, a ten-year flood) are less frequent than smaller-flow events (two- or five-year floods), the valley floods less frequently. The results are fourfold. First, there is a decrease in the water content of the sediments. This triggers a shift in vegetation type to more drought-tolerant species, such as grasses, forbs, and sagebrush. Second, the water table drops over time. The subsurface flow back to the stream decreases, and streams that once flowed year-round may become dry for part of the year. Third, bank vegetation shifts to less dense and less deeply rooted types, making the banks more vulnerable to stream erosion. And finally, as greater amounts of water remain in the channel, the erosive power of a flood increases, and greater bank erosion and property loss occur.

Factors Controlling Stream Channel Change

The rates and processes by which a stream channel changes vary over time and space. A growing body of research reveals a series of complex interactions between sediment, vegetation, stream bank characteristics, stream flow regime, valley floor characteristics, and human land use, such as cattle grazing.[2] These complex interactions are a function of the following factors:

- Composition and stratigraphy (layering) of the stream bank sediments
- Vegetation type and density along the stream banks and on the valley floor

Table 1. Comparison of Stream Characteristics as a Function of Stream Health

HEALTHY VALLEY STREAMS	DEGRADED VALLEY STREAMS
Narrow stream channel with well-vegetated, overhanging stream banks bound by deep root systems.	Wide stream channel with sparse stream bank vegetation with shallow root systems. No overhanging banks.
Distinct deep pools and shallow riffles or perhaps just deep water with little pool/riffle distinction. Varies depending on channel width.	Distinct pools absent. Channel bottom topography fairly uniform.
Well-sorted channel bottom sediments.	Unsorted channel bottom sediments (lots of fine sediment mixed in with gravels and cobbles).
Dense, lush riparian vegetation on the valley bottom indicating a high water table.	Dominance of drought-tolerant plants (sagebrush, grasses, forbs) on the valley bottom, indicating a low water table and infrequent flooding of the valley floor.
Sinuous channel—wanders back and forth across the valley bottom, often very tightly. May have multiple channels.	Single, nearly straight channel.
Few unvegetated gravel and sand bars.	Many bare gravel and sand bars.
Well-vegetated banks. Bank sediments rarely visible except at channel bends.	Minimal bank vegetation. Bank sediments are exposed and susceptible to stream erosion.

Note: This table applies only to streams that flow through wide, flat valleys and that do not have beaver.

- Ability of riparian vegetation to reestablish itself and grow after a disturbance (flood, grazing, or fire)
- Abundance of sediment
- Type and duration of disturbance acting on the stream system—long-term, chronic (seasonal cattle grazing) or infrequent (fire)
- Location within a stream channel reach (bend versus straight section)[3]
- Presence of stream channel features such as bedrock knobs, constrictions, debris jams, or islands.[4]

Bank composition, bank stratigraphy, and vegetation are important because they influence the way stream banks fail and control the amount and determine the type of sediment that enters the stream channel.[5] Bank composition and vegetation determine bank cohesion—that is, how well the bank sediments stick together. As cohesion increases, the ability of the stream banks to resist erosion by water or trampling also increases, and the channel's susceptibility to widening decreases. When vegetation is removed, as by cattle grazing, bank cohesion becomes dependent primarily on bank composition.

Stream Restoration

Restoration of streams degraded by cattle is complex because of the magnitude and nature of the stream channel changes that have occurred directly as a result of cattle grazing impacts and indirectly as a result of processes that cattle set in motion. However, restoration begins with the reestablishment of lush riparian vegetation on the stream banks and channel bars. Vegetation slows the water down, protects the banks from erosion, and captures fine sediment. Sediment, when deposited, provides sites for vegetation to become established, which in turn can trap more sediment.

A third component of stream systems, rarely mentioned but capable of accelerating restoration, is beaver and beaver ponds. Beaver ponds effectively capture fine sediment and raise the water level in the stream. The ponds help stabilize the valley water table and facilitate the recovery of vegetation. The reestablishment of vegetation, sediment deposition, and beaver ponds all lead to a hydrologic reconnection of the stream and its valley floor. How quickly that reconnection occurs varies depending on how quickly riparian vegetation becomes reestablished, the amount of sediment available, and the presence or absence of beaver. Of the three, acquiring the large volumes of sediment needed to replace what has been lost historically is the greatest challenge.

Current sources for the sediment in any given stream are limited to upstream banks and, occasionally, debris flows, landslides, and hillslope erosion, setting in motion frequent lose-lose situations for the watershed. Potential downstream gains in sediment are offset by real upstream bank and hillslope losses. In the absence of well-vegetated bars and banks downstream, the fine sediments contributed by the upstream sources are transported downstream and out of the watershed as suspended load, resulting in an overall increase in the channel area. To prevent further channel widening, it is critical that stream bank erosion be minimized. This is only possible when lush, densely rooted riparian vegetation is present on the banks.

Herein, however, lies a restoration dilemma. Well-vegetated banks (good) result in minimal sediment inputs (a problem). Where sediment inputs are low, channel narrowing is slow to nonexistent, regardless of the abundance of stream bank and channel bar vegetation,[6] and the valley and stream remain disconnected. It is in examining this dilemma that the significance of beaver becomes obvious.

Beaver ponds quickly reconnect the stream and valley floor by elevating water levels in the stream, thus stabilizing valley water tables and increasing the frequency of valley floor flooding.[7] However, beaver require abundant vegetation to build and maintain their dams successfully. Cattle use of the riparian zone must therefore be minimized through more frequent moves, fences, and changes in the season of use—or completely eliminated to allow vegetation to recover naturally or plantings to grow. The removal or reduction of cattle grazing should be accompanied by changes in trapping policies and efforts directed at changing public perceptions of beaver from one of varmint or mere furbearer to one of keystone species. Otherwise, beaver trapping will continue to set recovering systems back to zero or worse.[8]

Recovery Begins with Cattle Removal

Cattle are one of the dominant controls on the composition and distribution of vegetation on lands throughout the arid and semiarid West. Cattle impacts on vegetation are particularly significant in the riparian zone and on channel bars where constant trampling and consumption prevents the development of lush, densely rooted vegetation that would otherwise grow—vegetation needed to stabilize the stream banks and channel bars and restore hydrologic and vegetative complexity and stability to these ecosystems. Thus, streams and riparian zones remain in a degraded condition, one that has marginal value for fish and wildlife and for downstream communities concerned about water quality and flooding.

Any serious effort at stream ecosystem restoration must, therefore, address the impact of cattle grazing in the riparian zone. Cattle must either be removed or their numbers reduced and their use strictly monitored. Patience is necessary, as the amount of alteration and disturbance is high, and recovery rates will be slow in many places. However, by understanding how cattle influence the evolution of the stream ecosystems, the potential for restoration can be improved.

STINK WATER
Declining Water Quality Due to Livestock Production

John Carter, Ph.D.

Livestock production is a major cause of water quality impairment in the United States. Livestock wastes, primarily from cattle, are a significant source of pollutants such as phosphorous, nitrogen, and bacteria. Increased nutrient concentrations, decreased oxygen, and higher pathogen levels in water are common consequences of livestock use, including on public lands. In general, government agencies are currently doing a poor job of monitoring water quality.

John Carter *holds a Ph.D. degree in ecology, an M.B.A. degree, and a B.S. degree in mechanical engineering and is a registered professional engineer in Utah. His work has included research on heavy metals in aquatic and terrestrial systems, research on aquatic and wetland ecology, and oversight of complex remediation and reclamation projects. He writes and speaks frequently on water quality issues. He founded the nonprofit organization Willow Creek Ecology in 1996 to address the massive impacts of livestock production on western public lands, and is currently Utah director for the Western Watersheds Project. His home is in Mendon, Utah.*

Wilderness stream, or livestock sewer? This fouled creek flows through the Popo Agie Wilderness in the Wind River Range of Wyoming. While campers are required to pitch their tents more than 200 yards from any water body, and are importuned to bury their bodily waste, livestock wander and defecate at will. The Forest Service says its wilderness regulations for recreationists are intended to protect water quality and fragile wetland habitat, yet when one cow produces the same amount of waste each day as sixteen people, and a herd of livestock leaves a mountain meadow smelling like a stockyard, it's easy to infer that what's being protected is a rancher's grazing permit, and not the water resource.

The feeding, housing, and grazing of livestock throughout the United States is pervasive. Watershed and water quality degradation accompany this industry and affect nearly every major water body in the lower forty-eight states. Government regulation is inconsistent and ineffective at controlling these problems. This essay highlights water pollution caused by livestock wastes. Examples of livestock impacts on water quality in the Wasatch-Cache National Forest in Idaho and Utah plainly illustrate that when livestock have access to streams, water quality is degraded.

Livestock Wastes

Agriculture is the major source of water quality impairment in this country. Siltation, introduction of excessive "nutrient" materials (nitrogen and phosphorous are nutrients for growth in smaller amounts, but at higher levels, they interfere with the maintenance of healthy aquatic ecosystems), bacteria, proliferation of oxygen-depleting substances, and pesticides rank as the top causes of water quality decline in rivers, and agriculture—including livestock production—is linked to all of them.[1] Livestock waste alone is a major factor in the nutrient pollution of streams, increase of pathogenic bacteria in water supplies, and the decline of dissolved oxygen levels in rivers, lakes, and other water bodies.

The amount of animal manure and urine generated in the United States on an annual basis is staggering.[2] Table 1 provides a summary, by type of livestock, of the waste generated and the amounts of nitrogen and phosphorous contained in that waste. Table 2 gives a further summary of livestock waste in the eleven western states, broken down by type of livestock and state. Cattle are by far the largest generators of waste, producing about 3.5 tons per year for every man, woman, and child in the United States. Cattle waste exceeds all others in the eleven western states, and the total waste generated by all forms of livestock in the West makes up about 18 percent of the national livestock waste stream.

In 1990 and 1991, the condition of surface water for 18 percent of the nation's streams was assessed by the states and reported to the Environmental Protection Agency (EPA). Water quality in 38 percent of the stream-miles surveyed was impaired by crop and animal agriculture. Eighteen states, representing all regions of the country, reported on the sources of impairment by type of agricultural activity. In agriculturally impaired streams, the activities contributing to declines in water quality were irrigated cropland (42 percent), nonirrigated cropland (31 percent), feedlots (26 percent), and rangeland (25 percent). Manure accounted for significant percentages of the nitrogen and phosphorous inputs to watersheds across the country. For example, in the western United States, manure accounted for 39 percent of phosphorous and 53 percent of nitrogen input to watersheds. Statistical studies also indicated that increases in stream loading of these nutrients were correlated with increases in the concentration of livestock populations in the watersheds.[3]

Public concern has been raised by the occurrence of drinking water contamination, fish kills, shellfish contamination, swimming advisories, nuisance odors, and the links of these problems to agricultural practices. According to the EPA:

> AFO [animal feeding operation] activities can cause a range of environmental and public health problems, including oxygen depletion and disease transmission in surface water, pathogens and nutrient contamination in surface and ground water, methane emissions to the air, and excessive buildup of toxins, metals, and nutrients in soil. . . . AFOs have also been identified as substantial contributors of nutrients (e.g., nitrogen and phosphorous) in water bodies that have experienced severe anoxia (i.e., low levels of dissolved oxygen) or outbreaks of microbes, such as *Pfiesteria piscidia.*[4]

In 1991, a billion fish died from a *Pfiesteria* bloom in North Carolina's Neuse River Estuary.[5]

EPA efforts to address environmental and health concerns from AFOs and CAFOs (concentrated animal feeding operations, or feedlots with more than 300 animal units) began in the 1970s. These efforts have included issuing permits under the Clean Water Act and promoting voluntary efforts among livestock producers to limit pollution. These efforts have not worked; the problem persists and has intensified as the size and numbers of these operations have increased. According to the EPA:

> Evidence suggests that EPA's regulatory and voluntary efforts to date have been insufficient to solve the environmental and health problems associated with AFOs. Agricultural practices in the United States are estimated to contribute to the impairment of 60 percent of the nation's surveyed rivers and streams; 50 percent of the nation's surveyed lakes, ponds, and reservoirs; and 34 percent of the nation's surveyed estuaries.[6]

The EPA estimates that feedlots alone adversely impact 16 percent of impaired waters.[7] Land application of manure and grazing of livestock on private and public lands contribute a majority of this agricultural pollution. Although the Federal Water Pollution Control Act in 1972 designated feedlots as point sources (geographically focused pollution sources—the classic example is a factory waste pipe draining into a stream), amendments to the act, now known as the Clean Water Act, excluded from permit requirements agricultural storm water discharges and return flows from irrigated agriculture. Pastures and rangeland were also excluded, although a 1997 appellate court ruled that runoff from cropland used for disposal of manure from a facility designated as a point source was also a point source.[8]

After almost thirty years' delay, the EPA is now requiring states to establish total maximum daily loads (TMDLs) under Section 303D of the Clean Water Act. A TMDL is a calculation of the maximum amount of a pollutant that a water body can receive from both point and nonpoint sources (nonpoint being a large-area, or extensive type of pollution source, such as pesticide-treated cropland) and still meet water quality standards, and an allocation of that amount to the pollutant's sources. A potential danger lies within the TMDL process whereby streams with low levels of pollution may be allowed to become more polluted as long as they can still meet beneficial use criteria.

Between 1987 and 1992, the number of animal units in the United States increased by about 4.5 million, or 3 percent, with a decrease in small operations (AFOs) and an increase in the larger operations (CAFOs).[9] In 1995, there were about 6,600 CAFOs with more than 1,000 animal units in the United States.[10] Current EPA strategy to address this growing problem is the increased permitting of CAFOs and the associated land application of manure from permitted facilities. EPA efforts are concentrated on priority watersheds, determined according to the number of CAFOs, AFOs, and animal units; on revision of existing regulations; and on increased coordination with other government agencies and agriculture. In addition, the EPA is encouraging (and may financially aid) voluntary efforts among operators to implement best management practices.[11] During 1992–1994, for example, the EPA provided $89 million to farmers through these voluntary assistance programs. These efforts, however, leave the grazing of livestock on public and private rangelands largely unaddressed:

> According to EPA, many operations with more than 1,000 animal unit equivalents are not required to have point source permits because they do not discharge during most storm events; others should have permits but do not because of mistaken exemptions or limited federal or state resources for identifying operations needing permits.[12]

Excessive Nitrogen and Phosphorous Levels

Livestock wastes in water have detrimental effects for a number of reasons. Elements that are normally beneficial—thus called "nutrients"—become disruptive at high concentrations. Manure and urine contain the nutrient elements nitrogen and phosphorous, and when high numbers of livestock produce waste that gets into water supplies, the excessive nitrogen and phosphorous inputs harm aquatic ecosystems. Algal populations increase in response to the increased nutrients and result in lowered dissolved oxygen levels. In addition, levels of disease-causing organisms can increase as a direct result of the introduction of these wastes into watersheds and surface and ground water.

Nutrient concentrations in streams and lakes also increase as a result of erosion and runoff from disturbed watersheds. Nutrients from all sources become more concentrated as stream flows are reduced and reservoirs are drawn down by livestock production activities, such as irrigation of hayfields and feed crops.[13] In addition, trampling and removal of stream bank and watershed vegetation by livestock result in altered stream morphology and hydrology, leading to reductions in late-season flows.

One study documented increases in runoff from more heavily grazed pastures when compared with those subjected to less pressure.[14] The research suggested a linear relationship between runoff volume and soil nutrient loss. Another study of runoff from land application of dairy cattle wastes concluded that nutrient concentrations in runoff were directly related to the application rate of dairy wastes.[15] Researchers have also found that nutrient and chemical outputs were directly related to precipitation, stocking rate, hydrologic characteristics, and sediment content in runoff.[16]

Dynamics of nutrients in grazed watersheds are altered by livestock. Watersheds can be depleted of nutrients by livestock through removal of vegetation, soil disturbance, erosion, and leaching. For example, in one investigation, researchers noted that cattle (and other livestock) export nutrients out of the watershed when they are physically removed from the watershed. Approximately 17 percent of the nitrogen in the forage livestock consume leaves with those livestock—the nitrogen incorporated into their bodies—when they are rounded up and taken off-site.[17] The remaining nutrients consumed from forage are placed on the watershed in the form of feces and urine that can be more readily leached into water bodies. This net export of nutrients from grazing can be significant. Studies in the Wasatch-Cache National Forest by the author indicate that soil nitrogen in areas grazed by livestock is reduced to about half that of areas ungrazed by livestock, reflecting losses through export, erosion, and interruption of nutrient cycling by soil microorganisms.[18]

Adequate dissolved oxygen is necessary in maintaining healthy populations of fish and other aquatic life. Dissolved oxygen levels decline when water temperature rises because the oxygen holding capacity of water is less at higher temperatures. Water temperatures rise when livestock remove riparian vegetation and alter streams over time so they are wider and shallower and have reduced late-summer flows. In addition, algal blooms occur, resulting from nutrient enrichment of the water by livestock waste. These algal blooms deplete oxygen by respiration at night in the absence of photosynthesis. As the blooms die, decomposition of the algae as well as the organic matter in the animal waste consumes additional oxygen.

Increases in Disease-Causing Microorganisms

A 1997 review of major pathogens and health effects associated with dairy wastes noted that "aside from the problem of disease transmission among animals, more than 150 pathogens can cause zoonotic infections" (that is, infections transmitted from animals to humans).[19] Organisms causing health effects in humans, ranging from gasteroenteritis to death, included protozoan species such as *Cryptosporidium* and *Giardia*, as well as bacterial species such as

Table 1. Summary of Animal Wastes in the United States

Livestock Type	Number	Waste (tons/yr)	Nitrogen in Waste (tons/yr)	Phosphorous in Waste (tons/yr)
Hogs	57,450,288	110,000,000	650,000	225,000
Cattle	99,275,900	750,000,000	4,100,000	1,000,000
Poultry	1,316,425,230	50,000,000	650,000	205,000
Sheep	7,588,377	3,000,000	32,000	6,500
Totals	**1,480,739,795**	**913,000,000**	**5,432,000**	**1,436,500**

Source: Environmental Defense Fund, "Animal Waste—A National Overview," *Environmental Defense Fund Scorecard,* 15 January 2000.

Table 2. Livestock Waste Generated in the Eleven Western States

State	Cattle Waste (tons/yr)	Sheep Waste (tons/yr)	Hog Waste (tons/yr)	Poultry Waste (tons/yr)
Arizona	6,900,000	30,000	17,000	400
California	51,000,000	310,000	380,000	2,800,000
Colorado	19,000,000	230,000	850,000	1,600
Idaho	15,000,000	100,000	55,000	820
Montana	19,000,000	170,000	290,000	5,800
Nevada	4,100,000	36,000	2,300	210
New Mexico	13,000,000	120,000	11,000	540
Oregon	11,000,000	110,000	60,000	78,000
Utah	7,000,000	170,000	550,000	21,000
Washington	11,000,000	21,000	69,000	230,000
Wyoming	11,000,000	280,000	150,000	520
Totals	**168,000,000**	**1,577,000**	**2,434,300**	**3,138,890**

Source: Environmental Defense Fund, "Animal Waste—A National Overview," *Environmental Defense Fund Scorecard,* 15 January 2000.

Salmonella, *E. coli* O157:H7, *Brucella*, *Leptospira*, *Chlamydia*, *Rickettsia*, *Listeria*, and *Yersinia*.[20]

Cryptosporidium oocysts in the Milwaukee water supply affected 403,000 people in 1993.[21] *E. coli* O157:H7 is of concern because many outbreaks have been traced to ground beef and raw milk.[22] Waterborne sources of infection in humans have also been documented.[23] *E. coli* O157:H7 can lead to kidney failure and death in some individuals. A 2000 article in the *Salt Lake Tribune* reported from U.S. Department of Agriculture (USDA) findings that *E. coli* O157:H7 kills an estimated 60 Americans each year and sickens 70,000 more.[24] In addition, 28 percent of cattle entering Midwest slaughterhouses were found to be contaminated.[25] That same year, agriculture researchers studying the path of *E. coli* from farm to food stated: "At least 68 percent of the *E. coli* O157:H7 on meat traces back to a live animal of the same group of cattle."[26] They also found that, in summer, up to 75 percent of cattle feces contain *E. coli* O157:H7. As of September 2001, the USDA Food Safety and Inspection Meat and Poultry Product Recalls web site had listed twenty-one recalls of ground beef for possible *E. coli* O157:H7 contamination in 2001.[27]

Fecal coliform bacteria are a group of bacteria that reside in the intestinal tract of warm-blooded animals. They are used as indicators of water pollution related to waterborne diseases associated with many of the organisms described above.[28] A single cow excretes between 30 and 49 pounds of urine and between 29 to 70 pounds of feces per day, containing 5.4 billion fecal coliform bacteria and 31 billion fecal streptococcus bacteria. Since cattle spend a significant portion of their time in or near streams, lakes, and wetland areas, they can contribute significant numbers of these organisms to surface waters.[29] Because large numbers of cows are grazed on our public lands and their wastes are deposited in the watersheds and in streams and lakes, pathogenic contamination of water supplies is a major risk.

Livestock-Impaired Water Quality, Wasatch-Cache National Forest

Despite extensive documentation of watershed damage and water pollution by livestock, public land agencies and states almost universally protect the livestock industry. They largely ignore the negative effects of livestock on the landscape and our water supplies, as well as the economic costs of this pollution to taxpayers. When confronted with evidence of damage or pollution, government officials engage in denial and obfuscation regarding the cause.

The General Accounting Office has found that permittee and political pressure has prevented reduction of livestock grazing levels on some allotments identified as being very degraded.[30] The EPA admits that regulatory and voluntary efforts to date have been insufficient to solve the environmental and health problems associated with AFOs.[31] Many environmental advocates and scientists have worked on these issues, providing overwhelming documentation of the problems. The following examples provide plain evidence of the ease with which these problems can be documented, and are based on studies conducted by the author in the Wasatch-Cache National Forest in Idaho and Utah.

During 1997, fecal coliform bacteria were monitored in Spawn Creek, Utah, and its tributaries in relation to the presence or absence of cattle. Concentrations of fecal coliform bacteria in Spawn Creek upstream of the cattle varied from 0 to 16 fecal coliforms/100 ml, whereas downstream of the cattle, concentrations ranged up to 201 fecal coliforms/100 ml. Concentrations in small tributaries of Spawn Creek ranged up to 1,370 fecal coliforms/100 ml when cattle were present. Tributary inputs from an adjacent upland pasture with livestock kept fecal coliform levels elevated in Spawn Creek for months after the cattle were removed from the creek area itself. Fecal coliform levels were again elevated in the spring before the cattle returned, because of overland flow and leaching of cattle manure during spring snowmelt. This phenomenon has also been documented in studies by the EPA.[32]

A clear example of the dynamics of fecal coliform concentrations in relation to livestock presence is also provided by data collected in 1998 in Paris Creek, a stream in the Idaho portion of the Wasatch-Cache National Forest. Paris Creek arises as a spring flowing through an ungrazed portion of the forest into a cattle allotment and private grazing land, then downstream into private property where livestock are excluded. Data was collected on two dates, when cattle were present on the allotment and the private grazing land and after they were removed. During the time cattle were present, fecal coliform levels were elevated; after they were removed, fecal coliform numbers declined to near background. Upstream of the cattle, fecal coliforms were absent.[33]

The forest plan for the Wasatch-Cache National Forest states that water "leaving the Forest" in the Logan River meets water quality standards.[34] This statement implies that water quality within the national forest meets criteria, when in fact it does not, and is typical of the obfuscation tactics used by public lands agencies. The Utah Department of Water Quality, using data from a single station representing 106 miles of the Logan River, asserts that water quality meets criteria, implying that the single station represents the entire watershed.[35] Our site-specific studies demonstrate that when cattle are present, water quality is degraded, and in many cases pollutants exceed beneficial use criteria. These are violations of antidegradation standards as well as numeric criteria—violations that go undocumented by the Forest Service and the State of Utah.

Time to Monitor the Monitors

The impacts of livestock on watersheds and water quality are predictable and dire. However, public lands management agencies and many state agencies either ignore the problems or persist in denial, rarely monitor, and refuse to use current science as a basis for decision making. Because monitoring requirements are time consuming, expensive, and seldom carried out, public lands agencies are able to perpetuate the myth that waters on our public lands meet criteria. It is time to act on the basis of our knowledge, and to require land managers and livestock operators to meet ground cover and water quality standards, to have discharge permits, to conduct monitoring, and to keep livestock away from streams.

This dead cow is floating down the San Juan River, through BLM lands in southern Utah. Along with manure, urine, and excess sediment, the livestock industry occasionally contributes whole, rotting carcasses to the West's waterways. Public lands ranchers typically put out their cattle at the start of the grazing period and do not see them again until it's time to take them off the allotment. If an animal is injured or sickened and dies, usually there's no one around to do anything about it.

GUZZLING THE WEST'S WATER
Squandering a Public Resource at Public Expense

George Wuerthner

Livestock production, which includes the irrigation of livestock feed crops, accounts for the greatest consumption of water in the West. Such a water-intensive industry is poorly suited to the arid West. Dewatering of rivers and groundwater pumping for irrigation is a major cause of species decline throughout the region, and water development for agriculture is costly to taxpayers.

When people think of California and water, they often imagine sprawling cities dotted liberally with swimming pools and watered lawns; legions of vain auto owners washing their SUVs, sports cars, and minivans; and endless acres of verdant golf courses—all sucking down rivers both near and far. This image is partly correct—rivers are going dry. But the major reason is not direct consumption by humans—urbanites running sprinklers on their front yards and the like. In California, the major user of water is agriculture, and within agriculture, the thirstiest commodity is the cow.

Overall, agriculture accounts for 83 percent of all water used in California. It's true that California grows the majority of America's fruits and vegetables, so liberal use of water by its agricultural sector would not be unexpected. However, few people would suspect that growing feed for cattle is the predominant agricultural use of water in California. In 1997, 1.7 million acres of the state were planted to alfalfa alone. Irrigated pasture and hayfields consume more water than any other single crop in California—more than a third of all irrigation water.[1] Together, alfalfa and hay and pasturage account for approximately half of all water used in the state.

The story is similar in other western states. In Colorado, some 25 percent of all water consumed goes to alfalfa crops.[2] In Montana, agriculture takes 97 percent of all water used in the state, and just about the only irrigated crop there is hay and pasture forage; more than 5 million acres in the state are irrigated hay meadows.[3] In Nevada—the most arid state in the country—domestic water use amounted to 9.8 million gallons a day in 1993. By contrast, agriculture used 2.8 *billion* gallons of water per day.[4] Altogether, agriculture uses 83 percent of Nevada's water[5]—and the major crop is hay for cattle fodder. In Nevada, while cow pastures are flood irrigated, wetlands at wildlife refuges and the state's rivers often go bone-dry.[6]

Cows are poorly adapted to arid environments. They are profligate consumers of water. Beef production demands an estimated 3,430 gallons of water just to produce one steak![7] Most western rangelands simply don't provide enough forage alone—because the climate is too dry—to run livestock economically. Supplemental feed and irrigated pasture are also needed. Many of the ecological and health impacts of livestock production in the West are associated with the use and abuse of water: the livestock industry alters water quantity and quality and water flow regimes.

The removal of water from streams and aquifers for irrigation threatens many species with local extinction. Rivers and springs are often completely dewatered. According to the Montana Department of Fish, Wildlife and Parks, some 3,778 miles of river are dewatered in Montana annually.[8] Dewatering of streams is a major factor in the decline of many fish species across the West, including most native trout and many salmon stocks.

Dewatering leaves fish stranded in shallow pools, where they are more vulnerable to predators. Fish eggs can be left high and dry when water levels drop during the irrigation season. Many young fish are diverted, along with portions of their streams, into irrigation canals; they subsequently die when water ceases to flow down the canal. In one study in the Bitterroot Valley, Montana, up to 90 percent of the annual production of young westslope cutthroat trout—a species petitioned for listing under the Endangered Species Act—was lost out of some streams because of irrigation canals. Dewatering also leads to higher water temperatures in streams and concentrates pollutants—all to the detriment of native aquatic life.

Another problem is that irrigation exacerbates an already naturally high loss of water from the land into the atmosphere. Huge amounts of water evaporate from storage reservoirs or are transpired into the air by water-hogging crops, such as alfalfa.

Irrigation leads not only to the concentration of pollutants already in streams: new pollutants enter the water, thanks to the diversion of water onto fields and the subsequent return of some of that water to groundwater or surface water bodies. Contaminants picked up from fields include excess nitrogen and

Irrigation sprinkler delivers water to a crop of alfalfa near Mud Lake, Idaho.

minerals leached out of the irrigated soils. In Nevada, for instance, used irrigation water diverted back into the Stillwater National Wildlife Refuge is so full of mercury, selenium, and boron leached from agricultural fields that waterfowl and other wildlife at the refuge are being adversely affected.[9] A similar problem with polluted irrigation return water has been documented at Kesterson National Wildlife Refuge in California.[10]

Groundwater pumping has diminished or destroyed water sources for numerous species around the West. In southern Idaho, water is pumped out of the aquifer to grow hay; as a consequence of this activity, Bruneau Hot Springs—sole habitat for the Bruneau Hot Springsnail—has been drying up. The snail was listed as endangered by the U.S. Fish and Wildlife Service in 1993. Similar groundwater depletion once threatened a host of unique fish and snail species at Ash Meadows in Nevada. Fortunately for these species, the offending agricultural operations were purchased and retired to create Ash Meadows National Wildlife Refuge, but other aquatic species haven't been so lucky. Groundwater pumping for irrigation has already caused some desert fish to go extinct.[11]

Dewatering of streams and aquifers has led to a shrinkage in riparian vegetation and naturally subirrigated lands (such as valley bottom meadows) around the West. This reduction in riparian habitat has serious consequences for wildlife, since an estimated 70 to 80 percent of all western species—plants and animals—are dependent on these thin zones of moisture for survival.[12]

What is particularly ironic about livestock-caused stream dewatering is that it usually makes little economic sense. In much of the West, the value of leaving water in the river to sustain native fisheries or to provide for water-based recreation is often vastly greater than that of the beef produced with the same amount of water.[13] Leaving water in the river to support fishing may ultimately be far more beneficial to local economies than using it for irrigation.[14] Yet we regularly sacrifice the fish to produce beef—a commodity that is already produced more economically and with less environmental impact in other, naturally wetter, parts of the country.

In biology, it can be useful to categorize causative factors as either proximate or ultimate. In the arid West, livestock production is often the ultimate cause of species endangerment, though other factors, often more readily recognized, may be proximate causes. Thus, many dams in the West are proving to be ecological disasters, yet the dams themselves are only proximate causes of deteriorating aquatic ecosystems. Many dams would not have been built but for the demand for water storage for irrigation. Other uses, such as recreation or hydropower production, were often secondary rationalizations for dam construction. Without livestock production, it's likely that many fewer dams would exist in the West.

Dams fragment aquatic systems, preventing free movement of species such as salmon. Dams obviously flood habitat, too, making it unusable for many species. For instance, several species of Snake River snails are now listed under the Endangered Species Act because of dam construction and flooding of the river channel. The change in flow regime and water temperature occasioned by the construction of dams in the Colorado River system has led to the decline of the bonytail chub, the Colorado pikeminnow, the razorback sucker, and the humpback chub.[15] Since most of these dams were constructed for water

storage—with the bulk used for irrigation, mainly of livestock feed—partial blame for the decline in these native species lies with the livestock industry.

Taxpayers carry much of the burden for western water projects that benefit ranchers and the livestock industry. A review of Bureau of Reclamation water projects found that most western irrigation projects are subsidized in three ways.[16] First, irrigators often receive no-interest or extremely low-interest loans for project construction, with repayments scheduled over very long time periods—forty or fifty years or longer. Second, many project costs are forgiven and charged instead to taxpayers, since the projects are seen as having "public benefits," such as recreation. Third, Congress frequently legislates repayment relief.

In a 1996 review of 133 federally funded irrigation projects, the General Accounting Office found that for only 14 projects had irrigators paid, or were scheduled to pay, their entire allotted share of construction costs. In nearly 90 percent of the water projects, irrigation assistance and/or charge-offs accounted for payment or relief of some portion of the irrigators' repayment obligation.[17]

And in a 1988 study conducted for Congressman George Miller, irrigators on the Vernal Unit of the Central Utah Project paid only $3.68 per acre-foot for water that cost the government $204.60 per acre-foot to deliver.[18] Such discrepancies between the cost of water storage and delivery and what irrigators ultimately pay are widespread throughout the West.

Livestock also produce actual, not merely economic, waste. Agriculture is the greatest source for nonpoint water pollution (that deriving from an extensive area, such as a farm field or a grazed hillside, rather than a single site, such as the waste pipe of a factory) in the United States.[19] Pollutants from livestock agriculture include sediments as well as animal waste. Nationwide, the output of livestock manure is more than 130 times that of human waste—yet most of the livestock waste enters waterways and groundwater untreated. The Environmental Protection Agency has found that of all rivers it has identified as "impaired" in some way, agricultural runoff, including animal waste, is the culprit in 60 percent of the cases.[20] And cattle are the biggest producers of livestock manure: more than 1.2 billion tons per year.[21]

The heavy input of phosphorus and nitrogen from manure can elevate microbial activity in streams and lakes, consequently driving oxygen levels down and harming other aquatic organisms.[22] Animal wastes also carry diseases that can be transmitted to humans via water sources. Among the infectious diseases that can be acquired from livestock-contaminated water are salmonellosis, Johne's disease, leptospirosis, anthrax, listeriosis, tetanus, tularemia, erysipelas, and colibacillosis.[23] Again, neither the cost of treating domestic water supplies to make water safe for human consumption nor the cost of disease outbreaks caused by waterborne pathogens originating with the livestock industry are carried by livestock producers.

Livestock affect water quality and quantity through their impact on vegetation and soils. Trampling compacts soils, reducing water infiltration, which in turn leads to greater overland flow and flooding.[24] Trampling also can reduce late-season stream flows by as much as half. The removal and destruction of streamside vegetation by livestock increases bank erosion and allows the current to increase speed and downcutting capability.[25] Livestock damage to watersheds is the most serious cause of excess sedimentation (that is, above the levels of natural erosion) in much of the West. Sedimentation hurts trout and salmon, as well as many lesser known aquatic species, such as freshwater snails.[26]

According to a 1998 study, dams and other water developments are responsible for the endangerment of 30 percent of *all* species listed as threatened or endangered in the United States.[27] Therefore, it should not be surprising that half of the fish species found west of the Continental Divide are listed, or are candidates for listing, under the Endangered Species Act.[28] Of course, livestock production isn't the only reason for these declines, but it plays a major role in many cases. In addition to the livestock impacts to water and streams enumerated above, livestock can threaten fish species indirectly, by degrading habitat to the point that nonnative fish gain competitive advantage.

In the moisture-limited West, raising water-loving livestock makes about as much sense as raising oranges in Alaska. If you can get most of your costs subsidized, and if you and society are willing to ignore the environmental consequences, it can be done. However, as more of the true costs of western livestock production are realized, including the cost in precious water resources, society may want to reconsider this folly.

Most agricultural water [in the West] grows low-value crops. In California, for example, nearly 1 million acres of irrigated pasture requires about 4.2 million acre-feet of water per year—as much as an urban population of 23 million. Pasture, though it is the single largest water user in California, is an extremely low-value crop.
— Marc Reisner and Sarah Bates, *Overtapped Oasis*, 1990

THE SOIL'S LIVING SURFACE
Biological Crusts

George Wuerthner

Biological crusts are assemblages of microscopic organisms dwelling on the soil surface in arid regions.
They are important for retaining water, reducing erosion, cycling nutrients, and diminishing the invasion
of exotic plants. Range managers have typically disregarded the ecological role of biological crusts,
yet they are easily disturbed and destroyed by livestock, and recovery can take years.

The plants most people think of as characteristic of the arid West are the large, vascular types, such as various grasses, sagebrush, rabbitbrush, bitterbrush, cacti, and juniper. Few people are aware of one of the most important groups of plants found on arid lands: biological soil crusts. These are assemblages of tiny, often microscopic organisms, such as cyanobacteria, green algae, fungi, lichens, and mosses, living on or just beneath the soil surface in the spaces between the larger, more prominent vegetation. Although inconspicuous, biological crusts are critical to the productivity of many arid land ecosystems and in some places account for 70 percent of the living plant cover on soils.[1]

Unfortunately, the important role of biological crusts has been unnoticed or ignored by many people, including most range managers and livestock grazing proponents. Traditionally, only the impact of livestock grazing on vascular plants has been a concern in evaluations of rangeland health. Yet recent research suggests that even if vascular plant communities are not affected in any detectable way by livestock, there can be significant differences between grazed and ungrazed sites in the proportion of ground covered by biological crust.[2] And over time, livestock damage to biological crusts can lead to the declining health of the entire ecological system—from increased soil erosion, diminished water-holding capacity of the soil, and less favorable nutrient flows, to greater vulnerability to invasion by exotic plants.

Biological Crusts as Part of Arid Ecosystems

Biological crusts, perhaps in keeping with their rather hidden nature, are known by many terms, such as *microbiotic crusts*, *cryptogamic crusts*, and *cryptobiotic crusts*. They are particularly important components of arid ecosystems, such as those in the Great Basin, the Colorado Plateau, and the deserts of the Southwest, although they can be found in rangeland ecosystems from alpine areas to the Great Plains. Biological crusts are native elements of most western public lands.[3] As a group they are amazingly diverse and often account for a far greater number of species than the vascular plants with which they are associated.[4] For example, in southern Idaho, botanist Roger Rosentreter found 16 vascular plant species and 39 biological soil crust species in 140 plots placed throughout the rangeland plant community.[5]

Biological crusts help to hold the soil surface together and thus reduce soil erosion from wind and water.[6] They play an important role in reducing the impact of raindrops; on unprotected soils (lacking biological crusts), heavy rain breaks up soil aggregates, which leads to the clogging of soil pores and reduces water infiltration rates, sometimes as much as 90 percent.

The crusts also create small-scale roughness or depressions in the surface of the soil that catch water, allowing it to infiltrate, thus reducing sheet erosion.[7] Some biological crusts have microfilaments that weave soil particles together,[8] again anchoring the soil against erosion. Biological soil crusts also act as mulch, reducing evaporative water losses.

Some biological crusts capture and fix atmospheric nitrogen,[9] and all of them can contribute to carbon fixation,[10] providing an important source of carbon for microbial soil populations. Since nitrogen and carbon are both limiting factors in arid environments, maintaining normal nitrogen cycles and carbon deposition is critical to soil fertility and prevention of desertification.[11] Vascular plants growing in soils with intact biological crusts have been found to have a higher concentration of nitrogen than plants growing in soils lacking such crusts.[12]

By occupying the spaces between perennial plants, biological crusts also prevent the establishment and spread of exotic weeds. Most native perennials found in North American deserts tend to have seeds with self-burial mechanisms or that are cached by rodents—ensuring that they will be covered by soil or plant litter and will be able to germinate. However, the seeds of most exotic species, such as cheatgrass, do not use these strategies; rather, they germinate on the soil surface. Where biological crusts are intact, seeds of exotics

Area surrounded by livestock-inaccessible cliffs, Arch Canyon, Bureau of Land Management lands, Utah. A biological crust covers the soil surface on this desert site. These crusts are critical to the conservation and productivity of arid land soils. They hold the soil together, reduce water evaporation rates from the soil, and add nitrogen to the soil. They make it more difficult for weeds and exotic plants to become established. However, biological crusts are easily destroyed by trampling. Livestock are the major culprits in the decline of biological crusts throughout the West.

generally fail to germinate successfully. Indeed, the loss of crusts in the bunch-grass communities of the Intermountain West may be largely responsible for the widespread establishment of cheatgrass and other exotic annuals.[13]

Another unexpected positive aspect of intact biological crusts is their role in creating favorable microclimates. Most biological crusts are dark and can raise temperatures as much as 23 degrees Fahrenheit above that of adjacent surfaces.[14] Heightening soil temperatures can increase nutrient uptake and speed seed germination, photosynthetic rates, and nitrogenase activity for associated vascular plants. Ants, arthropods, reptiles, and small mammals are able to forage more effectively and more quickly with warmer soil temperatures, because they themselves are then warmer and more active.[15]

Higher temperatures may be critical in many desert environments since soil moisture is typically higher during the cooler fall, winter, and spring months, and biological activity may be dependent on favorable soil temperature and moisture. When the dark-colored biological soil crusts are eliminated, the result can be lowered biological activity, with green-up pushed back to later in the spring and early summer. This can negatively affect vascular plants, since they are usually limited by soil moisture, and soils generally dry out as the season progresses into the warmer months.

Finally, biological crusts play a role in moderating fire frequency and intensity. Native plants in the most arid parts of the West are naturally widely spaced, and fires usually do not carry far because of the discontinuous and patchy distribution of fuels. Biological crusts occupy the open spaces between the larger plants—impeding the establishment of exotics, such as cheatgrass, which allow fires to carry farther and also increase fire frequency. So long as the crusts help maintain these mini firebreaks, fires are slowed, and their intensity is decreased.[16] Furthermore, under low-intensity blazes, soil crusts remain intact, limiting potential erosion that may occur in the aftermath of a fire.[17]

Effects of Livestock Production

Various human activities can damage biological crusts, including use of off-road vehicles and even hiking. However, no human activity is as ubiquitous on western public lands as livestock grazing.

Livestock damage biological crusts primarily by trampling them. Except perhaps at the lightest stocking rates, the presence of livestock results in broken, degraded crusts. Livestock also tend to compact soils by walking on them repeatedly. Compaction can lead to changes in soil moisture and nutrient flow, which in turn can alter the species makeup of crusts. These changes may occur before differences in biological crust cover are apparent at the macroscopic level.[18]

Biological crusts need moisture for growth and reproduction. Livestock grazing in the spring, just prior to the beginning of hot, dry periods, limits opportunity for regrowth of crusts. The net effect of the loss of biological crusts is magnified in areas where high-intensity summer thunderstorms occur; heavy rains on unprotected soil surfaces lead to significant erosion.[19] Livestock grazing in summer and fall is also detrimental since biological crusts are particularly susceptible to breakage and fragmentation when dry.[20] Spring, summer, and fall are the primary seasons for livestock grazing on public lands.

Full recovery of badly trampled biological crusts typically requires more than a few years. Since most public rangelands are not allowed more than a season or two of rest, even under the best rest-rotation management plans, complete recovery is essentially precluded under any livestock grazing regime.[21] It is important to understand that biological crusts occur most prominently in ecosystems that did not evolve with large herds of grazing ungulates. Along with the grasses native to such areas as the Great Basin, the Colorado Plateau, and the Mojave, Chihuahuan, and Sonoran Deserts,[22] the biological crusts lack adaptations to the frequent presence of big-bodied herbivores. This fact helps explain why crusts are so vulnerable to damage in the face of livestock grazing.

The negative effects of livestock on biological crusts contribute to lower productivity, accelerated invasion of exotics—particularly cheatgrass—changes in fire regime, changes in soil structure, reduction in water infiltration, higher soil erosion from wind and rain, and changes in energy pathways. These impacts are nearly unavoidable when livestock are present, and thus the policy of allowing livestock grazing on public lands is in direct conflict with such goals as maintaining healthy ecosystems and limiting the occurrence of costly and ecologically damaging cheatgrass-fueled fires.

Bunchgrasses in Wupatki National Monument, Arizona. No livestock are allowed here.

COMRADES IN HARM

Livestock and Exotic Weeds in the Intermountain West

Joy Belsky, Ph.D., and Jonathan L. Gelbard

Exotic weed invasion is one of the greatest ecological threats to grass and shrub ecosystems in the
arid West, and livestock grazing is a leading cause of weed invasion. Livestock carry in weed seeds on their
coats and in their digestive systems; they weaken native plants by grazing them; and they disturb
the soil surface, thereby creating more favorable conditions for exotic invaders
and less favorable conditions for native plants.

Joy Belsky *was a well-respected grassland ecologist and outspoken critic
of traditional range management methods. At the time of her death in 2001,
she was staff ecologist at the Oregon Natural Desert Association, where
she reviewed federal resource management plans and worked to develop
scientific bases for ecosystem protection. She held a Ph.D. degree in plant
ecology from the University of Washington and published over forty-five peer-
reviewed scientific papers on North American as well as African rangelands.*

Jonathan L. Gelbard *holds a master's degree in environmental management
from Duke University and is currently a Ph.D. candidate at the University of
California at Davis. His work focuses on the science and management of
exotic plant invasions in the American West.*

In the midst of the vast expanses of sagebrush and bunchgrass that blanket
the public lands of the Great Basin, a hiker passes through a livestock allotment in which native grasses have been grazed to the ground. Only nonnative
plants, such as pink bull thistle, yellow leafy spurge, and brown curly dock,
remain standing tall. The weeds seem poised to invade bare soils that were
only recently vegetated by native bunchgrasses tall enough to reach a horse's
underside. Dozens of fresh cattle patties dot the area, fouling every breath
with the stench of fresh dung and revealing the cause of the damage. The hiker
stands witness as livestock initiate the invasion and replacement of native
grasslands by weeds.

Exotic weed invasions are possibly the greatest threat facing the grasslands and
shrublands of the arid and semiarid West today. Species-rich ecosystems are
being converted into monotonous "weedlands" as aggressive weeds replace
native plants and degrade habitat for native wildlife. Some of the most notori-
ous invaders—nonnative species such as cheatgrass, medusahead, knapweed,
yellow starthistle, and leafy spurge—have already spread over more than 100
million acres of western lands[1] and are invading new areas at the rate of 5,000
acres per day.[2]

During the past century, a large number of scientific studies have documented
that cattle and sheep are major causes of weed invasions into grasslands and
shrublands of the arid West.

First, livestock carry weed seeds on their coats and in their guts. Where these
seeds are brushed off the animals or excreted in dung, they can grow into
mature plants capable of producing hundreds to thousands of seeds. One study
in Alberta, for example, found that in a single growing season, one cow moved
270,000 viable weed seeds around a pasture.[3] It is clear that the millions of
cattle and sheep now grazing our western public lands are annually moving
tens of millions, if not hundreds of millions, of weed seeds from weed-infested
communities into uninfested areas, even on our most remote public lands.

Second, livestock weaken many native plants by grazing them, thus removing their leaves and flowering stems—that is, their photosynthetic and

Along the Salmon River, Idaho. It may look like one lovely
green sward, but cheatgrass and other weed species have
invaded the lower slope, while native bunchgrass still grows
on the upper slope.

reproductive organs. Grasses and other plants of the Intermountain West are especially vulnerable to grazing by large herbivores since they evolved in an environment that has not been home to many large grazers for the past ten thousand years. Bison are predominantly a Great Plains species, and only low densities of elk, deer, and pronghorn occupy the arid lands west of the Rocky Mountains.[4] As a result, Great Basin grasses and flowering plants evolved little tolerance of herbivory and are severely damaged by close and repeated grazing.[5] In addition, livestock frequently prefer native plants to weeds, which are often covered with spines or contain toxic and distasteful compounds.[6] Where they preferentially consume native grasses and wildflowers, they leave weeds to grow unharmed and with little competition for water and nutrients.[7] Consequently, weedy species grow large and increase in number while native species decline.[8]

Finally, livestock contribute to weed invasions by disturbing the soil surface. Several factors are involved:

- Livestock trample the soil, creating patches of bare ground that serve as natural seed beds for the germination of weed seeds.[9] Trampling also compacts the soil, damaging the roots of native plants and preventing them from acquiring sufficient water and nutrients for vigorous growth.[10]
- By reducing plant cover through grazing and disturbing the soil surface with their hooves, livestock enhance wind and water erosion.[11] Dislodged soil particles then bury the weed seeds, increasing their ability to germinate.[12]
- Livestock hooves destroy fragile biological crusts that blanket exposed soils in deserts, arid grasslands, and shrublands. These crusts, which are composed of algae, bacteria, lichens, and mosses, enrich the soil with nutrients, especially nitrogen, and increase the vigor of native plants.[13] They also stabilize the soil and act as physical barriers to weed invasions. As the hooves of livestock pulverize the biological crusts, they remove an important defense against the invasion of weeds.[14]
- Livestock trampling also reduces the number of soil mycorrhizae, the microscopic fungi that benefit native plants by transporting nutrients and water from the soil into plant roots. Many exotic weeds, such as Russian thistle and halogeton, do not require or benefit from these fungi. As trampling reduces concentrations of mycorrhizae in the soil, the ability of native grasses to acquire nutrients and water is reduced, giving the exotic weeds a competitive advantage over the native plants.[15]
- Livestock deposit nitrogen on the ground in their urine and feces. These nitrogen "hot spots" are concentrated where livestock congregate, especially near streams, water tanks, and salt licks. They intensify invasions by nitrogen-loving weeds, such as cheatgrass and medusahead.[16] Repeatedly, scientists have found that sites that are disturbed *and* also receive high concentrations of livestock waste are the most severely invaded.
- By reducing plant and litter cover and compacting the soil, livestock create warmer and drier soils, an impact especially severe in parched deserts, where plants are already highly stressed by lack of water.[17] These drier soils reduce the vigor of native plants, whereas annual weeds simply go dormant.

Most, but not all, exotic weed species require the type of disturbance and open space created by livestock to germinate and grow vigorously. A few species, however, are able to flourish in plant communities ungrazed by livestock, as can be seen in national parks and other natural areas. This is because vehicles, miners, native wildlife, hikers, wind, and flooding streams can also carry weed seeds into grasslands and disturb the soils. Rarely, however, are these other influences as numerous or as widely distributed as livestock. Studies have shown that in most cases, plant species that invade undisturbed natural areas are less dense inside the natural areas than outside[18]—with localized exceptions, such as sites near roads and trails, or sites disturbed by recreationists and wildlife.

Many in the livestock industry and in federal agencies such as the Forest Service and the Bureau of Land Management ignore the connection between livestock grazing and weed invasions. Since these agencies deny the role of livestock grazing, they seldom reduce the number of livestock allowed to graze public lands, even in areas where weeds are a major problem. Agency personnel prefer using herbicides and biocontrol agents to eradicate the weeds rather than trying to prevent the invasion of weeds in the first place.

Because federal agencies ignore a major cause of weed invasions—that is, livestock grazing, which is also the major land use in the western United States—their recent attempts to hold back the flood of exotic weeds onto public lands have been ineffective. By pouring toxic herbicides onto grasslands and shrublands, rather than working to prevent the invasions, they compound the problem, since herbicides kill beneficial species, poison soil ecosystems, and prepare soils for the next onslaught of weeds.

Preventing weed invasions by controlling livestock is the best tool we have, but unfortunately it is not being used.

The cause of the substitution is overgrazing. When the too-great herds and flocks chewed and trampled the hide off the foothills, something had to cover the raw eroding earth. Cheat did.
—Aldo Leopold, *A Sand County Almanac*, 1949

Frank Church/River of No Return Wilderness, Idaho. This open slope of ponderosa pine and native bunchgrass has never been grazed by domestic livestock. There is little evidence of weed invasion.

SILENT SPRINGS
Threats to Frog Habitat from Livestock Production

Janice Engle

Amphibians are key indicators of ecosystem health. They depend on both aquatic and terrestrial environments and are therefore vulnerable to habitat changes wrought by livestock production. Spring development and water diversion; soil compaction, erosion, and trampling; water pollution from livestock wastes; overgrazing; road building; and reservoir construction are some of the livestock-related causes of amphibian decline.

Janice Engle *holds master's degrees from Boise State University, where she completed her thesis on Columbia spotted frogs, and from Furman University, where she studied biology and education. She has worked for the Georgia Department of Natural Resources, the Idaho Department of Fish and Game, and the Bureau of Land Management and is currently an employee of the U.S. Fish and Wildlife Service. Her number one field assistant is her son, Robby, who has spent the last five summers in the high desert of southwest Idaho with her, capturing and tagging frogs.*

Amphibians are often considered to provide early warning signals of environmental stress because of their physiological and behavioral characteristics and their key positions in food webs of both aquatic and terrestrial systems.[1] They are very susceptible to subtle as well as dramatic changes in their environments and provide us with critical indications of the overall health of the ecosystem. The challenge is to recognize conditions that affect amphibians' survival over the long term and to follow through with sound management practices to restore and maintain stable amphibian populations, thus fostering overall ecosystem integrity.

Livestock impact amphibians in a number of ways. Primary threats to the Columbia spotted frog *(Rana luteiventris)* are habitat loss due to the long-term impact of overgrazing, spring development, irrigation, loss of beaver, road construction, and reservoir development.[2] Because of the Columbia spotted frog's status as a candidate for listing as a threatened or endangered species under the federal Endangered Species Act, research on this frog in southwest Idaho has intensified—to better understand the threats facing the population there, and to discover more about the population dynamics of this species throughout its range in high-desert sagebrush-steppe habitat.

Habitat Fragmentation

The diversion of water for livestock use has caused the complete drying of stream beds in areas known to have once supported spotted frog populations. Spotted frogs utilize riparian corridors for migratory movement between breeding, foraging, and hibernating sites.[3] Loss of moisture in these movement corridors results in an increase in distance between suitable habitat patches. Habitat fragmentation is a well-documented phenomenon that results in the decreased recolonization of isolated patches and an increased risk of extinction.[4] Although diversion of water is easily observable as a major culprit, the lowering of the water table due to compaction, erosion, and the loss of willows is less obvious.

Compaction and Trampling of the Microhabitat

Livestock tend to congregate in riparian areas for green vegetation, shade, level ground, and available water, disturbing the microhabitat preferred

Pacific tree frog. Frog populations are declining around the world for a host of reasons. But in the arid West—where the majority of wetlands, riparian areas, seeps, springs, and other moist habitat is either trampled by livestock or dried up for livestock needs—the silencing of the frogs is, in part, another disastrous consequence of livestock production.

by spotted frogs.[5] Bank-stabilizing rushes, sedges, and willows may be replaced by weedy species and less protective grasses.[6] The decrease in vegetative cover reduces the humidity at ground level and exposes frogs to predators and desiccation. Food (insect prey) is more abundant in vegetated as opposed to denuded habitat.

Spotted frogs have been observed to exhibit natal pond fidelity—that is, they return to their birth site to breed. Because of natal pond fidelity, spotted frogs lay their eggs in specific locations, even when other suitable sites are available. When cattle trample these breeding sites, entire egg masses may become stranded in cow hoofprints. Even if eggs do not freeze in these exposed four-inch hoof pools, they are likely to develop into tadpoles with no hope of transforming into frogs before the pitted shore of the pond or stream dries. In addition, trampled breeding sites increase the density of tadpoles in such a confined area that garter snakes and other predators may wipe out an entire cohort even before desiccation becomes a problem. The trampling of pond banks and seeps may also compress and close subterranean passageways to preferred hibernation sites.

It has long been believed that most frogs of the family Ranidae (to which Columbia spotted frogs belong) hibernate in deep reservoirs as protection from freezing, but researchers have found that green frogs (*Rana clamitans*) avoid overwintering in the sediment of shallow ponds and lakes if alternative sites are available.[7] Instead, these frogs choose overwintering sites that are moist, that remain unfrozen, and that have a higher oxygen content than would be available in breeding ponds. Adult ranid frogs cannot tolerate a lack of oxygen,[8] and waters with high concentrations of nitrogen and phosphorous have a reduced oxygen content, especially at the sediment/water interface, because of bacterial decomposition of organic material.[9] Each cow removes 15 to 20 gallons of water a day by drinking and replaces it with highly concentrated organic matter (in other words, manure and urine), often deposited directly into the water source.

Erosion

Overgrazing leads to exposed soils that are subjected to erosion by trampling, wind, and runoff.[10] Increased runoff and decreased water storage capacity lead to reduced infiltration of water from precipitation or standing water sources into the soil, and reduced deposition of sediment necessary for building stream banks, wet meadows, and floodplains.[11] Erosion causes an increase in water velocity, resulting in a loss of suitable frog breeding sites. Spotted frogs deposit egg masses in eddies of streams, backwash pools and oxbows, and "improved" spring ponds, but not in swiftly flowing waters.

Loss of Willows

Willows help retain the water table, create stable microhabitat conditions and subterranean overwintering sites for many animals, and provide shelter from predators. However, willows (and junipers) are frequently burned in riparian areas in southwestern Idaho to increase vegetative production for livestock. Once willows have been removed, it is difficult for them to reestablish in an area because livestock favor the tender shoots of seedlings and trample the tiny plants.

Willows afford protection for frogs along their migratory paths. If female frogs are unable to overwinter near their natal breeding site (because livestock-induced changes to travel corridors—such as removal of vegetation from riparian zones and severe diminishment of water flows—prevented them from making it back in the fall), they have to travel to their breeding sites in early spring, when temperature extremes are greatest and little vegetative cover is available. In southwest Idaho, the breeding period for spotted frogs is relatively short, from one day to one week in some locations. If females reach a breeding site too late, their eggs could go unfertilized because males cease breeding activities abruptly.

Other Deleterious Effects of Cattle Grazing

The development of springs for watering troughs alters the natural hydrology of watersheds, destroys critical overwintering sites for frogs, and creates new, artificial movement corridors. In 1999, the author found over a hundred newly metamorphosed froglets in a cattle trough. The frogs had migrated up the outflow in an attempt to reach the spring source for hibernation. Not only was this a migrational dead end, but bird droppings at the outflow pipe indicated that many of these recently metamorphosed frogs were funneled right up to waiting predatory birds.

In another incident, livestock trampled an old beaver dam that had provided backwater for critical breeding and overwintering habitat. This dam was lost in 1998, whereupon the frog population plummeted, from an estimated 165 individuals in 1995 to just one in 2000. The velocity of the water greatly increased, resulting in banks (that had previously been observed to contain hibernation sites) tumbling into the stream. The subsequent drop in the water table also left breeding sites dry and more accessible to livestock trampling.

At one site in southwest Idaho, a small pond was fenced to exclude livestock disturbance of a known hibernaculum (frog wintering site). Not only did livestock exclusion have a positive effect on vegetation, but close population monitoring revealed an immediate increase in frog numbers at this protected pond. It was the only pond among many to have an increase in numbers in 1999; in 2000, frog numbers increased or stabilized at the two ponds nearest the protected site as well. Apparently, animals choosing reproductive or foraging sites poorly in one year might be able to improve in following years.[12] Additionally, the pond from which livestock were excluded showed the highest degree of frog survivability across a winter. From 1998 to 2000, the protected pond showed stability in numbers, whereas the unprotected ponds and riparian corridor showed decreases in spotted frog numbers.

Precipitation decreased during the period of this study (1997–2000), but grazing practices were not modified to accommodate dryer conditions. If grazing had been reduced, a decline in frog survival might not have occurred, as demonstrated at the protected pond. Unfortunately, grazing management plans are commonly incorporated into public land leases that are not reviewed or modified annually on the basis of fluctuating environmental changes.

Water trough at Hays Spring, Prineville, Oregon, Bureau of Land Management lands. Water developments for cattle deplete and destroy natural springs and thus harm frog habitat.

NATIVE SNAILS
Indicators of Ecosystem Health

Terrence J. Frest, Ph.D.

Native snails and other mollusks are among the most sensitive indicators of disturbance in western ecosystems. The West also possesses a remarkably diverse mollusk fauna. In arid regions, livestock production is a primary cause of decline for many mollusk species. Livestock trample habitat and snail colonies, compact soils, remove vegetation, alter plant communities, and reduce water quality. Spring development, water diversion, and groundwater pumping associated with livestock production seriously threaten many species.

Terrence J. Frest *holds a Ph.D. degree in geology (paleontology and Quaternary studies) from the University of Iowa. He is a malacologist, specializing in land and freshwater mollusks of the Late Cenozoic to the modern era in western North America. He is senior partner with Deixis Consultants, based in Seattle, Washington.*

These tiny black dots are snails—specifically, endangered Bruneau Hot Springsnails. Like numerous other native mollusks in the West, this species has been placed in jeopardy because of livestock production. Groundwater pumping in Idaho's Snake River Plain has virtually dried up the once voluminous Bruneau Hot Springs—the snail's only home.

Most of us think of snails and slugs only as slimy garden pests. These "pests" are almost always nonnatives, often from western Europe. Native land mollusks, which include shelled snails and partly or wholly shell-less slugs, are not found in flowerbeds. They stick to wild habitats and are not very tolerant of disturbance. In fact, they and their freshwater counterparts (aquatic snails and clams) are among the most sensitive indicators of disturbance in many western ecosystems. Although there are numerous threats to the habitat of native western mollusks, livestock production is a primary factor in the more arid regions of the American West.

Grasping the scope of the problem, and the reasons why snails are vulnerable to the impacts of livestock, requires an understanding of the diversity of our native snails and their particular ecological requirements. Snail endangerment in the West is significant not only because of the potential loss of many mollusk species but also because it signals a general dismantling of key habitats (such as springs and riparian areas) and because terrestrial and freshwater mollusks are often important, if usually hidden, players in the ecosystems they occupy.

Diversity of Western Mollusks

Veteran hikers in the mountainous West commonly come across "bones" (dead shells, bleached white from exposure) of large (1-to-3-inch) land snails. In particular, *Oreohelix* shells are found over most of the western United States east of the Coast Range, especially in the Rocky Mountains. *Oreohelix* means "mountainsnail"; the 120 named forms are most often seen in such locales.[1] But they are equally at home in near-deserts, in open rock talus or outcrops, among grasses, or even in true deserts, such as in Baja California Sur. Others live only in moist mountain forests; a few range right up to the edges of Montana glaciers.

The soft, exposed body, even of shelled forms, makes it easy to see why mollusks might be disturbance-sensitive, so the wide range of habitats mollusks can successfully exploit is remarkable. Still, each species generally has narrow habitat tolerances, and many species are found only in small areas. Each is wonderfully adapted to local conditions, even those as harsh as a lava rock talus in eastern Oregon, for example. But all native mollusks can quickly die if their native habitat changes significantly or they are removed from it.

Western Washington fir forest snails cannot live in eastern Washington, and it is very easy to drown a desert land snail. Likewise, native freshwater snails of the western United States are quite well adapted to and dependent on the original, regionally prevalent pristine water conditions. For example, the Lancidae, a family of large freshwater limpets, lack both lungs and the gills typical of other freshwater snail groups. They respire only through their skin and thrive only in very cold, clear, fast-moving, oxygen-saturated waters. Lancidae are found only in the American West, where such habitat was once common, and now only at a few sites in a few river systems.

Characteristic of western fresh waters are the hydrobiids, or springsnails, once common from Mexico to Canada and from California to the Great Plains. The genus *Pyrgulopsis* has over 100 species of springsnails scattered over most of the West but especially in the Great Basin. These creatures are small, under one-half inch long, and respire with gills. A few live in warm springs, lakes, or streams, but most live only in cold springs, which must be permanent, quite chilly (fed by groundwater), and clear.

Springsnail sites generally have excellent water quality, free of pollution and major disturbance. Such habitats are of great importance all over the West, but especially in relatively dry regions. Western freshwater habitats differ from those of eastern North America in that larger streams in such arid areas as the Great Basin may dry seasonally. Hence, year-round springs are of paramount importance to native animals and plants, as well as to humans looking for dependable water supplies for their own usage or for crops or livestock. Springs are major foci for biodiversity in general.

The western United States displays a near endless variety of landforms, and the snail biota reflects this in its own diversity. As malacologists (mollusk specialists), a colleague and I recently determined that the northwestern United States has about 30 to 40 native genera of land snails and slugs and about an equal number of freshwater types. We recognized about 225 species of land snail, perhaps 36 slugs, and about 150 freshwater snail species. For the whole western United States, the figures would at least double. Many of the mollusks are undescribed (known to be there but not yet scientifically named or studied). We projected a total of 450 Northwest land snails, 72 slugs, and perhaps 300 freshwater snails still "out there." Thus, at least 50 percent of western mollusk species are not yet even named, though mollusks are among the best-known invertebrates.[2]

This situation is unlike that in the East, where most taxa (that is, species, or groups of species classified according to their evolutionary relationships) are known, and new ones are fairly unusual. In the West, even in comparatively well studied groups like mollusks, we face the real and tragic problem of centinelan species—that is, species unknown to science before their extinction and hence never recorded.[3] Currently there is little support or funding for finding and describing new species, and there are few taxonomists who can do so. Hence, we may lose a sizable part of our own native biodiversity with very little trace. Scientists are involved in a race against time to find and describe native biodiversity.

One example: After long neglect, the western U.S. spring biota has begun to attract scientific attention. In 1982, about 400 freshwater snail species were reported from the whole United States.[4] By 1999, the number had reached about 700. Most new ones are western.[5] Smithsonian malacologist Bob Hershler visited some 2,000 sites in his major Great Basin springsnail project. In the genus *Pyrgulopsis* alone, he discovered and described 58 new species.[6] However, by the time his study was published, 2 had already become extinct. At that rate of loss (2 species in three to four years), less than a century would be required to extirpate all. No new taxa will arise during the same period, and one can only wonder how many have already disappeared in the 150 years of western settlement and development preceding Hershler's study.

Should one be concerned about the extinction of springsnails? Basically, animals like mollusks represent what has been termed "the hidden 99.5%" of the diversity of life and "the little things that run the world."[7] Aside from the fact that they are fascinating animals (as every malacologist knows), many people feel that every species has an inherent right to exist and live out its evolutionary life span unmolested. More practically, mollusks are a major basal food chain member, usually at the herbivore (plant-eating) level, in much of the western United States. While not as diverse as some groups, notably insects, they make up in mass what they lack in species. In western forest habitats, they are major consumers of forest floor detritus and recyclers of animal and plant wastes. In stream environments, they are often the major basal food chain herbivore (major aquatic plant consumer and recycler) and serve as an important food resource for fish, amphibians and reptiles, and water-loving birds and mammals. On land, they are a major food item for amphibians, reptiles, some birds, and many small- to medium-sized mammals.

Ecosystem Health Indicators

Snails are exceptional forest and freshwater ecosystem health indicators on two levels. A typical mollusk fauna has about equal numbers of widely distributed and very local (endemic) species. The more cosmopolitan forms are useful for regional assessments of ecosystem health and function; the endemics are most useful at a local scale. Most snails are easy to see and find and have easily preserved shells that provide a record of occurrence even after severe disturbance. The number of taxa at a site is usually small, so identification is comparatively simple. The habitat specificity of many snails means that very fine microhabitats can be detected and monitored.

Mollusks are also effectively sessile—that is, they don't easily move far. This makes them more vulnerable to local disturbance, and also more useful in assessing local habitat stability. Part of a region's mollusk fauna is often relictual—that is, they reflect climate types and particular microhabitats that were once widespread but which today cover much smaller areas or have become disconnected from other fragments of similar habitat. Snail colonies can survive on very tiny patches of ground or in single springs, so long as basic habitat remains unchanged. Thus, endemics often reflect an area's geologic history and past climates on a scale of hundreds or millions of years, whereas generalists reflect today's ecology. This unique historical perspective cannot be gained from migratory groups, such as birds or mammals. And snails or clams make just as good or even better fossil indicators: for one thing, fossil mollusks are much more abundant and likely to be preserved than plants, insects, or vertebrate animals.

Western U.S. mollusks often have little relationship to those of the rest of the country. The mountainsnails, for example, date back at least to the time of the dinosaurs (Cretaceous). But despite (or may be because of!) their humble status, they have survived essentially unchanged, whereas the larger flora and fauna around them have vanished. Fossil land snails look much like living representatives and occur in about the same region, as do such western endemic freshwater groups as the freshwater limpet group, Lancidae, and the western springsnails *(Pyrgulopsis).* These, too, date back at least 65 million years and lived in much the same areas as now. The farthest east *Oreohelix* ever got was to the Mississippi River in one short-lived Ice Age incursion.[8] Recently, Hershler proved that western U.S. springsnails are substantially different from all of their eastern and central U.S. relatives.[9] This difference is one of the more striking aspects of the western biota generally.

Once established, mollusks are often persistent. Land snail colonies, for example, are known to inhabit the same site for 10,000 years or more.[10] Snail fossils of Upper Klamath Lake in Oregon indicate that several freshwater forms have lived in the same area for at least 5 million years.[11] These small animals may thus reflect aspects of the climate and ecology of a region and its history on a geologic scale, something not possible for many other groups.

Native snails may be able to survive natural catastrophes, even on the tremendous scale of volcanic eruptions and Bretz floods (enormous deluges that periodically swept eastern Washington during the Ice Age, following the collapse of natural dams blocking glacially formed lakes). Yet some features of their biology make them vulnerable to extirpation by human-mediated causes. Partly, this reflects the scale, pace of change, and repetitive nature of human activities. Livestock grazing, for example, affects 70 to 75 percent of the total acreage in the American West.[12] Impressive though something like the Mount St. Helen's eruption may have been, the area affected was tiny by comparison. Though effects linger, duration of the single, destructive event is short when compared with many human activities. Among many causes of snail and slug extirpation, two related ones—livestock grazing and spring modification—are preeminent.

There is a relative dearth of objective, quantitative scientific studies on the matter of native mollusks and impacts of livestock production. Many federal land managers have little interest or desire to know how activities such as livestock grazing affect freshwater snails and other species, so there has been slight incentive for investigation. Careful analysis of the existing literature, however, does lead to some fairly well established conclusions. Moreover, increased public interest in public lands management has recently led to a few more objective studies.[13] For specifics on snails, I draw especially on results of examination of over six thousand sites in the western United States since 1988.[14]

Livestock Grazing Impacts

Livestock grazing is a major factor causing extirpation or reduction of both land and freshwater mollusks. With up to 90 percent of all western federal lands allotted to use by livestock producers, grazing is an extremely severe problem, especially in sensitive habitats. Heavily grazed areas often lack land snails altogether or have only introduced pest taxa. Even moderately grazed sites generally have a small fauna, often composed only of generalist species. I know of no instances in which moderate to heavy grazing improved native terrestrial or freshwater mollusk diversity or abundance, but examples of reduction or extirpation are easy to find.

Direct trampling, soil compaction, and colony fragmentation are major problems, but the resulting vegetation changes and the usual reduction in plant biomass and effective cover and shelter are also significant. So are concomitant changes in moisture and insolation (the penetration of sunlight to the ground). Worldwide, perhaps the single most important factor causing land snail death is dessication,[15] which grazing exacerbates. Soil compaction and trampling extirpate snail colonies, tend to dry up springs and seeps, and induce plant community changes, such as simplification, diversity loss, or introduction of nonnative taxa.

The physical consumption of plants by livestock can result in elimination of some plant species (and the smaller animals that consume them) and give competitive advantage to other plants, particularly waste species (species living in highly disturbed terrain, often species that are introduced "nuisance" species) and heavily protected tough or toxic taxa. Grazing also tends to increase insolation, which warms and dries soil during the warm-weather months. Livestock trample and break up large woody debris and other shelter sites used by mollusks for aestivation (summer torpor), hibernation, seasonal protection from dryness, and egg laying.

Locally, manure and urine deposits can change edaphic (soil) conditions and degrade water quality. Changes to soil pH (domestic animal wastes are rich in ammonia and ureic acids) can affect biota in a drastic way. Also, soil erosion is generally increased in grazed areas. Litter or duff, a major source of food and shelter for mollusks, is often largely or totally absent in heavily used sites. Moreover, associated activities—such as rangeland "improvements" to springs, seeps, bogs, riparian areas, or other unique and uncommon microhabitats—have major deleterious effects. Even light grazing by domestic species seems to have substantial negative effects on land snail diversity and abundance.

Together with my colleagues, I have documented specific instances of land snail colony reduction or extirpation. Some of the most egregious involve seven Idaho land snails recommended for federal Endangered Species Act listing as threatened or endangered at the inception of the law in 1973.[16] The two localities of a new species described in 1975[17] had both been either greatly damaged or destroyed by 1995—one largely by cattle grazing and the other by domestic sheep grazing.[18] Fortunately, we were able to locate additional sites. This species is still not listed, and its present condition is unknown. We also noted instances in which a colony bisected by a fence either stopped completely or had only dead shells on the grazed side. Even if areas have been fenced off from livestock, fencing is rarely adequate, and its long-term usefulness is questionable.

It seems likely that native snail taxa are well adapted to light grazing by native herbivores. Indeed, many taxa have evidently survived such levels of herbivory for substantial time periods. Greater-intensity grazing (with its accompanying habitat changes), however, particularly by introduced herbivores, seems to present severe problems. For land snails and slugs in particular, there are severe effects that hit these animals in their biological weak spots.

Livestock grazing negatively affects a wide range of ecosystem functions. It simplifies the plant community (resulting in loss of forage species and possible modification of soil pH, microclimate, and composition); increases insolation; shrinks or removes cover, litter, hibernation, and shelter sites; decreases winter ground temperature; increases summer ground temperature; decreases effective available moisture and humidity; compacts soil; and physically destroys or fragments land snail individuals and colonies. It diminishes water quality and quantity, stream channel morphology, groundwater hydrology, and biodiversity.

Because livestock grazing is often concentrated in riparian areas and lowlands, it has disproportionate effects on the biologically richest parts of the landscape, increasing floodplain erosion and especially reducing aquatic and riparian wildlife. Livestock grazing tends to be heaviest around permanent water sources, such as springs and streams, hence exacerbating negative effects in some of the most important and irreplaceable parts of the ecosystem. Upland sites, often originally forested, are also quite commonly subject to grazing, so that forest floor invertebrates are often doubly challenged.

Multiple or synergistic impacts are typical. A common example in western forests is logging, followed by livestock grazing. Sagebrush removal is often followed by grass seeding and livestock grazing in semiarid lands. Logging in itself has negative effects, but common site preparation practices thereafter administer the coup de grâce. Effects of single practices may be detrimental in themselves, but bearable; combinations are especially destructive.

One side effect of livestock grazing is road building, which accompanies almost all human uses of western lands. The direct effect on an existing colony is extirpation in the roadway proper; site preparation often extends the effects. Road building also increases human traffic, including foot traffic; increases exposure, insolation, and effective ground temperature; changes the local plant community; often leads to introduction of disturbance plants and nonnative and noxious plant and animal species; and initiates damaging side effects, such as herbicide spraying. Even for larger animals, roads may be impassable migration barriers; they are even more so for smaller and soft-bodied animals like snails.

Spring Development and Destruction

A second major cause of mollusk extirpation deserves special treatment. Destruction of springs by livestock grazing, logging, and human exploitation (such as troughing, capping, or diverting for stock use, or appropriating for human water supplies) has already caused extinction of species throughout western North America. The Great Basin region has many such examples.[19] In some Bureau of Land Management (BLM) districts, 90 percent of all named springs have had their native mollusks completely extirpated owing to these causes. In certain areas, pumping of groundwater has depleted many natural springs. As population pressure in the West grows, competition for limited clean-water supplies will grow ever more serious. Native biota will increasingly be pitted against ranching and other agricultural interests as well as urban and industrial users.

Some notion of the importance of springs to native mollusks may be conveyed by the following example. In areas with undisturbed middle- to low-elevation springs, many springs have springsnails. In 1991–1994, we surveyed 501 aquatic mollusk sites, mostly surviving springs, in southeastern Idaho (362 in the upper Snake River drainage and 139 in the Bear River drainage). Nearly all had suffered some human impact. Only 115 (23 percent) retained springsnails. Indications are that about 12 described and undescribed species remain.[20]

Surveys by Hershler and others have led to the discovery of about 100 surviving species of *Pyrgulopsis* in the western United States. This phenomenal fauna is now at high risk of extinction, largely owing to direct and indirect human impacts on springs.[21] Examples of snail survival in springs modified for livestock are few and far between and are likely to be ephemeral. Few cold-water species can live in impoundments or tolerate eutrophification, a process in which dissolved oxygen levels decline because of a rise in nitrogen and phosphorous levels. Livestock wastes are often the major source of these excessive substances. Many western snails graze organic matter from stones and cobbles and require stable, hard substrates and high-quality water. Hence, siltation (such as that resulting from clearcutting and livestock grazing) generally means loss of habitat and at least local extirpation. Western snails often respond negatively to increases in water temperature, as occurs when removal of overhanging vegetation increases insolation.

Spring development generally results in loss of native freshwater mollusk species, particularly endemics. Effects include drying out of the original spring and spring meadow area; disruption of soil, rock, and vegetative cover; and increased stock visits, with accompanying trampling effects and accumulation of acidic manure and urine. Unless the source area is left intact and carefully protected, development can completely extirpate the native freshwater mollusks as well as reduce diversity in other animal and plant groups.

At least 3,500 springs have been "developed"—often at public expense—in Idaho and Montana alone.[22] Their biota are forever lost. Ironically, such development often fails in its primary function—to make water more available to stock. Piping, channeling, and similar activities may disturb the groundwater source or be so inexpertly done as to dry up the spring. Moreover, this type of modification tends to concentrate stock in an extremely limited area, thereby exaggerating the damaging impacts of livestock use.

In the drier parts of the West, springs are major focal points for plant and animal biodiversity. Spring development thus tends to eliminate selectively the relatively few rich islands of plant and animal diversity in arid regions. Even in better-watered locales, springs, seeps, and spring-fed small streams are high-diversity areas and seasonal and moisture refuges for much of the biota. Numerous plant taxa are restricted largely or completely to such areas, and they are sites of concentration and breeding for many insects, amphibians, and reptiles. Many of the more sensitive land mollusk species are restricted wholly or in part to springs and seeps or their borders.

Spring development is one of the most common "range improvements" done on the western public lands. Yet, for native flora and fauna, these actions are anything but improvements. This is well stated in a report by Public Employees for Environmental Responsibility, a federal agency watchdog organization:

> Range improvement projects are a major BLM management emphasis. These projects would be more accurately termed as livestock management facilities. One of the most common range improvement projects is to run water through a pipe from a natural spring to a watering trough. In Idaho

and Montana alone, the BLM and livestock permittees have developed over 3,500 springs on public lands. Some BLM Districts have developed all known springs. Yet in desert ecosystems, natural springs are critical areas for maintaining biological diversity.

The BLM often states that the purpose of these spring developments is to improve riparian area condition. Yet the BLM does not monitor the effectiveness of these projects for riparian improvements. Diverting spring water to a trough results in a dewatered wetland or spring riparian area, and a net loss of wetlands acreage. This effect is not only inconsistent with the Bureau's publicized goals for wetland/riparian improvement but is inconsistent with the national policy of no net loss of wetlands.[23]

There are other causes for spring development. Conversion for human domestic usage is significant, especially where surface streams are not dependable. In upland areas, even where forests and permanent streams are present, springs may be the most dependable water sources and thus are subject to conversion. Another use of upland springs is the "pump chance"—a dug-out area modified to provide a permanent pond as a source of water (ostensibly)

to fight forest fires. However, since many pump chances are virtually inaccessible, their value for fire suppression is nil. Instead, these often function as de facto livestock ponds in poorly watered areas. As with other types of spring development, the result is almost always extirpation of native biota, especially rare and strongly endemic forms.

Portents of Peril for Other Native Species

Livestock production has negative repercussions for nearly all native biota. Activities ancillary to livestock grazing, such as spring development and road building, tie directly into the negative aspects of livestock grazing but are not generally recognized as detrimental themselves, even when their impacts are even more significant than that of grazing per se.

It would tax even the most dedicated biophile to be personally concerned with every species assemblage that makes up the planet's biosphere. Yet, more detailed consideration of even one such group—the native snails of the western United States—can give a sense of our impacts on the many more, effectively anonymous taxa that comprise most of Earth's biodiversity.

Each shell, each crawling insect holds a rank
Important in the plan of Him who framed
This scale of beings; holds a rank, which lost
Would break the chain and leave behind a gap
Which Nature's self would rue.
—Benjamin Stillingfleet, British poet, 1702–1771

BIRDS AND BOVINES
Effects of Livestock Grazing on Birds in the West

Carl E. Bock, Ph.D.

Because livestock grazing is widespread, it has the potential to affect all bird species in western grassland, shrub/steppe, and riparian habitats. However, bird responses to livestock grazing are species- and habitat-specific, so that some populations benefit while others are harmed. Large, permanent livestock exclosures are needed throughout the West to provide critical habitat for the assortment of species that are negatively affected by livestock grazing.

Carl E. Bock *is a professor of biology at the University of Colorado at Boulder. His research interests include ecology and conservation biology, especially as related to grasslands, birds, livestock grazing, and fire. He is the author of over one hundred scientific articles and book chapters and recently published a book,* The View from Bald Hill: Thirty Years in an Arizona Grassland, *with his wife, Dr. Jane Bock, about their work at the Audubon Society's Research Ranch Sanctuary—ungrazed by cows since 1968.*

Hawk caught in a barbed wire fence.

Livestock grazing is the most widespread economic use of both public and private lands in western North America,[1] and it potentially affects all birds associated with grassland, shrub/steppe, and riparian habitats.[2] Birds may be especially responsive to livestock grazing, compared with other native plant and animal populations.[3] North American Breeding Bird Survey data indicate that grassland species are experiencing national declines that are more widespread and dramatic than in any other bird group.[4] Therefore, we have reason to be concerned about continuing negative effects of livestock grazing on birds, and we ought to plan and implement conservation strategies designed to offset these effects.

There is little doubt that domestic livestock frequently have a controlling influence over the structure and function of the ecosystems where they are grazed. Nevertheless, it is essential that we not overgeneralize negatively or positively about effects of grazing on bird populations.

There are three reasons for being careful and case-specific in making our conclusions. First, a variety of factors, in addition to direct effects of livestock grazing, have changed western ecosystems.[5] It is important that bird population changes be properly attributed among these possible causes. Agricultural cultivation of former grasslands, dewatering of wetlands and riparian habitats, fire suppression, introduction and spread of exotic vegetation, extirpation of native herbivores such as bison *(Bison bison)* and prairie dogs (*Cynomys* sp.), predator and pest control, and habitat fragmentation due to agriculture and urban sprawl all have impacted western bird populations. Some, but not all, of these are indirectly related to livestock production, as, for example, dewatering of riparian habitats for irrigation of livestock forage.

Second, ecological consequences of livestock grazing depend very much on the historical association of particular sorts of ecosystems with large herds of native grazers, especially bison. Rangelands of the Intermountain West and Southwest apparently supported few if any bison,[6] and so large herds of domestic grazers represent an alien ecological force that has drastically changed these ecosystems from what they once were. Birds doubtless have responded accordingly. By contrast, grasslands of the central and western Great Plains had a long evolutionary association with large herds of bison, and here livestock may hold the grassland ecosystems in something more like their prehistoric condition than would an absence of domestic grazers.[7]

Third, individual bird populations will respond differently to livestock grazing, depending on their particular habitat and resource needs. Some birds almost always respond negatively to effects of livestock, others usually respond positively, and still others vary in their responses, depending on the particular plant community and the type and intensity of grazing.

My objectives in this essay are four: to describe briefly the various ways that livestock can affect birds, either positively or negatively; to synthesize evidence from the literature about the direct effects of livestock grazing on bird populations in different western ecosystems; to review some case studies illustrating the range of responses of individual species, in particular habitats, to livestock grazing; and to make recommendations regarding livestock grazing and the conservation of bird populations in the American West in general, and on public rangelands in particular.

Livestock Grazing Impacts on Bird Populations

The brown-headed cowbird *(Molothrus ater)* is an obligate brood parasite—a bird that never builds its own nest and always lays its eggs in the nests of other bird species.[8] The host parents then incubate the cowbird eggs and feed the cowbird nestlings, often to the detriment of their own young. Cowbirds forage in shortgrass and edge habitats, usually following herds of grazing mammals that flush their insect foods. Brown-headed cowbirds originally followed bison herds and apparently were restricted to areas where bison were common. Cowbirds switched from bison to livestock when that option became available, greatly expanding cowbird numbers and range. They have since negatively impacted a wide variety of birds in places where avian nest parasites once were scarce or absent—for example, plumbeous vireos *(Vireo plumbeus)* in Colorado[9] and willow flycatchers *(Empidonax traillii)* in California.[10]

The principal means by which livestock grazing affects bird populations is by altering habitat structure and food availability. Grazing invariably reduces the height and ground cover of grasses, at least temporarily, thus depriving many birds of the cover they need, both as a refuge from predators and as a favorable thermal environment for roosting and nesting. Grazing-related loss of grass cover also decreases the frequency and intensity of fire in grassland and shrub/steppe ecosystems, thereby facilitating the spread of woody vegetation that otherwise might be outcompeted by grasses or killed by fire. These changes in habitat structure can significantly alter the avifauna by favoring species that use woody vegetation as cover, and by reducing species whose predator escape strategies require relatively open habitats.[11] Grazing also can reduce grass seed production, significantly affecting winter bird densities.[12]

In riparian habitats, livestock can widen stream channels by trampling the banks, and grazing then reduces or eliminates recruitment of trees and shrubs.[13] These changes in riparian plant community structure can significantly change avian diversity, abundance, community composition, and reproductive success.[14]

Literature Review

In 1993, three colleagues and I summarized the results from forty-two different studies published through the early 1990s that compared bird populations in grassland, shrub/steppe, and riparian habitats in the West that differed in terms of grazing history.[15] Individual birds species were categorized as responding positively, negatively, or in uncertain or mixed ways to the direct impacts of livestock grazing (see Table 1). An additional category that emerged just for grasslands was a group of nine species that appear to benefit from grazing in relatively lush mixed and tallgrass prairies of the central and eastern Great Plains while being negatively impacted in arid shortgrass, desert, and intermountain grasslands.

Relationships between birds and livestock grazing clearly are more complex than this simple summary suggests, since the comparisons involved study areas that differed in such variables as grazing intensity, season, and duration. In many cases there were no ungrazed areas available, and the comparisons involved lightly versus heavily grazed sites. Despite these confounding factors, the data summarized in Table 1 serve to make one point clear—that livestock grazing is neither universally beneficial nor harmful to bird populations, but that the responses are species- and habitat-specific. The data also suggest that livestock grazing has negatively impacted a higher proportion of species in riparian habitats than in grasslands, with shrub/steppe intermediate.

Case Studies

A few examples will serve to illustrate the species-specific nature of avian responses to livestock grazing and some of the complexities involved in considering such aspects as habitat type, direct versus indirect impacts, and historical versus contemporary effects.

The mountain plover *(Charadrius montanus)* breeds primarily in shortgrass prairies of the western Great Plains, where it requires disturbed and open patches of bare ground and very low stature vegetation, both of which are essential for foraging and for early visual detection of approaching predators.[16] Historically, this plover occurred in areas where bison and black-tailed prairie dogs *(Cynomys ludovicianus)* created these habitat conditions. The mountain plover has declined significantly in recent times, and the species has been proposed for federal listing as threatened. It has been argued that livestock are beneficial to this species because of its dependence on heavily grazed habitats,[17] and the plover is included in Table 1 as a species that usually benefits from livestock grazing. However, the situation is more complex, because of two negative impacts indirectly related to livestock production. First, livestock growers have eliminated prairie dogs from much of the historic range of the mountain plover. Second, one of the most significant negative effects on plovers is tilled agriculture,[18] and much of the product of that tilled agriculture is grain for livestock.

The grasshopper sparrow *(Ammodramus savannarum)* is a North American grassland specialist that has experienced widespread recent declines.[19] This species generally relies on intermediate levels of grass cover for nesting and foraging.[20] Grasshopper sparrows can virtually disappear from moderately to heavily grazed western grasslands[21] but may be dependent on grazing or other disturbances to create openings in tallgrass prairie.

The Botteri's sparrow *(Aimophila botterii)* is a southwestern tallgrass specialist that occurs north of Mexico only along the coastal plain of south Texas and in grasslands of southeastern Arizona.[22] So dependent is this species on heavy grass cover that it disappeared from Arizona between 1903 and 1932, following drought and overgrazing in the late 1890s. It remains dependent on stands of thick and tall grass cover, and today it occurs in Arizona only in such habitats. The related Cassin's sparrow *(Aimophila cassinii)* is another species that reaches highest densities only in those southwestern arid shrub/grasslands that are ungrazed or lightly grazed.[23] At the same time, this species requires some woody vegetation on its territories to serve as song perches. Historical grazing-related invasions of shrubs and mesquite into formerly pure grasslands may benefit the Cassin's sparrow, as long as adequate grass cover remains.

The Brewer's sparrow *(Spizella breweri)* likely has had a similar relationship to livestock grazing in shrub/steppe habitats of the Great Basin, profiting from early shrub increases but then suffering from continued grazing-related depletion of herbaceous ground cover.[24]

Relatively little is known about the impacts of livestock grazing on birds of prey, perhaps because their large home ranges frequently exceed the sizes of habitat patches with different grazing histories, especially the sizes of most livestock exclosures. Species such as the short-eared owl *(Asio flammeus)* and northern harrier *(Circus cyaneus)*, which both hunt and nest in relatively heavy ground cover, are likely to do best in areas with moderate to light or no grazing.[25] Species such as the ferruginous hawk *(Buteo regalis)* may benefit from a mosaic of habitat patches, including protected areas in which to place their ground nests, and grazed sites where prey may be more conspicuous and abundant.[26] At the same time, control of ground squirrels and prairie dogs in the name of enhancing livestock production will negatively affect this bird, as it will any raptor with a rodent prey base.

In riparian ecosystems with intact woodland canopy, the sorts of birds most negatively impacted by livestock grazing include many relatively uncommon ground and understory nesters,[27] such as the willow flycatcher *(Empidonax traillii)*, veery *(Catharus fuscescens)*, and yellow-breasted chat *(Icteria virens)*. Other common riparian species that forage on open ground, such as the killdeer *(Charadrius vociferus)* and American robin *(Turdus migratorius)*, may be more abundant in grazed sites. Canopy-nesting riparian birds may be little affected as long as the mature trees survive.

Recommendations

The effects of livestock grazing on wildlife in the American West are substantially scale-dependent. Moderate levels of grazing may increase avian diversity at a local scale, because the habitat needs of many species will be met, as, for example, in a mixture of shrubs and grasses in the desert Southwest.[28] However, at a larger scale, such a uniform management strategy would depress avian diversity, because the most grazing-sensitive species would have no refuge. Without data from replicated long-term livestock exclosures, we almost certainly have a distorted view of what the avifauna of many western North American ecosystems once was and might again become. There are scarcely any such exclosures around that are large enough to support self-sustaining bird populations.

Given the prevalence of livestock grazing in grassland, shrub/steppe, and riparian habitats, especially on western public lands, it is those bird species least tolerant of livestock activities that have the fewest secure places in which to live. It is on this group that our most concerted conservation efforts must be focused. I renew my call for establishment of a system of large, permanent livestock exclosures throughout the American West, to provide critical habitat for that assortment of species negatively impacted by livestock grazing.[29]

Table 1. Numbers of Bird Species Responding in Various Ways to Livestock Grazing in Grassland, Shrub/Steppe, and Riparian Habitats in the American West

	Grassland	Shrub/Steppe	Riparian
Species usually responding positively to grazing	10	12	20
Species usually responding negatively to grazing	7	12	31
Species usually responding positively in taller, more mesic grasslands but negatively in shorter, more arid grasslands	9	Not applicable	Not applicable
Species unresponsive, or showing mixed or uncertain responses	7	10	17

Source: V. A. Saab et al., "Livestock Grazing Effects in Western North America," in *Ecology and Management of Neotropical Migratory Birds,* edited by T. E. Martin and D. M. Finch (New York: Oxford University Press, 1995).

RANCHING IN BEAR COUNTRY
Conflict and Conservation

Brian L. Horejsi, Ph.D.

The arrival of livestock operations in the West changed the human-bear relationship from one of occasional, localized conflict to a generalized opposition to bear presence and survival. By the 1920s and 1930s, grizzly bears were extinct or extremely rare throughout the West. Today, livestock-related conflict is still a leading cause of bear mortality, and government agencies do not act effectively to reduce bear deaths. Livestock also have impacts on bear habitat, further stressing bear populations. Public land managers must prioritize bear survival and habitat needs over those of livestock.

Brian L. Horejsi *is a wildlife scientist with a Ph.D. degree from the University of Calgary and a B.S. degree in forestry from the University of Montana. He was formerly range management forester for the Alberta Forest Service. He has conducted field studies of grizzly and black bears, bighorn sheep, moose, and caribou, and for twenty years he has been a conservation biology consultant and an activist for the protection of native biological diversity. He is the founder of the nonprofit Speak Up for Wildlife Foundation.*

Recorded conflict between bears and humans and their livestock in North America dates back to the first appearance of Europeans. Early European occupants of western North America brought with them an attitude, based largely on fear and lack of knowledge, of little tolerance and no mercy for bears. Settlers killed bears indiscriminately,[1] whether the encounters were haphazard or whether they resulted from pursuit of bears for food and fur. Explorers and settlers had horses and small numbers of cattle for subsistence. Livestock-related conflict between humans and bears at this stage in the settlement of North America was localized and sporadic, with limited impact on bear populations outside inhabited regions.

Even though bear numbers were depleted locally by hunting for food, including market hunting, and in "defense" of settlement, a fundamental shift in the nature of the conflict between humans and bears in the West began when livestock proliferated with the appearance of large-scale commercial operations in the latter half of the nineteenth century. Three substantial changes ensued, resulting in a major escalation in human-bear conflicts. First, bears, which have always exploited every available food source, now routinely found themselves in contact with livestock and those who protect livestock; few safe havens remained. Second, poison and baiting were introduced, efficiently exploiting the basic survival strategy that had made bears so successful— their inclination to eat just about anything. And third, through the permanent occupation of land by residency or through permits issued by public land managers, stockmen conceived it their right to make the land safe for their possessions.

As ranchers settled in, they systematically set about cleansing the environment of "obnoxious" animals. They did so through direct intervention, through political pressure on government to hire hunters, and through private employment of bear hunters. Political reward and/or commercial gain for killing bears meant that pursuit was no longer a function of diminishing effort and diminishing return. With taxpayer support, bear hunting took on a life of its own and resulted in the destruction of the very last bears in much of the U.S. West.

Young grizzly bear.

In roughly one hundred years, these circumstances led to the extinction of the grizzly bear in the Southwest. The last grizzly in New Mexico was killed by stockmen in 1931. The last grizzly in Arizona was killed in 1935, in response to a depredation complaint. At that time it was estimated by the U.S. Forest Service that between 10 and 28 grizzly bears survived in other parts of the Southwest; yet even in those areas "there is an undercurrent of feeling among stockmen that bear should be taken off the protected list, due to their resentment of being forced to first secure a permit from the State Game Department before we will attempt to take bear."[2]

In western Canada, the story was not much different; by 1880, prairie grizzly bear populations had been exterminated by a combination of overhunting, human settlement, and protection of livestock and grain fields. Grizzlies remained only in the foothills and mountains—public lands occupied by tens of thousands of cattle, horses, and sheep—and they were in constant danger.

By 1920, from what had been more or less continuous distribution of bears across the U.S. West, there remained but forty-three disjunct grizzly bear populations left in the lower United States,[3] most of them proving to be nonviable relics—"the living dead" occupying isolated islands of habitat. By 1970, all but five populations (Greater Yellowstone, Northern Continental Divide, Cabinet-Yaak, Selkirk, and North Cascades)—or possibly seven if the Selway-Bitterroot and Kettle populations are included—had become extinct.

The timing was perhaps good fortune—combined, I suspect, with foresight—but the science of grizzly bear conservation dawned in 1957, when brothers John and Frank Craighead began their investigations of grizzly bear behavior and ecology in the Yellowstone ecosystem. These pioneering efforts led to realization of the precarious state of grizzly bear populations in the lower forty-eight states and eventually resulted in the recognition of grizzlies as a threatened species in 1975. So began what I term the "modern era" of livestock–grizzly bear conflict.

This essay focuses on what we currently know and where we must go to end, or to reduce to manageable levels, the conflict between public and private interests: bears and the people of the United States and Canada on one side, on the other, livestock owners and the land and wildlife management agencies, legislatures, and occasionally universities that support them.

We know that the level of mortality that bears can sustain is very low (3 to 7 percent) and that cumulative impacts as a consequence of livestock-related conflict are rarely sustainable. There is also substantial evidence that livestock have a wide array of direct and indirect impacts on the ecology of both black bears *(Ursus americanus)* and grizzly bears *(Ursus arctos)*, their habitat and behavior, and the dynamics and viability of their populations. Together the ecological and mortality impacts of livestock production on bears fall into two broad categories and various subcategories that are not mutually exclusive:

  I. Mortality impacts
    A. Direct killing of bears by livestock owners due to:
      1. Real and perceived depredation of livestock, resulting in intense bear "management"
      2. Threats to human safety when bears are in close contact with agricultural activities
      3. Bear attractants, such as bone yards, livestock concentrations, and garbage, which bring bears, often from long distances
    B. Relocation of bears (for the above reasons), which removes them from a population or results in their death elsewhere
  II. Ecological impacts
    A. Habitat degradation (changes in abundance, diversity, and distribution of vegetation), including:
      1. Introduction of nonnative plants
      2. Forage competition from livestock for important bear foods
      3. Reduction in cover and declining security
      4. Impacts on populations of mammals, fish, and birds that are or were a natural part of bear ecology
    B. Habitat destruction, in which natural features of use to bears are destroyed or altered
    C. Displacement of wary, free-ranging bears that avoid humans, leading to alienation of bears from suitable habitat

Losing Bears: Mortality and Relocation

The most obvious impact of having cattle, horses, and sheep in bear habitat is mortality and direct removal of bears from the ecosystem as a consequence of defense of livestock by humans. An extremely important aspect of this issue is that there is no predictable correlation between bear numbers and livestock conflicts or livestock operator complaints, except that there would be no livestock-related complaints if there were no livestock.

Complaints by livestock owners can lead promptly to very high levels of mortality or relocation of grizzly and black bears, even when a bear population's status is recognized as threatened. Intolerance of bears by livestock interests is the largest single source of reported conflict with bears in rural areas. For example, livestock concentrations represented 30 percent of the attractants leading to human-bear conflicts reported in one study[4] on the east side of the Rocky Mountains in Montana, but these conflicts accounted for 44 percent of reported bear complaints.[5] A summary of the source of conflicts between humans and grizzly bears in the ranching country on the east side of the increasingly urbanized Northern Continental Divide ecosystem in Montana between 1986 and 1990 revealed that 39 percent and 30 percent of 129 incidents involved cattle and sheep, respectively.[6]

A study of cattle-, sheep-, and swine-related conflicts with black bears in Alberta highlighted one of the erroneous and negative responses to bears that still persists: the long-standing practice of assuming that "missing" animals have been killed by bears. In this study, which covered conflicts in the period 1974–1979, 39 percent of the losses for which black bears had been blamed were actually stock that had simply "gone missing." When Fish and Wildlife officers investigated, they were able to associate black bears with only one-third of the claims made by stockmen.[7] Even on more remote public land leases, where 180,000 cattle grazed, only 28 head of livestock (14 percent of claims for lost animals) were confirmed as black bear kills.

These numbers reflect similar situations in the United States. In Wyoming and Montana, for example, 5.2 million cattle and calves grazed in 2000, but fewer than 200 were killed by bears.[8] Other causes of death accounted for 134,000 animals. These large losses should draw attention to the serious consequences of inadequate husbandry practice and caretaking. Management errors—such as inadequate herding, fencing, and salting; lack of veterinary care; and improper stocking levels, age classes, and dates of entry—lead to much higher levels of livestock mortality than do bears.[9]

Sheep and swine, by virtue of their smaller size and more limited mobility, are more susceptible to black bear predation than are cattle. They are highly social, and evidence of their presence is concentrated and readily detected by both black and grizzly bears. Protection of sheep on public lands is difficult: they must be herded daily to find fresh forage, but they must also be confined at night so they can be defended. Both sexes and all ages of sheep are susceptible to bear mortality.[10]

Retribution against bears by land users and wildlife and land managers for real or suspected livestock deaths can be severe. An estimated 62 black bears were killed in the Yellowstone ecosystem on Wyoming national forests over an eight-year period in the 1970s because of perceived livestock conflict.[11] From 1996 to 1998, sheep- and cattle-related incidents in national forests in Wyoming resulted in capture and relocation of 8 adult grizzly bears (2 females, 6 males).[12] On Montana's Rocky Mountain East Front between 1986 and 1990, 77 of 100 so-called nuisance incidents involved livestock and grizzly bears; 7 bears were relocated and 4 were destroyed.[13]

In a small Alberta corner of the same ecosystem, a very high level of mortality and relocation of grizzly bears has persisted for decades, resulting often from alleged conflict between cattle and bears. In the 1970s and 1980s, at least 77 grizzly bears and an unknown number of black bears were removed from this international population.[14] Livestock conflicts continue to deplete this population: 37 grizzly bears were captured and relocated between 1990 and 1997; of these relocations, 27 were due to livestock-related incidents.[15] In British Columbia, I estimate that a minimum of 85 black bears and 5 grizzly bears were destroyed annually between 1992 and 1996 because of livestock-related conflicts.

Bear destruction and relocation by livestock owners and wildlife managers is often indiscriminate. Individual bears are not easily identified and are difficult to "target." In northwest Alberta, between 1974 and 1979, Fish and Wildlife officers killed 60 black bears as presumed predators, but only 18 of these fit criteria defining them as offenders.[16] Even under cases of intensive "management," the impact on bears can be substantial. Few jurisdictions have, or can be expected to have, adequate knowledge of individual bears. In the context of a combined "protection of private property" and "liability reduction" strategy, there is intense pressure to remove "problem" bears. Under these circumstances, nonoffender casualties can be expected to be high.

Extreme reaction in defense of livestock is not uncommon. In response to the suspected killing of four cows (only one confirmed kill) by a grizzly bear, the Alberta Fish and Wildlife Division spent twenty-three days over two years, engaged up to four conservation officers for at least seventy-six workdays, employed helicopter assistance, imposed two area closures on public activity, laid thirteen legal charges against the public in relation to the closures, and handled three nontarget black bears.[17] I estimate taxpayers paid over $20,000 for this extravaganza.

Indirect and progressive effects on bears involved in livestock conflicts cannot be discounted. For example, young bears learn food habits and foraging behavior from their mother. In recent depredations in Wyoming, one female grizzly was accompanied by four cubs (two litters).[18] We can expect that these young bears learned a great deal about, and probably adopted at least some aspects of, this artificial lifestyle. Consequently, the prospect that they will live long and natural lives is dim.

In some cases, few bears participate in livestock depredation, and among those who do, the majority do so only occasionally. Of 7 radio-collared grizzly bears recently monitored in Wyoming, 2 adult males were suspected of sixteen depredations each, 1 was suspected of five depredations, and 4 were suspected of only one depredation each.[19]

The prospects that a "culture of depredation" on livestock can be created within bear populations cannot be dismissed. In areas where individual grizzly bears are not marked and the composition, size, and dynamics of the population is poorly known, bears that are not now preying and that may never have preyed on livestock are often killed or captured and relocated. Those who survive become inappropriately labeled as "management" bears, often negatively changing the way they and their young are perceived and dealt with from that point onward.

As is to be expected from animals whose young accompany their mother for any length of time, livestock allotments become part of the offsprings' home range, and some young bears will learn to associate with and prey on livestock, sharply increasing their risk of lethal contact with humans. If females with young seek a degree of security by moving into areas from which males have been removed[20] for alleged or known depredations, heavy-handed management of male bears may intensify conflicts between females and livestock.

Domestic livestock use of public lands occurs in seasons when native wildlife carcasses, most of which result from late winter mortality, are rarely available; thus, the presence of domestic livestock carcasses creates an unnatural distribution and concentration of bears, leading to increased mortality rates and greater danger to long-term population viability. The unnatural presence of livestock and livestock carcasses beginning in spring and continuing through fall exposes young cubs to scavenging, and perhaps preying on, livestock.[21]

The impacts of livestock on vegetation further marginalize bear habitats, some of which are not highly productive even under natural conditions. With decreased foraging opportunities, bears are forced to search for alternative foods, to exploit livestock or garbage, and to expose themselves and their young to risk in the process. In addition to the direct impact the decrease of foraging opportunities has on bears, it also contributes to the inappropriate labeling of bears as "nuisance." The connotations of this label degrade the public's vision of free-ranging and wild bears.

The cumulative effect of all the preceding livestock-related impacts is to push bears and ecosystems to a state I refer to as ecological overload.

In 1927, the biologist who first directed the U.S. Predatory Animal and Rodent Control Branch, Stokely Ligon, observed:

> As a group, bears are not livestock killers but individuals, especially among the grizzlies, *occasionally* vary from traditional habits and become

serious destroyers of sheep and cattle. Poverty stricken ranges, as result of *excessive range utilization*, and drought often render their usual food so scanty that out of *need* bears become killers; hence, as respects losses from bears, forage conservation would result in increased saving of cattle and sheep.[22]

This observation has been with us a long time, but it is one that present-day management consistently overlooks. Many years later, a study of the feeding ecology of Yellowstone bears noted "the apparent fidelity between vegetatively deficient habitats, predation, and use of garbage."[23] It is this human-influenced relationship, established in areas with a long history of livestock presence and management intervention—including disproportionate reliance on data from "marked" bears, many of which are also "management" bears—that has produced observations that a high proportion of the bear population is involved in livestock conflicts.

Killing or removing bears that are, or are perceived to be, predators of livestock creates a mortality sink for the regional bear population. Even occasional removal of bears at localized sources of conflict, a common management practice, can have far-reaching negative implications for endangered or threatened populations.

Bears in the Yellowstone ecosystem are a perfect example. Recovery to a size and distribution necessary to assure long-term population persistence is a legal obligation under the Endangered Species Act, but that process is hampered by a legally defined "recovery area" that is inadequate in the context of existing human use in and near that zone. The state of Wyoming's "new" guidelines, which allow "lethal take" of grizzly bears[24] on public lands outside the existing recovery area, continues the long-standing practice of dealing with grizzly bears in the face of domestic livestock demands as "nuisance" bears.[25]

In the absence of a determined strategy to relocate or remove livestock from public lands, the "nuisance bear" reaction by managers and livestock operators appears to be but a variation of the reaction that exterminated grizzly bears in the U.S. Southwest over seventy years ago.

Destruction of a bear that occasionally preys on livestock can lead to immigration (replacement) by a more aggressive or less wary bear and a subsequent sustained or higher rate of predation.[26] A relatively sophisticated knowledge of individual bears and of the seasonality of conflicts is important if managers are to deal with bears in a scientifically sound manner; this is information rarely available to management agencies.

In high-priority, public grizzly bear habitat, the continued presence of cattle and sheep is not compatible with grizzly bear population viability or recovery. The susceptibility of calves and sheep to predation, with consequences for bears of subsequent management actions, appears also to be a sound and necessary reason to exclude these classes of livestock from most, if not all, grizzly bear habitat.

Ecological Impacts of Livestock on Bear Habitat

Riparian habitats have historically provided travel corridors and security and feeding areas for grizzly bears.[27] In much of the West, use of these areas by bears has declined, partly because vegetation has been dramatically impacted,

and partly because many of these habitats are now in private ownership, a situation that has led to intensive and often year-long use by humans and livestock. Paradoxically, in some areas, bears are attracted to riparian zones by heavy saturation of livestock odors and remnants,[28] even when these habitats are ecologically degraded.

On Montana's Rocky Mountain East Front, grizzly bears appear to have increased use of low-elevation "shoestring" habitats, many of which are on private land.[29] In these situations, grizzlies are more nocturnal than bears on adjacent wilderness and publicly owned habitat.[30] In addition, the presence of domestic livestock alters the distribution and grazing preferences of these bears, which display a preference for ungrazed areas. Only 10.8 percent of the area was ungrazed, but that area received 20.2 percent of use.[31] These figures indicate that, at times, some bears are avoiding humans and livestock and are altering their habitat use and activity in doing so. Research with black bears indicates they rely "greatly on sight to locate and obtain food," particularly such food as berries, nuts, and insects.[32] Bears, particularly grizzlies, have historically been active during daylight hours because it was beneficial to do so. Consequently, it is probable that bears feeding at night and under pressure of disturbance are less energetically efficient.

Cattle and sheep compete with grizzly bears for forage, particularly in riparian areas. All select the most nutritious and digestible forage available, but through sheer numbers and extended presence, livestock deplete the standing crop of the most nutritious plants. Prolonged use also results in a loss of productivity and diversity, with preferred species the first to be affected. The absence of succulent forbs such as *Heracleum* spp., *Angelica* spp., and *Osmorhiza* spp. in the diet of grizzly bears in Montana's Rocky Mountain Front, as a consequence of overuse by livestock, was termed "alarming" by bear researchers, who concluded that "[t]he denudation of riparian plant communities, and the resultant inabilities of grizzly bears to utilize them, could be a major factor limiting grizzly bear communities in the lower 48 United States."[33]

In another study, long-term sheep grazing over more than a hundred years eliminated preferred species and made aspen communities on the Rocky Mountain Front unproductive of black bear foods by converting understory vegetation to unpalatable species. Remaining fruit-bearing shrubs were stunted and unproductive. These shrubs were considered "influential to long-term population health, reproduction, and survival of black bears."[34]

A significant element in the altered ecology of riparian and wetland vegetation is trampling of vegetation and soils by livestock.[35] While compaction of upper soil layers is a serious problem in a hydrological context, it is also an issue in grizzly bear feeding ecology. Bears are known to select root feeding sites where soils are more porous and thus more easily dug.[36] Compaction of soil by livestock can also degrade site characteristics conducive to the existence of important bear foods. For example, cow parsnip *(Heracleum lanatum)* is often found in moist, depositionally disturbed sites[37] characterized by permeability and water retention—qualities that deteriorate with use by cattle.

Livestock indirectly affect bears through their influence on the number and distribution of native ungulates, such as deer, elk, moose, and bighorn sheep. Bear populations that have a higher proportion of meat in their diet, be it through predation on young ungulates or scavenging of winter-killed carcasses, are more productive and occur at higher densities.[38] Livestock are routinely allocated a considerable portion of the forage on public lands, including crucial winter ranges.[39] The implications are reduced wild ungulate numbers and altered distribution and movement dynamics,[40] all potentially important aspects of bear ecology. The consequences for bears may be dramatic.

In summary, intensive and/or long-term grazing by domestic animals has altered the diversity and productivity of native plant communities and habitats,[41] suppressed regeneration of shrubs and trees important for bear food and security, depressed native ungulate populations and altered their distribution and movement dynamics, and created additional risk of negative and often lethal interaction between bears, livestock owners, and public land managers across western North America.

The Public Interest and Regulatory Solutions

Public lands need fundamental protection through legislation to protect biodiversity. With this essential foundation land managers and the public could assess livestock production practices in their proper and complete context, which would include the recovery and viability of grizzly and black bear populations. Solutions will almost certainly have to conform largely to the following practices:

1. *Removal of domestic sheep from publicly owned bear habitat.* Evidence shows clearly that the presence of domestic sheep is not compatible with bear population viability.
2. *Expansion of grizzly bear recovery areas.* The continued presence of cattle in recovery areas for threatened and endangered grizzly bear populations is not compatible with population viability or recovery. Existing recovery areas appear to be inadequate and require expansion to increase probability that long-term recovery will occur. It will be necessary to incorporate buffer areas and potential grizzly bear habitat on adjacent public lands. Management of livestock on these lands must be consistent with recovery objectives.
3. *Routine evaluation of livestock use of grizzly bear habitat.* Where grizzly bear populations are reduced in density and distribution from historical levels, but are not known to be endangered or threatened, livestock use must be routinely evaluated to eliminate unsustainable bear mortality and habitat degradation. Livestock use of less productive or substantially impacted bear habitats may be something bear populations cannot afford.
4. *Exclusion of calves from grizzly bear habitat.* The susceptibility of calves to predation, with subsequent bear management actions, is a sound reason to exclude calves from grizzly bear habitat.
5. *No bear relocation or destruction.* As a condition of entry to public lands, livestock permittees will be required to consent that livestock losses to wild carnivores will be on a "live with it or don't use it" basis. With very infrequent exceptions, there should be no bear relocation or destruction.
6. *Adaptive management of agricultural activity on public lands.* As bear populations recover, it will be necessary to modify existing agricultural activities on public lands in favor of bears and their habitat (adaptive management). For example, strategies to reduce or eliminate the impact of grazing in riparian zones will be necessary.
7. *Protection of native ungulate populations.* Livestock use of native ungulate habitat will have to change to allow protection and management of ungulate populations as significant and natural components of bear habitat.
8. *Reduction of bear attractants on private lands.* Intensive management on private lands is needed to reduce bear attractants. Livestock carcasses are high-intensity focal points for bears; proactive solutions are necessary.
   a. Bone yards on private lands should be fenced with electrically powered bear deterrent fencing—or, alternatively,
   b. Livestock carcasses should not be available to bears; a carcass removal program on public and private lands is critical. Carcasses should be hauled to a central bone yard or landfill that is bear-proof.
9. *Intensive management at the interface of public and private lands.* Private lands on which livestock are raised attract bears. An essential element of any bear conservation strategy must be to moderate the friction that develops between public and private interests in these areas. Full-time biologists with a strong public interest mandate, specializing in conflict management and backed by strong agency support and resources, will be required for the long term.

There are those who say that the grizzly had to go, that his presence in the Southwest was incompatible with man's. There is some truth in this. Certainly there was no compatibility between the grizzly and the livestock industry—not at Southwest stocking levels. Like the wolf, the opportunistic grizzly was not about to forgo a new and readily available food source—not when this new-found prey had depleted the grizzly's natural food supplies.

— David E. Brown, *The Grizzly in the Southwest*, 1985

A WEST WITHOUT WOLVES
The Livestock Industry Hamstrings Wolf Recovery

Michael J. Robinson

In the early twentieth century, the livestock industry lobbied for a government-sponsored campaign
to eliminate wolves from the West. Today, the livestock industry is the major obstacle to wolf recovery.
Cases in the northern Rockies and the Southwest illustrate how wolf management remains highly biased
in favor of stock growers, even on public lands. Wolf predation was once a significant
ecological force in many western ecosystems; public lands livestock grazing is at
odds both with full wolf recovery and with ecosystem restoration.

Michael J. Robinson *works on predator recovery for the Center for Biological Diversity, headquartered in Tucson, Arizona. He holds a master of arts degree in literature from the University of Colorado at Boulder and has authored dozens of articles and opinion pieces on conservation issues that have appeared in publications ranging from* High Country News *to the* New York Times. *He is currently finishing work on a book about the wolf extermination campaign of the late nineteenth and early twentieth centuries. He lives in Pinos Altos, New Mexico, next to the Mexican wolf recovery area.*

Gray wolf, Montana.

Wolves were exterminated from the American West by a concerted campaign mounted by federal hunters and funded with local, state, and federal revenues. Using poison, traps, and bullets, the government pursued each wolf with the avowed goal of wiping the species off the face of the Earth.

The livestock industry was the sole beneficiary of, and the greatest political impetus for, this campaign. Today, the livestock industry stands at the heart of the opposition to wolf recovery and has blocked, hampered, and sabotaged reintroduction programs throughout the West. Unfortunately, the industry's political clout has profoundly shaped wolf recovery programs that are supposed to be guided by science.

The Northern Rockies

Wolf reintroduction in the northern Rocky Mountains of Yellowstone National Park and central Idaho was contested by the livestock industry and its supporters in Congress for over two decades. Under the Endangered Species Act, critical habitat for a listed species is supposed to be designated, and the species protected from being killed—whether it is reintroduced or recovering through natural recolonization. However, because of the power of the livestock industry, the plan to reintroduce wolves to parts of Idaho and Wyoming resulted in a compromise that designated the wolves as an "experimental, nonessential" population. This designation meant there would be no special protections for wolf habitat and that wolves that preyed on livestock would be killed or removed from the wild. Provisions were even made to allow ranchers themselves legally to kill wolves rather than waiting for government agents to show up and do the job.

The fact that cattle require huge quantities of water means they will always be vulnerable to wolves in the American West. For in this largely arid region, water and water-loving vegetation are so scarce, and scattered over such wide areas, that cattle must be similarly spread out, and that makes protecting them from wolves uneconomical; thus, as their forebears did, ranchers rely on federal agents to kill or remove wolves. Domestic sheep, much less numerous in the West than cattle, are even more vulnerable to predators, especially when flocks are not well protected. Thus, although wolves are a federally listed endangered species, their containment and control by the federal government constitutes one more subsidy that taxpayers provide the livestock industry in the West. (Some ranchers would no doubt happily dispense with this subsidy, as long as they were free to kill wolves at will, including putting out poison baits for them, as was common in the nineteenth century.)

Since gray wolves were released into Idaho and Wyoming in 1995, the federal government's "Wildlife Services" has executed numerous "control actions" because of wolf-livestock conflicts, killing a few dozen wolves either known or suspected of attacking cows or sheep.[1] Particularly egregious has been the capture or "lethal control" of wolves on public lands. Privately owned livestock grazing on public lands clearly take priority over endangered gray wolves, restored at public expense. In addition, somewhere between ten and twenty wolves have been killed illegally in the reintroduction areas.[2] In most of these cases, the perpetrator was never identified or charged.

Gray wolves that migrate naturally from Canada into Glacier National Park and surrounding areas of northwestern Montana—animals that were supposed to have complete protection under the Endangered Species Act—have likewise suffered at the hands of federal government hunters whenever livestock have been killed. As a result, their numbers are not likely to reach the threshold of ten breeding pairs set out in the official wolf recovery plan. So the U.S. Fish and Wildlife Service now proposes lowering the bar on wolf recovery by counting wolves in Idaho and Yellowstone toward recovery objectives in Montana, to remove the species from federal protection entirely and assuage the intense opposition of the livestock industry.

The Southwest

In the Southwest, Mexican wolf reintroduction began in 1998, almost two decades after the last five individuals were removed from the wild for an emergency captive breeding program. The Mexican wolf, a separate subspecies from the gray wolf inhabiting regions to the north, originally roamed throughout Arizona, New Mexico, and Texas, as well as northern Mexico. It, too, was extirpated from the United States by the federal government. Although the Mexican wolf is the most imperiled mammal in North America, it was designated "experimental, nonessential" like its kin in Idaho and the Yellowstone region, in an attempt to buy off livestock industry support for reintroduction.

It didn't work. Soon after the first eleven wolves were released, five were shot, two disappeared, and the remainder were recaptured for their own protection. The livestock industry cheered the killings, and the New Mexico Farm Bureau and Cattle Growers Association filed suit to remove the wolves but were rebuffed in court.

Over the next two years, government management of the Mexican wolves in conformance with their diminished protected status did even more damage than had the poachers. In 1999, the first released Mexican wolves to reproduce successfully in the wild were recaptured from the Apache National Forest in Arizona after they killed a couple of cows on national forest lands. In the course of that recapturing, three of the wild-born pups died from parvovirus. According to the veterinarian who necropsied them, the pups were already in the process of overcoming the disease at the time of capture, but the stress of that event likely caused them to succumb. After the survivors were rereleased

into the Gila National Forest in New Mexico, two of the surviving pups dispersed from the pack at a younger age than is normal for wolves, and one is missing and presumed dead. Biologists do not know whether their period of captivity altered their behavior.

Another pack of Mexican wolves also preyed on cattle on the Apache National Forest, but in this case the cattle were illegally present, having been ordered out by the Forest Service because of severe overgrazing. There was so little forage present that deer and javelina had already been displaced. The rancher failed to remove his cattle, and Forest Service officials failed to enforce their own order—which they later rescinded. Meanwhile, the U.S. Fish and Wildlife Service, unable to force the Forest Service to uphold its own decisions, managed to draw the wolves away to another (overgrazed) allotment on the Gila National Forest. But the wolves had become habituated to cattle, and a week after they discovered and scavenged on a dead cow in the Gila, they began killing cattle again. As a result, seven wolves were trapped, and one pup and a yearling disappeared; both likely died.

A third family of wolves didn't kill livestock at all. But they were also recaptured after scavenging on a dead cow and horse left out on the forest. It was feared that the wolves might learn to prey on livestock after they had tasted beef. In the course of the government's trapping effort, the adult female's leg was injured in a leghold trap and had to be amputated. The pack was rereleased into the Gila, but again, a previously tight family unit broke apart soon after. Two pups were subsequently trapped and returned to cages.

Stock Growers' Appropriation of Western Ecosystems

The conflict between the livestock industry and wolf recovery is more deeply rooted than the seemingly simple question of how to protect stock from predators. For even though a handful of ranchers—representing a tiny minority of the industry as a whole—have forsworn killing wolves and pledged themselves to living with the species, their cattle still displace elk, deer, and other native prey animals. Each blade of grass eaten by a cow means that much less for elk, and each cow shipped to market represents the removal from the ecosystem of hundreds of pounds of biomass that would otherwise take the form of deer, elk, moose, or pronghorn—all of which wolves might otherwise eat.

And when wolves prey on any cattle or domestic sheep, whether they belong to the most recalcitrant predator hater or to a "New Age" rancher, the government's response is the same: removal or killing of the wolves.

In all too many wild places, however, politics precludes recovery efforts even before such conflicts may arise. The livestock industry has so far successfully delimited not only the terms of wolf recovery but also where wolves will be allowed to roam. Thus, the southern Rockies of northern New Mexico, Colorado, and southern Wyoming have been excluded from wolf recovery consideration because the Colorado Wildlife Commission, an appointed body dominated by ranchers, browbeat the U.S. Fish and Wildlife Service into omitting this region. (Activists then persuaded Congress to mandate a habitat evaluation study of Colorado, which revealed that the state could support over 1,100 wolves, but even so the federal agency will not act on its own study and propose recovery.) As a result, wolf recovery on the limited terms proposed by the government will resemble small islands of predators surrounded by lethal "rangelands" dominated by the livestock industry.

The ongoing toll on predatory mammals from the livestock industry's federal killing campaign has skewed one of the fundamental relationships that shapes ecosystems. Wolves provide carrion for other animals, including bears, eagles, crows, magpies, raccoons, skunks, and wolverines. As scavengers, all of these species were decimated in the original, completely uncontrolled wolf poisoning campaign and its follow-up iteration as a coyote poisoning campaign. Today, with poisoning more limited but with wolves still absent from almost all western landscapes, these carrion eaters have lost one of their more reliable original providers.

Wolves kill coyotes, evidently regarding them as competitors. Coyotes similarly kill foxes. The diminutive kit fox of the western deserts and plains is imperiled today partly because of coyote predation. Biologists have speculated that the absence of wolves allows excessive coyote exploitation of kit foxes.

Predators, of course, also influence their prey species. As poet Robinson Jeffers noted, "What but the wolf's tooth whittled so fine / The fleet limbs of the antelope?" Indeed, the speed and keen eyesight of the pronghorn antelope, along with the fortitude of elk and moose, the sense of balance of bighorn sheep on mountain ledges, and the alertness of deer, evolved through predators' culling from the gene pool animals without such survival attributes. Today, with the absence of many predators and the diminishment of others, that predatory force has been profoundly altered. Will the deer, elk, and bighorn of future centuries sport the same traits that for millennia have helped define their very beauty to our species?

What we need are vast, wild landscapes in each type of western ecosystem, free over long periods of time from cattle and sheep, while well stocked with every native predator, including—and perhaps especially—wolves. Without such areas as scientific controls, many of the complex ecological relationships on western landscapes may remain forever closed to our perception and understanding.

In the face of the vast damage done to the American West by livestock production, predators would serve to help heal the natural landscape, to bring ecosystems back toward homeostasis. The systematic killing of predators keeps our otherwise wild places forever artifacts of our own civilization.

On public lands in the great western ecosystem, livestock will not have priority. The grazing of livestock will and must be subordinated to the natural order of the bison and the predator.
—Former secretary of the interior Bruce Babbitt, speaking at
Yellowstone National Park, Wyoming, January 2001

PRAIRIE DOG GONE

Myth, Persecution, and Preservation of a Keystone Species

Lauren McCain, Richard P. Reading, Ph.D., and Brian J. Miller, Ph.D.

Prairie dogs have declined drastically since European settlement of the Great Plains because of human persecution and habitat destruction. Traditionally, prairie dogs have been viewed as competitors for forage with livestock. Today, an emerging scientific view of prairie dogs as "keystone" species indicates a need to alter attitudes toward and management of these important grassland animals.

Lauren McCain *is president of the Southern Plains Land Trust, which works for conservation of native shortgrass prairie, and is a doctoral candidate in policy sciences at the University of Colorado.*

Richard P. Reading *is director of conservation biology at the Denver Zoological Foundation and associate research professor at the University of Denver. He received a Ph.D. degree in wildlife ecology from Yale University.*

Brian J. Miller *is a conservation biologist at the Denver Zoological Foundation. He received his Ph.D. degree from the University of Wyoming, where he studied the ecology of black-footed ferrets.*

Prairie dog (*Cynomys* sp.) decline began soon after Europeans began settling the Great Plains in the mid-nineteenth century. The twentieth century began with an estimated 41 million hectares of black-tailed prairie dog *(C. ludovicianus)* colonies spread across the western grasslands and grassland deserts of North America.[1] By 1960, that number had been reduced to about 600,000 hectares[2]—a decline of 98.5 percent in sixty years.

This trend continues. In 1995, approximately 540,000 hectares of black-tailed prairie dogs remained in the United States,[3] but by 1998 that number had dropped to 280,000–320,000 hectares.[4] This is a loss of 41–49 percent in three years and an overall decline of over 99 percent since 1900. Similarly, the area occupied by black-tailed prairie dogs in Mexico declined nearly 80 percent between 1988 and 1996.[5] The other species of prairie dogs have experienced similar declines. That prairie dog populations have been decimated throughout the western grasslands is undisputed among scientists.

Despite the dramatic decline of all five prairie dog species over the last century, it is only very recently that prairie dog management policy has taken into account any aim other than extermination, largely through poisoning. The Utah prairie dog *(C. parvidens)* did receive early protection in 1973 as an endangered species under the Endangered Species Act, but in 1983 the U.S. Fish and Wildlife Service downgraded its status to threatened, and over the past few years the species has experienced severe declines. As few as 1,500 remain today.[6] The Fish and Wildlife Service listed the Mexican prairie dog *(C. mexicanus)* as endangered in 1991[7] and began considering the black-tailed prairie dog as threatened in March 1999. The white-tailed *(C. leucurus)* and Gunnison's *(C. gunnisoni)* prairie dogs may also warrant listing.

The loss of prairie dogs can be attributed to the activities of one species—humans. People have caused tremendous damage to the prairie dog ecosystem.[8] Prairie dogs once existed in a matrix of colonies and off-colony habitat that shifted in space and time across the landscape of western grasslands of North America. As European settlers began colonizing the region in the late 1800s, they began converting prairies to farmland. Farther west, ranchers noted that prairie dogs ate grass, and in the early 1900s ranchers began

Black-tailed prairie dog.

poisoning prairie dogs in an attempt to provide more forage for livestock.[9] A few visionaries, such as naturalist Ernest Thompson Seton,[10] predicted catastrophic outcomes from the massive prairie dog poisoning campaigns, but such voices went largely unheeded. In addition, at the turn of the century, people accidentally introduced plague *(Yersinia pestis)* to San Francisco. With no natural immunity to this new disease, prairie dogs began dying in large numbers as plague spread across the West.[11]

Today, real estate developers destroy prairie dog colonies and replace them with houses, businesses, and parking lots as western cities swell. Still, ranching and other agricultural interests—the major industries overlapping prairie dog habitat—continue to have the greatest influence on both prairie dog ecosystems and policy, on both private and public lands. Thus, as a result of poisoning and other forms of persecution, plague, and habitat destruction, the landscape matrix of prairie dogs has been destroyed. Today we are left with only small, isolated fragments of a once-vast system.

The loss of prairie dogs is also linked to the decline of several other species. The most prominent example is the black-footed ferret *(Mustela nigripes)*, one of the most endangered mammals in North America.[12] Ferrets depend on prairie dogs for about 90 percent of their diet and require prairie dog burrows for shelter and rearing young.[13] In addition, the swift fox *(Vulpes velox)*, a predator closely associated with prairie dogs, was recognized as warranting listing in 1995.[14] The ferret and fox join the ferruginous hawk *(Buteo regalis)*, the mountain plover *(Charadrius montanus)*, and the burrowing owl *(Speotyto cunicularia)* as species that have all experienced severe declines and that are all closely tied to the prairie dog.[15] Should prairie dogs reach a nonviable population level, a wave of secondary extinctions would likely follow.[16]

A range of federal and state government policies run counter to long-term viability of prairie dogs. Poisoning still occurs throughout the increasingly fragmented range of prairie dogs on both public and private land.[17] Government agencies at all levels permit and often encourage prairie dog shooting on public lands. The federal land management agencies administering public lands—the Bureau of Land Management (BLM), the U.S. Forest Service, the Fish and Wildlife Service, and the National Park Service—have all conducted poisoning to control prairie dogs.[18]

Although in 1999 the Forest Service issued a moratorium on poisoning black-tailed prairie dogs, agency policy mandated limiting prairie dog acreage on the national grasslands and national forests to 1 percent or less.[19] On BLM lands, poisoning can occur at the discretion of the U.S. Department of Agriculture Animal and Plant Health Inspection Service Wildlife Services (formerly known as Animal Damage Control) without the knowledge of BLM administrators.[20] Federal bureaucracies, including the National Park Service, carry out so-called "good neighbor" policies, whereby prairie dogs are poisoned to provide a buffer to adjacent private landholders. Prairie dog shooting is permitted on most national grasslands and BLM lands.

Most states within prairie dog range designate prairie dogs as pest species, and some try simultaneously to manage them as wildlife. Colorado, Kansas, Montana, New Mexico, North Dakota, Oklahoma, South Dakota, Texas, and Wyoming all have laws that mandate the poisoning of prairie dogs. Only Colorado restricts shooting; yet that restriction—which sets a daily bag limit of five animals—applies only to contest shoots.

The Cowboy Myth

To understand our continued persecution of prairie dogs since European settlement of North America, we must examine the origin and elements of anti-prairie dog sentiment. The dominant attitude held by traditional power holders in the policy process—ranchers—is that of intolerance, even hatred, of these animals.[21] This attitude persists despite substantial and growing evidence of the importance of prairie dogs to the ecosystems they inhabit and the relatively small impact they impose on ranching operations.

Ranchers argue that prairie dogs compete with cattle for forage, injure livestock that step in their burrows, pose a public health threat, and cause environmental damage. They further suggest that prairie dogs are abundant—even suffering from overpopulation—and deserve no legal protection. We can view these deeply entrenched beliefs as an outgrowth of a larger myth, "the cowboy myth," that governs the human relationship to the land and treatment of the environment.

A myth is rooted in fundamental assumptions, regardless of their truth, that are believed by a community to the extent that they no longer appear to be myths.[22] Myths are supported by powerful symbols. Myths help people understand and relate to a world far too complicated to understand in its entirety, they promote solidarity, and they are utilized by power holders who manipulate key symbols to explain and justify their use of power.

Generally, rancher attitudes about prairie dogs and other wildlife can be understood as an outgrowth of the cowboy myth that espouses human dominion over other living beings—the philosophy that guided European settlement of the West.[23] This myth is deeply rooted in Christian ethics,[24] as well as in liberal political and economic philosophy.[25] According to the myth, nonhuman animals are either commodified or controlled to minimize interference with human economic activity.

With some exceptions, members of the ranching community tend to view many wildlife species as potential economic threats and have little to no tolerance for potential loss due to depredation (for example, coyotes, *Canis latrans*, and gray wolves, *C. lupus*[26]); the risk of disease transmission from wild to domestic animals (e.g., bison, *Bison bison*,[27] and bighorn sheep, *Ovis canadensis*); or competition for forage (e.g., elk, *Cervus canadensis*, and prairie dogs[28])—this, despite the high affinity of ranchers for open space and some types of wildlife,[29] especially nonthreatening species at low densities, such as modest numbers of deer and grouse.

Proponents of the cowboy myth, especially western ranchers and agency personnel, generally reject data demonstrating that fears about prairie dogs are exaggerated.[30] In the 1970s, scientists began quantifying the impact of prairie dogs on livestock operations. Studies found that cattle averaged no significant changes in body weight when grazed on prairie dog towns.[31] In addition, the annual cost of maintaining control of prairie dogs through poisoning exceeds the annual value in forage gained.[32] Most research finds that total vegetative cover decreases after prairie dogs abandon the land.[33] Furthermore, one extensive review of the literature revealed that although plant biomass in patches created by prairie dogs was lower, it was of higher nutritive value than plant biomass on uncolonized prairie.[34] In other words, the loss in forage quantity was almost fully compensated by a large increase in forage quality.

In addition, public health data indicate that plague poses a more significant danger to prairie dogs than it does to humans. The incidence of plague in humans is negligible. For example, the Colorado Department of Public Health and Environment has documented forty-three cases of plague between 1957 and 1998; only five of those cases were linked to prairie dogs.[35]

The Keystone Role of the Prairie Dog

Despite ranchers' claims that prairie dogs cause environmental damage, a growing body of data suggests just the opposite: that prairie dogs are keystone species. As we use the concept, keystone species are those that enrich ecosystem function uniquely and significantly through their activities and whose impact is larger than predicted by their numerical abundance.[36] Evidence is mounting that prairie dogs fulfill these requirements.[37]

The changes induced by prairie dogs lead to the creation of a unique ecological system referred to as the prairie dog ecosystem.[38] Over 200 vertebrate species have been observed on prairie dog colonies.[39] Some of these species appear to depend on prairie dog colonies for their survival, and many appear to benefit, at least seasonally or opportunistically.[40]

Prairie dogs and the other animals inhabiting their colonies represent a rich prey patch for a large number of predators, including prairie rattlesnakes *(Crotalus viridis)*, golden eagles *(Aquila chrysaetos)*, great horned owls *(Bubo virginianus)*, long-tailed weasels *(Mustela frenata)*, bobcats *(Lynx rufus)*, and coyotes.[41] Some predators, such as black-footed ferrets, are dependent on prairie dogs specifically.[42] Other species, such as badgers *(Taxidae taxus)*, swift foxes, and ferruginous hawks, have been shown to derive substantial benefits from the presence of prairie dogs as prey and eat other prey species in prairie dog colonies as well.[43]

The benefits from prairie dogs extend well beyond simply providing food for predators.[44] Since prairie dogs excavate more burrows than they regularly utilize, they create homes for many animals, such as cottontails *(Sylvilagus* spp.), burrowing owls, and several species of reptiles and amphibians.[45] These species and more also use the burrows as refugia from predators or temperature extremes. As a result, researchers have found that desert cottontails *(S. audonbonii)*, thirteen-lined ground squirrels *(Spermophilis tridecemlineatus)*, and northern grasshopper mice *(Onychomys leucogaster)* exist in higher numbers on prairie dog colonies than in surrounding grasslands.[46] Similarly, studies in Mexico found higher rodent species richness, density, and diversity and higher avian species richness on prairie dog colonies compared with surrounding grasslands in Chihuahua, Mexico.[47] Most of the work to date has focused on birds and mammals, with considerably less on reptiles and amphibians. Similarly, little is known about prairie invertebrates, yet the burrows in a prairie dog colony should offer habitat advantages to invertebrates as well.

Prairie dogs also have a large effect on vegetation structure, productivity, nutrient cycling, and ecosystem processes.[48] The activities of prairie dogs, especially their grazing and clipping of tall vegetation, result in changes in plant composition.[49] In general, the vegetation on prairie dog colonies is characterized by lower biomass (smaller quantity), a greater preponderance of annual forbs (broad-leaved, nonwoody plants such as wildflowers) and short grasses than tall grasses and shrubs, and higher nitrogen content than plants from surrounding areas.[50] Prairie dogs negatively impact some plant species,

reducing the prevalence and controlling the spread of taller grasses and several shrubs, such as mesquite *(Prosopis* spp.), sagebrush *(Artemesia* spp.), and *Ephedra trifurca*.[51] Ironically, prairie dogs are poisoned for livestock interests, but these shrubs preempt grass from cattle, and mesquite makes roundups more difficult.[52]

Prairie dog burrowing activities modify ecosystem processes such as water, mineral, and nutrient cycling. Prairie dogs turn over approximately 225 kilograms of soil per burrow system, which translates to several tons of soil per hectare.[53] By mixing in nutrient-rich urine and manure, prairie dog digging can change soil composition, chemistry, and microclimate, facilitate belowground herbivory, increase porosity of soil to permit deeper penetration of precipitation, and increase the incorporation of organic materials into the soil.[54] As a result, prairie dog colonies support higher numbers of nematodes and higher levels of soil nitrogen.[55] All of these processes contribute to aboveground plants with a higher nutritional content, higher digestibility, and a greater live-plant to dead-plant ratio, creating favorable feeding habitat for other herbivores.[56] Indeed, pronghorn *(Antilocapra americana)* and bison preferentially graze on prairie dog colonies.[57] Scientific models predict that bison can gain weight faster by grazing on a prairie dog colony than on grasslands without prairie dogs.[58]

Prairie dog researchers have concluded that collectively these functions are large, not wholly duplicated by other species (either in form or extent), and that the loss of prairie dogs would lead to "substantial erosion of biological diversity and landscape heterogeneity across the prairie."[59] The prairie dog therefore fulfills the definition of keystone.

Toward an Integrated Preservation Strategy

Reversing the trend of prairie dog decline demands urgent attention. Without changes in prairie dog management, all five species will face extinction. We must now work to reduce prairie dog mortality, recover dwindling populations, and protect habitat across the range of these species. Yet, policy changes to achieve these biological goals are unlikely without addressing the social and political processes that now govern prairie dog management. This task will be one of the most formidable challenges to prairie dog preservation. Initiating effective policy reform means exposing and countering the myths that engender the negative attitudes and values so ingrained in the western agrarian community and among the land and wildlife managers responsible for prairie dogs. Instilling more tolerant attitudes toward prairie dogs may be the key to recovery and long-term viability of their ecosystem. By overturning or altering dominant myths that hold prairie dogs as pests, we can begin developing a new set of myths conducive to sustaining the prairie dog ecosystem.

Shifting policy toward prairie dog preservation will first require enforcing already-existing relevant laws, terminating policies aimed at reducing prairie dog numbers, and developing a more effective legal framework directed toward protecting the prairie dog ecosystem. Currently, three of the five species of prairie dogs enjoy no federal legal protection and only limited protection in a few states.

Despite the mixed success to date, federal and state legal protection for all species is crucial. But laws must be adequately implemented and enforced to achieve their intended effect. Policy does not end once laws and regulations are promulgated.[60] Implementation is equally, if not more, important.[61] For example, properly enforcing prairie dog protection measures means devoting resources toward detecting and stopping illegal poisoning. Unauthorized poisoning has contributed to the decline of Mexican, black-tailed, and probably Utah prairie dogs.[62] We must also work to induce public land and wildlife managers to stop poisoning prairie dogs and start initiating proactive preservation programs. This will not be easy, as they are often proponents of continuing poisoning programs.

Along with legal measures, government agencies should initiate other actions to help reduce prairie dog mortality, recover populations, and protect prairie dog habitat. Agencies should support research focused on preventing outbreaks of plague and reducing mortality rates from plague infestations on prairie dog colonies.[63] Public lands within prairie dog range, such as national grasslands, wildlife refuges, and BLM land, should permit expansion of prairie dog colonies. A somewhat radical proposal, but one that is gaining support among certain members of the public, entails the eventual replacement of livestock with native ungulates, especially bison, on public lands.[64]

Another step toward prairie dog and prairie dog ecosystem preservation entails reconceptualizing and managing prairie dogs as keystone species. Keystone species conservation can be a sound basis for conserving entire natural areas efficiently and effectively, because the keystone helps regulate the entire system.[65] Thus, effectively managing prairie dogs as keystone species would aid efforts to switch from a single species strategy to an ecosystem approach.[66]

Focusing on prairie dog preservation makes not only biological sense but economic sense as well. Protecting prairie dogs and prairie dog habitat is a cost-effective means of protecting other species dependent on prairie dogs or otherwise associated with the prairie dog ecosystem. The black-footed ferret recovery program illustrates this point. The U.S. government is spending millions of dollars on black-footed ferret captive breeding and reintroduction. The Fish and Wildlife Service alone spent $1.5 million in 1991.[67] Ensuring an adequate number of prairie dogs by protecting prairie dog complexes remains a condition of success for black-footed ferret recovery.[68] Therefore, the greatest challenge facing ferret recovery is the lack of sufficient prairie dog populations to support even modest populations of ferrets.[69] Some of the resources now supporting single-species approaches to conservation and protection could be redirected to prairie dog preservation, as prairie dog habitat preservation equates to protection for many other species.

The social and political constraints facing the initiation of a comprehensive prairie dog preservation program are formidable. More integrated, interdisciplinary approaches are desperately needed. A more comprehensive program should identify all of the key stakeholder groups and devise separate strategies for each. In particular, local, state, and federal wildlife, land management, and agriculture agency personnel have demonstrated little concern for prairie dog preservation,[70] yet these individuals exert powerful influence on all phases of the policy process.[71] Broader public relations programs are unlikely to succeed in the face of opposition by agency personnel, especially local officials.

Similarly, the ranchers likely will resist any outside proposal that does not simultaneously bolster their own interests. Yet the economic and social crises of many ranching communities today may be a door to change. As family ranches become a thing of the past and the once "self-reliant" and libertarian rancher increasingly looks for government assistance to stay in business, prairie dogs serve as convenient scapegoats for the deeper, more complex problems of the livestock community.

On the other hand, in times of stress, such as economic downturns, communities are often most likely to seek out new ideas and embrace new myths.[72] Old myths may no longer work to justify traditional practices, including the control of native wildlife. Thus, some ranchers are indeed seeking out new ways, even trying to adopt more environmentally conscientious practices. If such innovators are successful, we may see others open to alternative techniques. Such a transformation is unlikely to occur among those who have grown up with the belief that prairie dogs are pests; however, younger generations may indeed reach out to new ideas. And if ranchers as a group are to be persuaded to adopt more tolerant attitudes toward prairie dogs, the best chance is for individuals among them to initiate and promote the change.

Outside the ranching industry, it will be up to prairie dog supporters—preservationists, wildlife biologists, and policymakers working with prairie dog-friendly ranchers—to promote symbols that will help forge more tolerant attitudes toward prairie wildlife. The symbols most likely to resonate with the ranching community are symbols that already hold meaning within this group. Preserving prairie dog habitat also means maintaining open space, practicing land stewardship, and retaining a sense of wildness to the western plains—all ideas that are already embraced by the traditional ranching community.

Beyond confronting the major myths that have helped reduce the prairie dog ecosystem from a once-rich network of biodiversity to a fragmented patchwork, American grasslands and the prairie dog ecosystems they include need a national public relations campaign. Unlike wolves and whales, prairie dogs do not muster the great charisma that the large mammals and predators enjoy. Moreover, grasslands as wild places lack resonance in the public's consciousness. Yet the Great Plains were once teeming with wildlife, rivaling the Serengeti Plains of Africa.[73] Along with the roving herds of bison, America's grasslands were also rife with wolves, elk, pronghorn, and grizzly bears. We must work to promote this image of the western plains as a place for wild nature, not just cowboys.

Black-tailed prairie dog pup.

SAGE GROUSE
Imperiled Icon of the Sagebrush Sea

Randy Webb, Ph.D., and Mark Salvo

Survival of the sage grouse is inherently tied to the health of sagebrush habitat. The range and numbers of this charismatic bird have shrunk dramatically over past decades owing to numerous factors, with livestock grazing and associated activities such as fencing, water developments, "brush" clearing, and herbicide spraying of sagebrush habitat chief among them. Indirect impacts of grazing on sage grouse include juniper invasion and the spread of exotic weeds, such as cheatgrass.

Randy Webb *is certified as a senior ecologist by the Ecological Society of America and is also an attorney. He holds a Ph.D. degree from Washington State University and a J.D. degree from the University of Oregon Law School. He has been a professor at Indiana University, Pennsylvania State University, and the University of Wyoming and has acted as a consultant to the U.S. Department of Defense and the Nature Conservancy. He has published numerous scientific papers, including several reports on the impacts of livestock grazing on sage grouse. He notes that he "lives in North America and stores his tent in Eugene, Oregon."*

Mark Salvo *serves as grasslands and deserts advocate for the American Lands Alliance in Portland, Oregon, where he coordinates American Lands' campaign to protect the sage grouse, the "spotted owl of the desert."*

Grazing of domestic livestock has affected the entire range of the sage grouse. Livestock production, along with its associated activities, is probably the number one threat to sage grouse survival. The historic range of this large, charismatic bird closely conformed to the distribution of the sagebrush steppe, covering much of sixteen western states and three Canadian provinces. Between 1 and 2 million sage grouse (and perhaps as many as 10 million) lived in the West when Lewis and Clark saw them in 1806, and huge flocks were reported still to darken the sky nearly a century later.[1] However, since 1900 the distribution of sage grouse has been shrinking dramatically, with complete extirpation of populations in Arizona, British Columbia, Kansas, Nebraska, New Mexico, and Oklahoma.[2]

The sage grouse is probably best known for its fascinating mating ritual. In early spring, male grouse congregate at leks, ancestral strutting grounds that are clear of large sagebrush and tall debris. "To attract a hen, cocks strut, fan their tail feathers, and swell their breasts to reveal bright yellow air sacs. The progression of wing movements and inflating and deflating air sacs elicits an acoustic *swish-swish-coo-oo-poink!*"[3]

Sage grouse must have sagebrush to survive, as they derive food, shelter, and cover from the shrub. But the birds need more than sagebrush: forbs (broad-leaved, nonwoody plants, such as wildflowers) and insects are essential food items. Tall grasses are also required for cover from predators.

Sage grouse prefer different habitats through the year. Ideal nesting habitat has a sagebrush overstory and a thick grass/forb understory.[4] Both provide food, shelter from the wind and sun, and cover from ground predators and raptors.[5] As chicks grow, they follow their mothers to sagebrush stands and forb-rich areas, including wet meadows and riparian areas.[6] Forbs and insects are required foods of sage grouse chicks. Good winter range will provide sage grouse with access to sagebrush under all snow conditions, as the grouse eat only sagebrush during the winter. During the year, sage grouse may range over one hundred miles of terrain to meet their seasonal needs.[7] Thus, sage grouse survival depends on vast expanses of healthy sagebrush habitat and functioning hydrologic systems.

Male sage grouse in courting display.

The sagebrush steppe is a little-loved landscape and has received minimal conservation attention in the past. Sagebrush habitat has been fragmented, damaged, and destroyed by livestock production, as well as agricultural conversion, suburbanization, mining, and off-road vehicles, among other uses. Altogether, the decimation of sagebrush habitat has reduced the size of the sage grouse breeding population to an estimated 140,000 individuals in Canada and eleven western states.[8] (A separate species, the Gunnison sage grouse, has also declined throughout its relatively small range in Colorado and southeastern Utah.)

The Impacts of Livestock

The geographical scale alone on which western livestock production occurs guarantees it will significantly affect sage grouse. Ungrazed sagebrush steppe in the Intermountain West has declined by 98 percent or more since European settlement.[9] The U.S. Department of the Interior has noted that "although only about 10% of the sagebrush steppe that dominates the Intermountain West has been converted to anthropologic habitats, more than 90% of this community is degraded by livestock grazing."[10]

Livestock grazing has multiple negative effects in arid ecosystems. Grazing changes habitat structure and species composition in both upland and riparian sites, spreads exotic species, and causes erosion and shrub encroachment into riparian areas.[11] Even light grazing is known to put stress on the herbaceous plants favored by livestock and required by sage grouse.[12]

Livestock grazing reduces water infiltration rates, reduces cover of herbaceous plants and litter, disturbs and compacts soils (creating microsites for invasion of exotics), and increases soil erosion, which reduces the productivity of vegetation.[13] Compacted soils may take centuries to recover in arid climates.

Grazing retards vegetative recovery from fires, from grazing itself, or from other disturbances. It can permanently diminish the productivity of an ecosystem, lower vegetative cover, reduce biomass and biodiversity, increase soil deterioration, and encourage other aspects of desertification.[14]

Even when livestock are not present at the time sage grouse use an area, they remove and stress sage grouse food and cover plants. These plants may not regrow at all or may not regrow to sufficient height, density, or nutrient composition in time for sage grouse to use them.

Hens with broods avoid meadows where livestock grazing has created steep, eroded stream banks, dense shrub cover, and low forb availability.[15] Riparian areas, critical for brood rearing, are heavily grazed in the arid lands that form the range of the sage grouse.[16] Livestock grazing has also caused plant communities to shift to species that tolerate heavy grazing, such as Canada thistle, Scotch thistle, and stinging nettle.[17] Although the U.S. Fish and Wildlife Service recognizes that "high-intensity grazing is incompatible with nest success" needed to ensure sage grouse population viability,[18] efforts to protect the bird from such impacts have been minimal to date.

Trampling

Trampling is possible on nests and eggs. However, slow-moving, dim-witted livestock are unlikely to present much of a threat to chicks or adults. A greater problem is the trampling of wet meadow areas needed by juvenile sage grouse.

Livestock trampling of such areas—often exacerbated by "development" of springs and seeps—typically transforms these naturally highly productive areas into little more than mudholes filled with cattle excrement.[19]

A common strategy used by public land agencies to protect stream areas is to move livestock use into uplands by fencing, herding, or placing "attractants" in the uplands, such as salt blocks or stock tanks. However, uplands are also sage grouse habitat, and moist upland sites such as streams, wet meadows, and springs provide important summer and fall habitat for the birds, especially in arid areas.[20]

Livestock are also known to severely degrade cryptogamic crusts[21]—the assemblages of tiny, often microscopic, organisms, such as cyanobacteria, green algae, fungi, lichens, and mosses, living on or just beneath the soil surface—which are important in providing favorable sites for the germination of vascular plants,[22] regulating soil hydrology,[23] stabilizing soil against wind and water erosion, retaining soil moisture, and promoting equable soil temperature regimes.[24] Intact cryptogamic crusts prevent invasion by cheatgrass and similar exotic species. These crusts may require fifty to one hundred years to recover from livestock trampling.[25]

Removal of Plants

Forbs and other understory plants are critical as food to sage grouse chicks and to insects, which are also key components of the chicks' diets during their early development. Even light grazing tends to remove preferred food plants of sage grouse, and heavy grazing can create barren spaces between sagebrush plants.[26] As the cryptogamic crust and forb and grass understory are destroyed, sagebrush plants increase in size and abundance, and canopy closure often occurs.[27] This can prevent reestablishment of the forb and grass understory, even after livestock have been removed. Also, many areas are so depleted of forbs that there is no seed source to recover the site, even if livestock grazing ceases.

Forbs and grasses near the nest provide shelter from wind and sun, and visual concealment from predators.[28] Livestock grazing directly harms sage grouse by removing sheltering plants near the nest.[29] Such absence of sheltering plants is known to diminish both nesting success and chick survival.[30] Sage grouse avoid grazed sagebrush steppe during the nesting season[31]—one reason many areas in Washington grazed by livestock no longer support sage grouse.[32]

Ecosystem Processes and Invasion of Exotics

Livestock grazing affects plant community composition by favoring plants tolerant of grazing over those less tolerant. Some scientists suggest that these alterations in the competitive balance in the plant community may be one of the most important pathways of habitat degradation in the sagebrush steppe.[33]

Livestock grazing also increases the vulnerability of plant communities to invasion by alien species.[34] Wandering livestock spread weed seeds attached to their bodies or dropped out in feces. Basin big sagebrush and Wyoming big sagebrush communities are highly susceptible to the incursion of cheatgrass.[35] Cheatgrass increases fire frequency, becoming the dominant plant species over vast areas and destroying sagebrush—effectively eliminating sagebrush habitat. Once invaded in this way, sagebrush communities are nearly impossible to restore. Consequently, the entire Great Basin is now "an endangered landscape."[36]

Fences

Fencing is often used for livestock operations and to delineate property boundaries. Fencing may vary from a few strands of barbed wire to a woven mesh. Sage grouse are killed by being caught on stranded barbed wire fence.[37] Even worse is a woven mesh fence that does not permit the birds to pass through by walking, their preferred mode of locomotion.

Livestock and other uses, including maintenance, often create trails along the fenceline. Such trails provide travel corridors for predators, increasing the risk to sage grouse populations.[38] Since fence posts also provide perches for avian predators,[39] sage grouse are known to avoid them.[40] Fences thus fragment habitat and populations.

Conversion of Habitat

It is not just the grazing and trampling effects of livestock that endanger sage grouse. Livestock production in the West requires a vast agroindustrial infrastructure, which consists of power lines, fences, watering facilities, dams, buildings, and winter pasturage for livestock. This infrastructure itself constitutes a complex of serious and ongoing threats to the survival of sage grouse.

Conversion of habitat, such as to agriculture or housing, completely eliminates the land's utility for sage grouse.[41] Habitat conversion is the only threat that is likely to be greater than livestock grazing,[42] particularly in Colorado, Washington, and perhaps in Idaho. Most of these converted acres go into agricultural production, such as wheat in Washington and potatoes in Idaho, as well as hay and alfalfa, to serve as feed for livestock. To a lesser degree, habitat conversion occurs for housing and industrial development.[43] In many areas, large reservoirs have been created, frequently providing irrigation water for the production of livestock feed and fodder. This further destroys and fragments grouse habitat. Throughout the Columbia River Basin and Snake River Plains, dryland farming, made economically viable by subsidized irrigation water delivery from dams, has resulted in the destruction of immense expanses of sage grouse habitat.

The infrastructure that ties rural and suburban areas to urban centers—networks of roads, power lines, pipelines, and so forth—also affects sage grouse by directly destroying, degrading, and fragmenting habitat.

Mining, particularly strip mining, is another type of habitat conversion. Mining operations release a diversity of pollutants, create additional roads, and add traffic to existing roads.

"Rangeland Improvements"

Treatment of so-called rangelands (an anthropocentric term, as if their only proper use was for the ranging of livestock) is often a matter of killing or controlling sagebrush to increase the amount of grass for domestic cattle and sheep. "Improving" rangeland can entail the use of defoliants and other herbicides and pesticides, blading (bulldozing of sagebrush), chaining (dragging a heavy chain between two vehicles to remove sagebrush mechanically), and fire.[44] The Bureau of Land Management alone has "treated" (destroyed) sagebrush on over 1.8 million hectares.[45]

Often, sagebrush is removed to allow for the growing of crested wheatgrass (an exotic) for livestock forage.[46] All studies of the impact of crested wheatgrass plantings on sage grouse show this activity to be detrimental.[47] The monocultural plantings usurp the place of native plants that sage grouse use for food, shelter, and concealment.[48] The alien grass also alters natural fire regimes. As one scientist stated, "Crested wheatgrass has no nutritive value to sage grouse . . . attracts few insects that can be used by sage grouse . . . [and provides] little cover value or structure."[49] The grass is too short even when mature to hide birds. Vulnerability to predation is exacerbated by typical straight-row plantings, in which predators gain long lines of sight for spotting grouse. The time required for habitat recovery can be seven to thirty years or more.[50] The fact that sage grouse occasionally use areas planted to crested wheatgrass as lekking grounds is attributable to the high site-tenacity of this bird. It is unlikely that sage grouse prefer crested wheatgrass for lekking or for any other need.

Juniper and Pinyon Pine Invasion

In many areas, native tree species have invaded former sagebrush steppe because of fire suppression and cattle grazing.[51] Juniper, in particular, is outcompeting formerly dominant grass species. Trees, along with fences and power lines, serve as raptor perches and create "killing zones" up to one square mile in area. Sage grouse avoid these areas, and hence the proliferation of such perches for raptors has meant a decline in available grouse habitat.[52] Under winter conditions, when access to food and other resources may be severely limited, sage grouse can be forced to come within forty meters of trees. However, sage grouse attempt to stay a half mile from trees, if they are within visual range.[53] Thus, cattle grazing can change one vegetative community into an entirely different one. These indirect effects can completely eliminate sage grouse from vast areas.

Endangered Species Act Listing Needed

Sage grouse are the proverbial canaries in the coal mine. Where the sage grouse struggle to survive, the landscape has suffered serious degradation. Because the United States lacks an ecosystem protection act, conservationists must address the symptoms of degraded ecosystems through the Endangered Species Act. Listing the sage grouse as a threatened or endangered species may be the necessary first step to reverse their decline and restore the sagebrush ecosystem.

WHERE BISON ONCE ROAMED
The Impacts of Cattle and Sheep on Native Herbivores

Bill Willers

Livestock grazing seriously impacts wild ungulates such as elk, bighorn sheep, and pronghorn through forage competition, disease transmission, social displacement, habitat degradation, and plant community alteration. On the majority of public lands, more forage is allotted to livestock than to native large herbivores. Whereas native species are an integral part of the ecosystems in which they have evolved, alien, domestic animals represent a denial and violation of ecological integrity.

Bill Willers *is emeritus professor of biology at the University of Wisconsin, Oshkosh, and editor of two recent anthologies,* Learning to Listen to the Land *(1991) and* Unmanaged Landscapes *(1999). He is founder and current president of Superior Wilderness Action Network (SWAN). For many years he has summered on family land along the North Fork Shoshone River, just east of Yellowstone National Park.*

Domestic sheep. On many public lands, if not most, livestock far outnumber the large native grazers, such as elk, deer, and bighorn sheep.

Consider Yellowstone, not simply a region harboring the crown jewel of America's parks but, for many, a standard for their very idea of wilderness—such a treasure as to have captured the imagination of millions of people around the world who may never hope to see it. Now consider the fact that in the Yellowstone ecosystem as a whole, the ratio of domestic livestock to all wild ungulates combined (elk, moose, mule deer, white-tailed deer, pronghorn, bighorn sheep, mountain goat, and bison) is greater than 2:1. For every wild hoofed mammal there are two domestic ones.[1]

Particularly startling is the ratio of domestic sheep to wild bighorn sheep, which succumbed in huge numbers from diseases and parasites introduced through domestic herds. In the Yellowstone ecosystem, the most optimistic figure for bighorns is 7,800, compared with more than 265,000 domestic sheep grazed there, a ratio of 1:34.

Actually, the Yellowstone ecosystem (the national park and surrounding lands—an area of some 16 million acres), beautiful and beloved as it is, and 76 percent of which is federal land, is managed with a great deal more care and attention to natural conditions than is the rest of the public domain. Not far to the east, for example, in the Bighorn Mountains, the creatures for which the range was named are gone, save perhaps for a few introduced individuals, some with identifying collars hanging about their necks. In their place are grazed some 52,500 of the sheep industry's stock. On warm summer days in the Bighorns, a visitor might experience surprise at seeing what at first appear in the distance to be banks of snow, only to have them be revealed as densely packed herds of domestic sheep.

Negative consequences as a result of livestock grazing, past and present, have been reported for "virtually all wild ungulates."[2] Joseph Townsend and Robert Smith minced no words: "Livestock grazing is the single most important factor limiting wildlife production in the West. It has been and continues to be administered without adequate consideration for wildlife, especially on federally owned lands."[3] Likewise, Frederic Wagner called livestock "the most ubiquitous influence for change in the West."[4]

The lush, heavily vegetated riparian areas adjacent to watercourses are of great biological richness, so the destructive impacts on them by cattle are particularly important in the arid West, where riparian areas constitute but 1 percent of federal rangelands. According to a 1990 publication by the Environmental Protection Agency, "Extensive field observations in the late 1980s suggest riparian areas throughout much of the West were in the worst condition in history."[5] Moreover, a study done by the General Accounting Office for the House Committee on Interior and Insular Affairs concluded that degradation of riparian areas on public lands is "largely a result of poorly managed livestock grazing [because] livestock tend to congregate in the riparian areas for extended periods, eat most of the vegetation, and trample the stream banks."[6]

The common argument that cattle are the ecological equivalents of bison is erroneous. Bison, being wanderers, are less likely to regraze a given site in a single season than are cattle. Bison can use drier, rougher forage than cattle and can forage more effectively in deep snow. And whereas cattle are well known for their ability to lay waste to riparian areas, bison typically go to water only once a day.[7]

Some of the comparisons of bison with cattle are done from a strictly managerial perspective—that is, how specific traits can "be more effectively exploited in land management."[8] But Glenn Plumb and Jerrold Dodd, who studied bison and cattle in a fenced "natural area," did admit that "bison reflect a greater degree of evolutionary context to a grassland natural area [and that] differences between the influence of free-roaming bison on pristine grasslands and semi-free-roaming bison on a fenced natural area must be much greater than those of the latter and domestic cattle." This admission is not only a concession to the importance of scale but also an invitation to question the use of "natural" in their fenced "natural areas." Others also have alluded to issues of scale and freedom of movement when they acknowledged that the change from "nomadic bison to resident cattle herds" coincided with subdivision of the land into fenced areas with managed watering and feeding situations, thus altering the spatial and temporal patterns of grazing and its impacts on vegetation.[9]

Wild, grazing herbivores in expansive, unfenced grazing ecosystems tend not to remain stationary, migrating instead toward optimal grazing conditions. Defoliation caused by native grazers passing through an area removes older plant tissues and promotes the growth of new shoots—a coevolved situation allowing for rapid recovery of habitat. Fenced lands and management for maximum yields of sedentary domestic livestock inhibit such recovery. Confining native grazers in such fenced-in situations has a similar effect.[10]

Andrew Isenberg, reporting on the history of bison, which were ultimately confined to a series of small sanctuaries, writes that "the creation of refuges both limited the ability of bison to seek new grasses and cemented the fragmentation of the bison population." He went on to comment about the genetic ramifications of this on the creature: "The fragmentation of the herds had a deleterious impact on the genetic diversity of the bison. All bison in North America are descended from the roughly five hundred survivors of the commercial slaughter of the nineteenth century, a so-called 'bottleneck' in the transfer of genes."[11]

Vegetational changes wrought by livestock may have a profound effect on wild species. For example, where cattle graze on grasslands so heavily that there is conversion to shrubland, they reduce the suitability of the land for grazing competitors such as bison and bighorn. In his 1978 essay, Frederic Wagner reported specific accounts of cattle-bighorn, cattle-elk, and domestic sheep-pronghorn competition.[12]

Impacts on vegetation may be purely mechanical. Where cattle are grazed, most of the vegetation loss may be due to "trampling effects."[13] Loss of shrubs and grass amounts to a loss of cover for deer and antelope fawns and elk calves, and a coinciding increase in their vulnerability to predation and weather.

Animals that feed on grasses and forbs (broad-leaved, nonwoody plants, such as wildflowers) are referred to as "grazers," whereas those tending toward shrubs and trees are called "browsers." These are broad generalizations, because a given species may have a wide preference range that would place it in both categories. A lack of preferred vegetation may also drive creatures to

Figure 1. Feeding Niches of Wild Ungulates and Domestic Herbivores, and Potential Competition Among Them

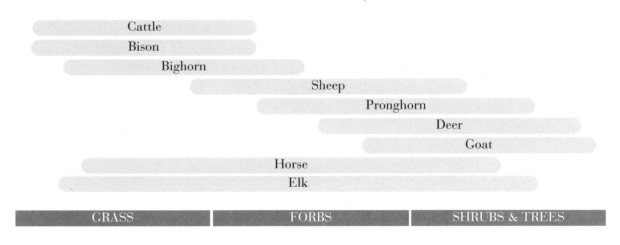

Source: Redrawn from F. Wagner, "Livestock Grazing and the Livestock Industry," in *Wildlife and America: Contributions to an Understanding of American Wildlife and its Conservation,* edited by H. P. Brokaw (Washington, D.C.: Council of Environmental Quality, 1978).

feed on plants they would normally pass up, thereby placing them in competition with species they wouldn't compete with under more typical conditions. In addition, feeding patterns may vary with season or with geographic location. Northern bighorns, for example, are primarily grazers, whereas desert bighorns tend to browse. Nevertheless, the norms for various species may be compared to allow for a general estimate of the potential for competition between and among species (see Figure 1).

Wild ungulates, having coevolved with native vegetation, tend to have more narrowly defined food preferences and more fixed feeding patterns than do domestic animals, in which selective breeding has apparently conferred an ability to shift diets with less stress: "As vegetation composition is altered through grazing, wild species may be affected detrimentally by slight or subtle changes while the range may still be in quite favorable condition for domestic animals."[14] Even if livestock use certain plants only in limited amounts, the impact may be magnified to the extent that those plants are specifically preferred by wild species.[15]

Livestock may have "psychological" effects on wild species. There are reports of avoidance by moose, deer, and elk of areas used by domestic stock. Such effects are in addition to any "operational impacts" that exist in the form of roads, fencing, brush control, pesticides, and disease transmission that come along with the livestock industry. One study found that 13 percent of deer mortalities being reported were due to "fence kills."[16]

The graph that appeared in Frederic Wagner's 1978 essay (see Figure 2), and which is still being widely reproduced and distributed, is an estimate of the shift from wild to domestic animals in the rangelands of the eleven western states since the middle of the nineteenth century. The unit used is an "animal unit month" (AUM), the amount of forage consumed in a month by a cow and her calf. The graph therefore shows the relative amounts of forage removed rather than sheer numbers of animals, but it does amount to a concise estimate of what we have lost in terms of wildlife on western lands managed for an industry now yielding less than 3 percent of the beef produced by the nation. Wagner estimated that between 1890 and 1940, the number of live-

stock on western ranges may have been close to twice the values for wild ungulates prior to European-American settlement. Livestock grazing pressure may have been half again as high as that exerted by wild ungulates. [17]

Another way of estimating impacts of domestic livestock on native creatures may be seen in governmental tables expressing "forage ratios." Although they are approximations that may vary one way or another, depending on a host of variables (such as quality of forage, variations in daily intake, species preferences, condition of animals, and the like), they give insight into the cost of the

Table 1. AUM Equivalents Based on 650 Pounds of Forage per Month (USDA)

Domestic cow with calf	1.00
Domestic bull	1.25
Horse	1.25
White-tailed deer	0.15
Mule deer	0.20
Pronghorn	0.20
Bison	1.25
Bighorn sheep	0.20

Source: U.S. Department of Agriculture, *National Range Handbook*, 1976.

Table 2. Wild Species Equivalents to 1 Cow-plus-Calf AUM (Wyoming)

Bighorn sheep	6.9
Pronghorn	10.8
Mule deer	7.8
Elk	2.1
Moose	1.2

Source: Wyoming Game and Fish Department, internal data, 1998.

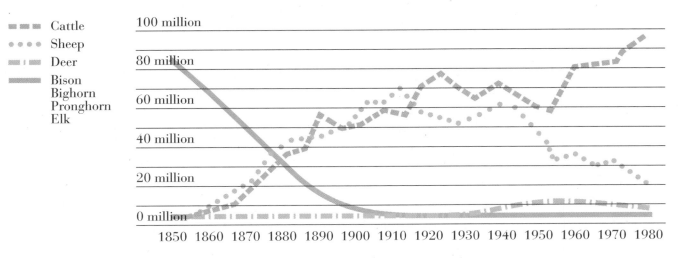

Figure 2. Conjectured AUMs of Wild and Domestic Grazing Pressures on Rangelands of the Eleven Western States

- – – Cattle
- · · · Sheep
- – · Deer
- ▬ Bison Bighorn Pronghorn Elk

Source: Redrawn from F. Wagner, "Livestock Grazing and the Livestock Industry," in *Wildlife and America: Contributions to an Understanding of American Wildlife and its Conservation*, edited by H. P. Brokaw (Washington, D.C.: Council of Environmental Quality, 1978).

livestock industry in terms of wildlife lost. The amounts of forage required by native game animals may be expressed as fractions of AUMs (as in the U.S. Department of Agriculture guide in Table 1) or simply in cow-game ratios (as in the State of Wyoming guide in Table 2).

There is lack of uniformity in these estimates. For example, an AUM is listed as the equivalent of 5 mule deer in the federal table (a mule deer is 0.20 AUM), but as 7.8 mule deer in the Wyoming table. Likewise, the federal table estimates an AUM to be the equivalent of 5 bighorns, as opposed to 6.9 bighorns in the Wyoming table. Through it all, though, there emerges a powerful message: the cost in lost wildlife from management of the nation's public lands for livestock is immense. And the fact that some 70 percent of the land area of the eleven western states is subject to grazing at least on a seasonal basis makes the public lands all the more valuable as safe havens for wild native species.

America's public lands, though, are managed primarily for the livestock industry. Big wild mammals there have become outnumbered or replaced by domestic beasts that are foreign to the landscape, unsuited to it, destructive of its habitats, but nevertheless maintained for the economic benefit of a small but vocal and politically powerful minority of individuals who believe that the country has an obligation to maintain their "way of life."

It is one of the great ironies of life in America that the 30,000 or so wealthy ranchers who wax fat at public expense detest "big government" as they do, for it's hard to imagine a group that receives more support from the rest of the nation's tax-paying citizens—this handed to them by the federal government. Looking at it another way, and since the population of the country is now at about 270 million, one can say that, on average, 9,000 American citizens pay to support each livestock baron operating on public lands. Ironic also is the fact that so many of these welfare recipients see themselves as paragons of rugged independence.

Biologically, it is significant that public lands are now managed for creatures not only alien to the environment but also selectively bred according to human values and whims. Paul Shepard, who wrote extensively on the subject of domestication, had this to say:

"To domesticate" means to change genetically. . . . Man substitutes controlled breeding for natural selection; animals are selected for special traits like milk production or passivity, at the expense of overall fitness and nature-wide relationships. . . . The animals become crude pawns in the farmer's breeding game, shorn of finesse and the exquisite detail so characteristic of wild forms. The animal departs from the hard-won species type. For man the animal ceases to be an adequate representation of a natural life form. Its debased behavior and appearance mislead us and miseducate us in fundamental perceptions of the rhythms of continuity and discontinuity, and of the specific patterns of the multiplicity of nature. . . . Civilization . . . has loosed a horde of "goofies."[18]

The issue of domesticity, as it pertains to the public lands of the West, is not trivial. It is a recognition of the fact that native animals, having been forged over countless generations as wild creatures, have evolved as integral parts of their environments. This goes to the very heart of biological integrity.

The failure to see native creatures as integral parts of larger entities—the ecosystems of which they are a part—has been our overriding problem all along. The reductionist worldview, having permeated all aspects of our culture, allowed for the reduction of western landscapes into parts that have never been allowed to come together again. We have failed to look at the larger systems as entities to be regarded, respected, and maintained.

If unmolested in the Intermountain West, "bison, elk, deer, antelope, mountain goats, bighorn, javelina, wolves, bear, and jaguar [would have] proliferated . . . on a scale unimaginable in terms of historical observations."[19] Moreover, "a drastic reduction in livestock numbers on western ranges and a return to large wildlife populations would be the most conservative [option] ecologically."[20] If citizens will work for removal of livestock and the return of native animals to a land that belongs to all citizens and to the descendants to whom they will bequeath it, they can become part of the move toward the reality. This is no pipe dream; it is very much in the realm of possibility. Those magnificent creatures know how to mate and to raise their young in the lands in which they are true natives. Give them the space and the habitat they require, and they will do the rest.

Livestock, ranchers say, benefit wildlife. If a mob of strangers tromped into your house, ate all your food; busted up your walls, windows, and furniture; and camped out in your living room, would you deem this of benefit to you?

A WAR AGAINST PREDATORS
The Killing of Wildlife Funded by Taxpayers

Brooks Fahy and Cheri Briggs

The federal Wildlife Services agency spends millions of taxpayer dollars annually in an inhumane and ineffective effort to protect private livestock by killing predators on both private and public lands. Western coyotes are the major target. Approximately 70,000 per year have been killed over the last decade. Methods of killing include leghold traps, aerial gunning, neck snares, sodium cyanide–loaded devices called M-44s, and denning (killing pups in dens).

Brooks Fahy *is a longtime wildlife activist with extensive experience in rescue and rehabilitation of injured and orphaned wildlife. He is executive director and cofounder of the Predator Defense Institute in Eugene, Oregon, which campaigns against predator control, trapping and trophy hunting of carnivores, and fiscally and ecologically irresponsible wildlife management agencies.*

Cheri Briggs *is cofounder of the Predator Defense Institute. Along with her work as a wildlife rehabilitator, she has been involved in campaigns to end aerial gunning of Alaskan wolves, to ban trapping, and to educate the public on the importance of protecting carnivores and their habitat.*

Coyote in trap.

Wildlife Services is the euphemistic name this arm of the U.S. Department of Agriculture chose in 1997 to replace "Animal Damage Control." Pleasantly named or not, "Wildlife Services" (WS) spends millions of taxpayer dollars every year to kill predators, primarily in the West, on both public and private lands, for the benefit of a small percentage of livestock producers. Ranchers are not required to change their management practices to reduce livestock/predator conflicts or to pay directly for the services they receive.

The latest available government figures show that in fiscal year 1999, more than $10 million in federal moneys was spent directly on killing predators in the United States for "livestock protection." The seventeen western states (Kansas, Nebraska, North and South Dakota, Oklahoma, and Texas, as well as the eleven states to the west) account for more than two-thirds of WS expenditures nationwide (67 percent), with most of this money going to livestock protection (71 percent). Taxpayers in western states pay twice, sometimes thrice over, as state and county governments also contribute to WS control activities. (Ranchers pay a modest portion of the total costs of government predator control—approximately 25 percent or less, sometimes much less—through county assessments on stock or through the contributions of livestock associations.) In the late 1990s, the cost of killing predators in the western states exceeded reported livestock losses to predators by a ratio of three to one.[1]

More than 95,000 predators were slaughtered by WS in 1999—a number that has varied little over the past decade. In the seventeen western states alone, 85,262 coyotes were killed, but the death toll also included bears, bobcats, foxes, mountain lions, and badgers (see Table 1).[2]

Typical killing methods include leghold traps, neck snares (strangulation), sodium cyanide–loaded M-44s, aerial gunning, call and shoot, and denning (killing pups in dens by using grappling hooks or incendiary devices, or digging them out and smashing them with shovels). Neither aerial gunning nor M-44s select for depredating individuals only—*all* coyotes in a specified area are vulnerable to such measures. Furthermore, nontarget animals are subject to WS's lethal devices. In 1998, *known* nontarget kills by M-44s totaled 1,984, including 1,277 foxes, 267 domestic dogs, 253 raccoons, and 1 gray wolf.[3]

Since animals do not always die immediately in the vicinity of the M-44, undoubtedly other nontarget animals were not included in the tally.

Table 1. Predators Killed by Wildlife Services in Fiscal Year 1999

	17 Western States	United States
Badgers	589	601
Bear	342	347
Bobcats	2,419	2,435
Coyotes	85,262	85,938
Fox	5,531	6,182
Mountain lion	359	359

Source: Predator Conservation Alliance, *"Wildlife Services"? A Presentation and Analysis of the USDA Wildlife Services Program's Expenditures and Kill Figures for Fiscal Year 1999* (Bozeman, Mont.: PCA, 2001); Wildlife Services, *Annual Report* (Washington, D.C.: USDA/APHIS Wildlife Services, 1999), www.aphis.usda.gov/ws.

Although nonlethal methods of predator control have been found to be more effective in protecting livestock, a 1995 U.S. General Accounting Office report revealed that WS uses lethal methods in almost all instances.[4] In addition, WS routinely launches lethal predator control programs *before* there are any confirmed livestock losses. Aerial gunning typifies this indiscriminate approach: shooters take whatever animals are unfortunate enough to come into their sights in the targeted area. In 1999, 390 bobcats, 30,875 coyotes, 231 foxes, and 3 badgers were gunned down from helicopters and fixed-wing aircraft.[5] The costs associated with aerial kills range between $200 and $800 per coyote.[6]

In addition to being inhumane and expensive, lethal control of predators like coyotes does not work. Current research shows that killing adult coyotes causes the overall reproductive rate to go up. Coyotes produce young at earlier ages in persecuted populations, and litters tend to be larger. Inexperienced parents with lots of mouths to feed resort to taking livestock much more readily than coyotes in more stable populations.[7]

Most predators never attack livestock; most livestock deaths (that is, prior to reaching the slaughterhouse) are not caused by predators. The National Agricultural Statistics Service found that in 1995, for example, coyotes caused 1.6 percent of all cattle and calf deaths, and predators overall only caused 2.7 percent.[8] Meanwhile, other causes of death were much more significant: respiratory problems (27.5 percent); digestive problems (19.7 percent); unknown causes (15.2 percent); birthing (14.8 percent); weather (9.5 percent); and other (9.1 percent). Only poison (1.1 percent) and theft (0.4 percent) were smaller problems for cattle growers than predation was.[9]

Predator take of sheep is higher, proportionally, than for cattle but is still minor when considered in the context of the population of domestic sheep as a whole. In 1999 alone, 742,900 sheep died in the United States, with 273,000 of these killed by predators. Altogether, predator losses accounted for 3.9 percent of the total domestic sheep population (70.3 million) in 1999.[10]

Wildlife Service's supporters and beneficiaries have resisted repeated attempts at legislative reform. For example, in June 1998, the House approved an amendment to eliminate the federal predator control program. Unfortunately,

on the following day, the House voted to reverse its decision. A critical player in the switch was Joe Skeen (R-NM), chair of the Appropriations Subcommittee on Agriculture and a fervent proponent of predator control. Documents requested through the Freedom of Information Act show that between 1991 and 1996, WS agents made ninety-nine visits to Skeen's New Mexico ranch, for a total of 315 hours of staff time. Reportedly, a total of three coyotes were killed during this period.

Government-sponsored predator killing has other powerful allies: ABC newsman Sam Donaldson, who also owns a ranch in New Mexico, has requested WS to protect his sheep numerous times. Between 1991 and 1996, this work occupied 178 staff days, and agents called on his ranch 412 times. Donaldson reported losing sixty-two animals, worth a total of $3,100.[11] Needless to say, neither the TV journalist nor the representative depends exclusively on ranch income to survive.

Repeated attempts by wildlife activists to cut funds from WS's livestock protection program have been largely unsuccessful to date because the power of the livestock lobby. Nevertheless, both the industry and WS understand that the slaughter of thousands of animals annually is not popular with the general public and will continue to come under harsh criticism. In response, WS is exhibiting greater interest in "nonlethal" methods of wildlife control. Researchers at WS facilities are currently investigating sterilization techniques for coyotes that can be practicably exported to the field.[12]

Although future mass sterilization of coyotes and other species offers the possibility of lower body counts, the ramifications for predator populations and the ecosystems of which they are a part are unclear. For example, significant reductions in coyote populations—whether from direct killing or contraceptive control—may allow some prey populations to expand dramatically. A study in Texas found that following the removal of coyotes, the number of rodents increased, but rodent biodiversity declined, with the number of rodent *species* decreasing from eleven to one. Black-tailed jackrabbits increased from three to eighteen times over pretreatment numbers. Numbers of small predators, such as skunks, also went up dramatically.[13]

The long-term ecological implications of permanently suppressing coyote populations through contraceptive control are not known, though it is possible that not only small mammals but also plant species could eventually be affected by alterations in predator densities and species composition. Agriculture itself could conceivably face new "pest" problems if predators that tend to keep rodents and other prey species in check are largely eliminated. Yet so far, wildlife contraception researchers proceed with little opposition or even skepticism from agricultural interests and conservationists alike.

The extensive manipulation of wild carnivores that has characterized WS's mission to date will not end with the development of new technologies but is likely to take a more silent and insidious form. "Biological warfare" may be the new frontier in predator control. It remains for Americans to decide if this is the type of wildlife management they want on their public lands, and also, if they wish to pay for it.

Clockwise, from upper left: Coyote pup pulled from its den with grappling hook—a common method of coyote killing, known as "denning"; skinned coyotes; pile of dead mountain lion heads, Arizona; coyote hung on a fence, Oregon; coyotes shot by government trappers; dead coyote in trap, Wyoming.

A HEAVY TOLL
Native Animals Harmed by Livestock Production

The animals listed here are negatively affected by livestock production. "Livestock production" includes all activities and inputs necessary to raise livestock in the arid West. The following activities and impacts are among the causes of species' decline.

- Livestock grazing per se, spurring changes in plant communities and competition for forage
- Trampling of vegetation and stream banks by livestock
- Cumulative effects of grazing and trampling, such as the destruction of riparian zones
- Predator "control" (killing)
- Pest "control" (as with prairie dogs or grasshoppers)
- Water diversion and impoundment for livestock watering or irrigation to produce livestock forage crops
- Conversion of habitat for pasture or for the growing of livestock forage crops, such as hay or corn
- Changes in natural fire regimes due to fire suppression or to vegetative changes resulting from livestock grazing
- Transmission of diseases
- Habitat fragmentation, as from fencing, roads, croplands, and pasture
- Water pollution from livestock wastes, from sedimentation caused by livestock grazing, and from other livestock-related activities

Not every impact or activity will affect every species. Also, there are often multiple reasons for a species to decline, and livestock production may be only one factor out of several. In some cases, a species may be harmed by livestock production in one region but be unaffected in another. Or the intensity of livestock production may be key to the nature of the effect. For instance, mule deer may increase under moderate amounts of livestock grazing because of a shift in the plant community to a greater number of browse species. However, even mule deer numbers will decline if grazing intensity is heavy enough to reduce browse availability.

The criterion for inclusion in the following list is sufficient scientific evidence and agreement that, at least in some places, livestock production is implicated in the decline of, or hinders the full recovery of, a particular species.

Some of the species listed here are not officially endangered or even candidates for listing. Yet, although a certain species may be abundant in a relative sense, if livestock production precludes it from reaching its biological potential in population size or geographic distribution, it was eligible for this list as well. For example, pronghorn antelope are relatively numerous; indeed, there are more today than at the turn of the nineteenth century, when overhunting wiped them out over much of their former range. Nevertheless, there is good evidence that livestock production can significantly diminish pronghorn numbers through forage competition, fencing that blocks migration, sagebrush eradication programs (carried out for the purpose of producing more livestock forage), and expropriation of water. Even though pronghorns are not endangered, their numbers are still short of their biological potential over large parts of their range, and the leading cause is livestock production. Moreover, the U.S. Endangered Species list includes only a small fraction of the many thousands of species that are in fact imperiled but do not have a constituency strong enough to win their protection.

In some cases, livestock production is the ultimate cause of species decline, even if it is not the proximate cause. For example, a reservoir, or water withdrawal for irrigation, may change the water quality and flow to favor an exotic fish species. With a competitive advantage over the native fish, the exotic gradually becomes dominant, and the native species dwindles. The immediate cause of the native's decline: an alien species. The larger, overarching reason for the native's downturn: changes to water quality and/or water flow regime, prompted by the demands of livestock production. Some species affected in such indirect ways are also listed here.

Cows in riparian zone, North Fork Big Lost River,
Challis National Forest, Idaho.

Insects

Idaho pointheaded grasshopper
Pinaleno monkey grasshopper
St. Anthony Sand Dunes tiger beetle
● Myrtle's silverspot butterfly
Great Basin silverspot butterfly
● Bay checkerspot butterfly
● Uncompahgre fritillary butterfly
● Pawnee montane skipper
Regal fritillary butterfly
Baking Powder Flat blue butterfly
● Sacramento Mountains checkerspot
butterfly
Colorado burrowing mayfly

Mollusks

● Banbury Springs limpet
▼ Pecos assiminea snail
▼ Page springsnail
Fossil springsnail
Montezuma Well springsnail
Kingman springsnail
Grand Wash springsnail
▼ Huachuca springsnail
▼ Three Forks springsnail
Brown springsnail
● Socorro springsnail
● Idaho springsnail
Pecos springsnail
▼ Roswell springsnail
▼ Chupadera springsnail
● Bruneau Hot Springsnail
Moapa pebblesnail
▼ Koster's tryonia snail
Virile Amargosa snail
● Bliss Rapids snail
● Snake River physa snail
● Kanab ambersnail
● Utah valvata snail

Fish

Pacific lamprey
Goose Lake lamprey
Snake River white sturgeon
Bonneville cisco
▼ Fluvial arctic grayling
● Bull trout
● Chinook salmon
● Sockeye salmon
● Gila trout
● Apache trout
▼ Bonneville cutthroat trout
▼ Westslope cutthroat trout
▼ Yellowstone cutthroat trout
▼ Colorado River cutthroat trout
▼ Lahontan cutthroat trout
● Paiute cutthroat trout
● Greenback cutthroat trout
▼ Rio Grande cutthroat trout
▼ Interior redband trout
● Little Kern golden trout
▼ Volcano Creek golden trout
Mexican tetra
● Colorado pikeminnow
Lahontan redside
California Pit roach
▼ Gila chub
● Sonora chub
● Yaqui chub
Leatherside chub
Alvord chub
● Borax Lake chub
● Chihuahua chub
● Humpback chub
Roundtail chub
● Pahranagat roundtail chub
Moapa roundtail chub
Gila roundtail chub
● Virgin River chub
● Bonytail chub
● Hutton tui chub
● Mohave tui chub
Goose Lake tui chub
Sheldon tui chub
Catlow tui chub
Summer Basin tui chub
Warner Basin tui chub
● Cowhead Lake tui chub
● Moapa dace
Relict dace
● Desert dace
● Little Colorado spinedace
Virgin spinedace
● White River spinedace
● Big Spring spinedace
● Spikedace
● Woundfin
Southern redbelly dace
Mexican stoneroller
● Ash Meadows speckled dace
● Clover Valley speckled dace
● Foskett speckled dace
● Independence Valley speckled dace
● Kendall Warm Springs speckled dace
Moapa speckled dace
Pahranagat speckled dace
● Loach minnow
▼ Sturgeon chub
● Beautiful shiner
✖ Phantom shiner
✖ Rio Grande bluntnose shiner
● Pecos bluntnose shiner
Blue sucker
● June sucker
● Cui-ui
● Shortnose sucker
● Modoc sucker
Desert sucker
● Razorback sucker
Tahoe sucker
● Lost River sucker
● Warner sucker
Zuni mountain sucker
Wall Canyon sucker
Goose Lake sucker
Gray redhorse
Jenny Creek sucker
● Yaqui catfish
● Pecos gambusia
● Pahrump poolfish
● Gila topminnow
● Desert pupfish
● Devil's Hole pupfish
● Owens pupfish
Pecos pupfish
White Sands pupfish
● Ash Meadows Amargosa pupfish
● Warm Springs pupfish
Cottonball Marsh pupfish
Palomas pupfish
● Railroad Valley springfish
● Hiko White River springfish
Threespine stickleback
● Unarmored threespine stickleback
Shoshone sculpin
Wood River sculpin
Bear Lake sculpin
Rough sculpin
Malheur mottled sculpin
Shorthead sculpin
Margined sculpin
Pit sculpin
▼ Arkansas darter
Greenthroat darter
Bigscale logperch

Amphibians and Reptiles

Sacramento Mountain salamander
- Desert slender salamander
Tehachapi slender salamander
Jemez Mountain salamander
- Sonoran tiger salamander
✖ Tarahumara frog
- California red-legged frog
▼ Oregon spotted frog
▼ Columbia spotted frog
▼ Mountain yellow-legged frog
- Chiricahua leopard frog
Lowland leopard frog
Yavapai leopard frog
▼ Yosemite toad
▼ Western boreal toad
Amargosa toad
Black toad
- Wyoming toad
Arizona toad
Gila monster
- Blunt-nosed leopard lizard
Bunch grass lizard
Cowles fringe-toed lizard
▼ Flat-tailed horned lizard
Texas horned lizard
Sagebrush lizard
Canyon spotted whiptail
Arizona Gilbert's skink
Mountain skink
Southwestern pond turtle
- Desert tortoise
Green rat snake
Mexican garter snake
- San Francisco garter snake
Plain-bellied water snake

Birds

Greater sandhill crane
- Whooping crane
White-faced ibis
▼ Trumpeter swan
- Yuma clapper rail
Yellow rail
- Mountain plover
- Piping plover
- Western snowy plover
Long-billed curlew

Black tern
- Least tern
▼ Northern goshawk
Ferruginous hawk
Swainson's hawk
Northern gray hawk
Golden eagle
- Bald eagle
Peregrine falcon
- California condor
Greater prairie-chicken
- Attwater's greater prairie-chicken
▼ Lesser prairie-chicken
▼ Columbian sharp-tailed grouse
Plains sharp-tailed grouse
▼ Greater sage grouse
▼ Gunnison sage grouse
▼ Western sage grouse
- Masked bobwhite
Gambel's quail
▼ Mountain quail
Elegant trogon
▼ Western yellow-billed cuckoo
Black-billed cuckoo
Short-eared owl
Great gray owl
- Mexican spotted owl
- Cactus ferruginous pygmy-owl
Burrowing owl
Gila woodpecker
Lewis' woodpecker
- Southwestern willow flycatcher
Cactus wren
Loggerhead shrike
Crissal thrasher
- Coastal California gnatcatcher
- Least Bell's vireo
Yellow warbler
Common yellowthroat
Abert's towhee
- Inyo California towhee
Grasshopper sparrow
▼ Baird's sparrow
LeConte's sparrow
Belding's savannah sparrow
Large-billed savannah sparrow
Dickcissel
Lark bunting

Mammals

Arizona shrew
Preble's shrew
Pygmy shrew
Spotted bat
- Mexican long-nosed bat
- Lesser long-nosed bat
Southern yellow bat
- Pygmy rabbit
Sierra Nevada snowshoe hare
White-tailed jackrabbit
- Point Arena mountain beaver
▼ Black-tailed prairie dog
▼ White-tailed prairie dog
- Utah prairie dog
Richardson's ground squirrel
Mohave ground squirrel
▼ Washington ground squirrel
- Northern Idaho ground squirrel
▼ Southern Idaho ground squirrel
San Joaquin antelope squirrel
Allen's thirteen-lined ground squirrel
Santa Catalina Mountains squirrel
Palmer's chipmunk
Uinta chipmunk
Fish Spring pocket gopher
San Antonio pocket gopher
Botta's pocket gopher
Northern pocket gopher
Hualapai southern pocket gopher
Prospect Valley pocket gopher
Pajarito southern pocket gopher
Searchlight southern pocket gopher
Harquahala southern pocket gopher
- Tipton kangaroo rat
- Morro Bay kangaroo rat
Merriam's kangaroo rat
Marble Canyon kangaroo rat
- Giant kangaroo rat
Fletcher dark kangaroo mouse
Desert Valley kangaroo mouse
Little pocket mouse
Silky pocket mouse
Wupatki Arizona pocket mouse
Yavapai Arizona pocket mouse
Coconino Arizona pocket mouse
Black Mountain pocket mouse
Yellow-nosed cotton rat
Yuma hispid cotton rat
Colorado River cotton rat
Yavapai Arizona cotton rat
Santa Catalina Mountains woodrat

- Riparian woodrat
Northern grasshopper mouse
Pinacate cactus mouse
Black Mountain cactus mouse
Water vole
Montane vole
Ash Meadows montane vole
- Amargosa vole
- Hualapai Mexican vole
Navaho Mountain Mexican vole
New Mexican jumping mouse
- Preble's meadow jumping mouse
- Gray wolf
- Mexican wolf
Kit fox
- San Joaquin kit fox
Swift fox
Sierra Nevada red fox
- Jaguar
- Black-footed ferret
Badger
Southwestern otter
- Grizzly bear
Bison
- Desert bighorn sheep
- California bighorn sheep
Rocky Mountain bighorn sheep
Pronghorn
- Sonoran pronghorn
Rocky Mountain elk
Mule deer

One species—cow—versus many species. (Occasionally, two species, domestic cattle and domestic sheep, versus many native species.) The livestock industry has nearly succeeded in substituting an animal ill suited to the arid West, brought from another continent, for a profusion of creatures, each keenly honed by evolution to its own particular niche. For more than a century, the livestock industry has been transforming the rich and diverse fauna of the West into a near-monoculture of cows.

1. Gila chub
2. Gila topminnow
3. Chinook salmon
4. Greenback cutthroat trout
5. Fluvial arctic grayling
6. Bonytail chub
7. Apache trout
8. Gila trout
9. Yaqui chub
10. Yellowstone cutthroat trout
11. Colorado pikeminnow
12. Desert tortoise
13. Sandhill crane
14. Golden trout
15. Swallowtail butterfly
16. River otter
17. Green-winged teal

Although ranchers pay almost nothing for public lands forage, there is no free lunch. It's a heavy price the native animals of the West pay for livestock production—they pay with their health, their lives, and perhaps the future of their species.

1. Mountain lion
2. Golden eagle
3. Yellow-bellied marmot
4. Hooded oriole
5. Masked bobwhite
6. Black-tailed prairie dog
7. Gray wolf
8. Burrowing owl
9. Bison
10. Sage grouse
11. Grizzly bear
12. Kit fox
13. Badger
14. Prairie falcon
15. Southwestern willow flycatcher
16. Pronghorn
17. Bighorn sheep

RANCHING
AND
SUBSIDIES

ECONOMICS LIVESTOCK

The True Cost of a Hamburger

RANGE
RESTORATION
Please use
Access Route
300 Yards West
of Here.

PART V

Many people assume, since most of the western landscape is given over to livestock production, that ranching must be economically important. But, as economist Thomas Power points out in the opening essay of this section, the livestock industry contributes almost nothing to western economies, even at the local level.

Despite the cowboy's image as a rugged, independent individual, a host of government subsidies keep him propped up in the saddle. The western rancher is dependent on what is, in essence, a welfare program. The much-publicized low fees paid by ranchers to graze federal lands are only the beginning. Other subsidies include taxpayer-supported research at western land grant universities and agricultural exemptions that lower property taxes paid by ranchers. There are handouts to help with nearly every problem: drought relief, low-interest agricultural loans, emergency livestock feed programs, emergency grazing on Conservation Reserve Program lands, to name a few. Even many of the fences crisscrossing the West's "open" spaces are paid for by American taxpayers.

And this is not all. Ranchers are literally mortgaging the public's resources for their private benefit. As Mark Salvo explains in his essay on the connection between the banking industry and public lands ranching, ranchers are able to take out loans based on the "value" of their grazing permits. This questionable arrangement forces government officials to consider the status of a rancher's debt when making range management decisions, rather than focusing on what is best for the land.

Beyond the economic subsidies are the health, social, and environmental costs of the animal agriculture industry in general—the larger context within which public lands livestock grazing is properly viewed. Ills such as heart disease, cancer, kidney disease, and hypertension may seem quite unrelated to ranching on western public lands, just as food security, loss of arable land, desertification, tropical deforestation, urban overcrowding, and poverty may appear unconnected to problems of ecosystem degradation in the arid West. Yet, all these difficulties are linked—directly or indirectly—to an international system of meat production and an increasingly global pattern of meat consumption. Western ranching is a part of these destructive worldwide trends.

Thus, Virginia Kisch Messina addresses the subject of meat eating and health, while Richard Schwartz and Mollie Matteson discuss the connections between industrialized animal agriculture around the globe and a variety of environmental and social dilemmas. In considering the impacts of public lands ranching, we should understand that these do not occur in isolation from the rest of the country or planet, nor are they disconnected from the most personal and serious aspects of our own lives—our individual health and that of our communities, and the well-being and stability of the world we leave behind for our children.

258–259: Range restoration, Arizona style.

Opposite: Cattle exclosure on Road Creek, Bureau of Land Management lands, Idaho. Most fencing and other developments on public lands are paid for in part, or in full, by taxpayers. These are expenses that, collectively, we would not have to bear if livestock were absent from our lands.

TAKING STOCK OF PUBLIC LANDS GRAZING
An Economic Analysis

Thomas M. Power, Ph.D.

Livestock grazing on federal lands is generally unimportant to local economies and even less so to state and regional economies. In terms of income and numbers of jobs provided, the contribution of federal lands grazing is less than 0.1 percent across the West. Farm and ranch operations are increasingly reliant on nonfarm income sources to be financially feasible, while livestock grazing competes with other uses of public lands—such as clean water, recreation, and wildlife habitat—that contribute to the ongoing vitality of western economies.

Thomas M. Power *is chair of the Economics Department at the University of Montana, where he has taught since 1968. He received his Ph.D. degree from Princeton University and specializes in natural resources and regional economic development issues. His books include* Post-Cowboy Economics: Pay and Prosperity in the New American West *(2001),* Lost Landscapes and Failed Economies: The Search for a Value of Place *(1996), and* Environmental Protection and Economic Well-Being: The Economic Pursuit of Quality *(1996).*

For the last decade and a half, one of the more emotional public policy issues in the western states has been the level of grazing on federal lands and the appropriate fee to charge for those domestic animals that are allowed to graze. Much of the emotion is tied to the perception that most ranching operations in the West rely on these federal lands, and that without access to the forage these federal lands provide, many western ranches would cease to be economically viable. Since, it is usually assumed, ranching is the economic backbone of the western economies, such a loss, it is concluded, would have a devastating impact on the western states. Arguments of this sort successfully blocked almost all the significant reforms of federal grazing policies attempted between 1975 and 1999 and fueled the political assertion that environmentalists and the Clinton administration were waging a "war on the West."

In this essay, I analyze these economic claims for the entire eleven-western-state region (Arizona, California, Colorado, Idaho, Montana, Nevada, New Mexico, Oregon, Utah, Washington, Wyoming), for the individual states, and for individual counties in two case-study areas. My empirical analysis demonstrates that grazing on federal lands contributes only a tiny sliver of economic activity to the local economies—usually a small fraction of 1 percent of total income and employment, and rarely more than 1 percent. During the 1990s, local economies in the West grew by this amount every few weeks. The ongoing rapid economic growth has been heavily fueled by families and businesses relocating in the pursuit of higher-quality living environments. Protecting the environmental integrity of public lands contributes to this ongoing economic vitality and almost certainly offsets any losses in the livestock sectors that may be associated with changes in livestock use of federal lands.

Measuring the Relative Economic Importance of Grazing on Federal Lands

Claims about the relative importance of federal grazing to the economies of the western states can be simply analyzed by answering the following four questions:

1. What portion of the value produced by cattle and sheep operations is associated with the feed used?

Peruvian sheepherder, Idaho. The livestock industry provides very few jobs relative to the economy as a whole, and ranch hands, sheepherders, and the like make very low wages. Some jobs pay so little that foreign workers have to be brought in to take them.

2. What portion of the feed for those cattle and sheep operations comes from grazing on federal lands?
3. What portion of the total agricultural activity involves raising cattle and sheep?
4. What part of the total economy is represented by agriculture?

Although it is easy to argue that without livestock feed there cannot be livestock produced, the same argument holds for all other inputs to livestock production. Without water; without trace elements, vitamins, and medicines; without land; without machinery; without fuel for the machinery; and importantly, without labor and management efforts, there would be no livestock produced, or much, much less. Clearly all of the inputs play an economic role, not just feed. According to the U.S. Department of Commerce, in 1992–1996, purchased feed—not including feed grown by the rancher—made up a fifth to a quarter of the total value of livestock sold.[1] In any case, feed—whether purchased, leased as pasture from private owners or public land agencies, or raised by the rancher—is not the only important input to livestock production.

For many of the western states, federal lands provide only a small percentage of the total feed needed to support cattle and sheep herds. California, Washington, and Montana, for instance, obtain less than 10 percent of their cattle and sheep feed from federal lands. Colorado, Oregon, and Wyoming obtain 20 percent or less of the feed for their livestock herds from this source. Overall, the eleven western states obtain only about a fifth of the feed needed to support their beef cattle and sheep herds from federal lands (see Table 1). From a national perspective, the reliance on western federal lands is dramati-

cally lower: only 4 percent of the feed consumed by beef cattle is provided by grazing federal lands.[2]

In many parts of the West, cattle raising is not the dominant agricultural activity. In Montana, dryland wheat operations are the source of about half of agricultural sales. In other areas of the West, irrigated crop production often is the dominant agricultural activity and includes everything from potatoes to cotton to grain. In still other areas, fruit or nut production is most important. In the Southwest (including Texas), livestock represents about two-thirds of the value of agricultural production. In the Rocky Mountain region, livestock represents about 60 percent of agricultural production. In the far West, livestock makes up about 30 percent of total agricultural production. Thus, agriculture in the West does not necessarily mean livestock production.[3]

It should also be kept in mind that "livestock" is not synonymous with "cattle" in the West. In many western states, poultry raising is the dominant form of "livestock" production. For instance, in California, the livestock workforce is not primarily "cowboys," but chicken or hog raisers. About 75 percent of California livestock marketings are not cattle or sheep. In Washington, Arizona, and Utah, only about half of the "livestock" are cattle or sheep. At the opposite extreme are Montana and Wyoming, where about 90 percent of livestock sales are cattle and sheep. In the eleven western states as a whole, only 53 percent of the livestock activity is associated with cattle or sheep.[4]

Next, it is important to realize that agriculture makes up only a tiny and decreasing fraction of the overall economic activity in the West. The West, like

Table 1. The Relative Importance of Federal Lands Grazing as a Source of Jobs and Income, 1997

State	Arizona	Calif.	Colorado	Idaho	Montana	Nevada
Agriculture as a source of income	0.8%	1.0%	0.8%	3.9%	2.7%	0.2%
Agriculture as a source of jobs	0.9%	1.6%	1.6%	5.5%	5.6%	0.5%
Livestock's share of agriculture	42.8%	27.3%	70.4%	46.7%	49.5%	64.5%
Cattle/sheep's share of livestock	53.2%	24.3%	83.4%	59.8%	89.5%	71.1%
Federal forage's share of total cattle/sheep feed	47.5%	7.3%	12.0%	26.5%	9.3%	50.3%
% of income derived from federal forage	0.09%	0.00%	0.06%	0.29%	0.11%	0.04%
% of jobs derived from federal forage	0.10%	0.01%	0.11%	0.41%	0.23%	0.11%
Days of real income growth to replace federal grazing	9	2	6	31	17	3
Days of job growth to replace federal grazing	10	5	13	45	30	9

State	New Mexico	Oregon	Utah	Wash.	Wyoming	11 W. States
Agriculture as a source of income	1.5%	1.1%	0.7%	1.2%	1.5%	1.0%
Agriculture as a source of jobs	2.4%	3.5%	1.6%	2.5%	4.2%	1.9%
Livestock's share of agriculture	68.9%	30.0%	74.9%	34.9%	77.5%	39.2%
Cattle/sheep's share of livestock	57.5%	59.2%	46.1%	42.9%	93.5%	52.8%
Federal forage's share of total cattle/sheep feed	32.2%	16.3%	31.7%	2.6%	21.1%	18.6%
% of income derived from federal forage	0.19%	0.03%	0.08%	0.00%	0.24%	0.04%
% of jobs derived from federal forage	0.30%	0.10%	0.18%	0.01%	0.64%	0.07%
Days of real income growth to replace federal grazing	23	4	7	1	54	8
Days of job growth to replace federal grazing	43	14	17	2	120	16

Sources: U.S. Department of Agriculture, Forest Service, Range Management, *Grazing Statistical Summary, FY 1997* (Washington, D.C.: Superintendent of Documents, 1998); U.S. Department of Agriculture, National Agricultural Statistics Service, *1997 Census of Agriculture*, vol. 1, *Geographic Area Series*, www.nass.usda.gov/census; U.S. Department of Commerce, Bureau of Economic Analysis, *Regional Economic Information System, 1996*, CD-ROM; U. S. Department of the Interior, *Public Land Statistics, Vol. 183, Statistical Appendix to the Annual Report of the Director, Bureau of Land Management, to the Secretary of the Interior* (Washington, D.C.: Superintendent of Documents, 1998).

most of the rest of the nation, is largely urban, and its economies are largely nonagricultural. This is true even of the nonmetropolitan areas. Using a five-year average to smooth out fluctuations in agricultural earnings, for the 1992–1996 period, agriculture was directly the source of only 1 percent of total income in the eleven western states. For the nonmetropolitan areas of those states, agricultural earnings represented about 3 percent of total income. Idaho and Montana were the most dependent on agriculture for income, at around 4 and 3 percent, respectively. In the nonmetropolitan parts of those two states, agriculture was directly the source of about 6 and 4 percent of total income, respectively. (See Table 2.)

Agricultural interests often seek to inflate the relative importance of agriculture by including in agriculture all food- and fiber-related activities. For instance, food stores and restaurants are included as part of agriculture because they sell food products. In addition, food processors, such as flour mills and meat packing and canning facilities, are included in the agricultural total. Following this logic, cotton and wool clothing manufacturing and sales could also be counted in the total. There is no limit to such creative calculations. Since without food we would all die, one could claim that all economic activity ultimately can be traced back to agriculture. Similarly, all products that contain mineral materials could be traced back to mining.

The problem with all these efforts to exaggerate the relative importance of a particular industry for political purposes is that they ignore the actual incremental contribution a particular activity makes to the creation of economic value. Forage on federal land is just one type of livestock feed, and livestock feed is just one of the many inputs that go into our ultimate enjoyment of a beef steak or hamburger. When competitive markets are functioning, the rewards to various inputs reflect their relative contribution to the production of economic value. That is the reason we focus here on actual farm earnings as a measure of economic importance.

Given that agriculture is the direct source of only a small fraction of total economic activity, that livestock grazing is only a fraction of total agricultural activity, that federal forage is only a fraction of the total feed required by western livestock, and that feed is only one source of the value created in livestock production, it should not be surprising if that federal forage supports only a very small fraction of total economic activity in the West.

Our calculation of the contribution that federal grazing makes to the ranching economy is based on the percentage of total feed needed to support the cattle and sheep herds. This direct calculation of the degree of ranching "dependence" on federal lands is to be contrasted with the method used by livestock interests in their efforts to make the case for protecting the status quo on western public lands. Often a ranch is labeled "dependent" on federal lands if any cow ever makes use of federal lands for forage. No matter how small the contribution to total feed needs, the ranch is considered dependent on that federal grazing. Alternatively, some studies label any ranch that obtains more than 5 percent of its cattle feed from federal lands as "dependent" on that federal grazing.

Such "dependence" is a rhetorical device intended to exaggerate ranches' reliance on federal grazing. Ranchers almost always have to supplement the forage they obtain from federal lands with other sources of feed. Ranchers rely on a mix of feed sources, depending on local supply and cost. They raise hay and feed grains themselves on their own or privately leased land. They graze cattle on their own or other private land. They purchase protein supplements

Table 2. Agriculture Earnings as a Percentage of Total Income (Average, 1992–1996)

State	% Total Income	% Nonmetro Income
Arizona	0.8%	0.9%
California	1.0%	4.1%
Colorado	0.8%	4.1%
Idaho	3.9%	5.9%
Montana	2.7%	3.5%
Nevada	0.2%	1.2%
New Mexico	1.5%	3.1%
Oregon	1.1%	1.8%
Utah	0.7%	2.6%
Washington	1.2%	4.1%
Wyoming	1.5%	2.1%
11 western states	1.0%	3.2%

Source: U.S. Department of Commerce, Bureau of Economic Analysis, *Regional Economic Information System, 1996*, CD-ROM.

and other feed for their cattle. Federal forage is only one source of feed for western livestock. Ranches are no more dependent on it than they are on other sources. Among those sources of feed, there are considerable opportunities for substitution, depending on the cost of each.

The contribution federal grazing makes to the economies of the eleven western states can be calculated by multiplying four percentages together. The percentage reliance of cattle and sheep on federal forage for feed is multiplied by the percentage of agricultural economic activity represented by livestock, which in turn is multiplied by the percentage of livestock sales that are from cattle and sheep operations. Finally, this product is multiplied by the direct contribution agriculture makes to the economy in terms of both income and employment. (See Table 1.)

For instance, for Montana, federal forage represents 9.3 percent of total cattle/sheep feed. Cattle and sheep operations are responsible for about 90 percent of livestock sales, which in turn are about 50 percent of the dollar value of all agricultural sales. Finally, agriculture is directly responsible for about 2.7 percent of total personal income in the state. As a result, federal grazing is responsible for about 0.11 percent of all income in Montana ($0.093 \times 0.895 \times 0.495 \times 0.027 = 0.0011$]).[5]

Note that in this calculation we attribute *all* the value of the livestock production to the feed used. Clearly this exaggerates the role of feed by a significant factor since the economic role of other inputs—such as management, labor, capital, equipment, buildings, fuel, water, and land—is ignored.

Table 1 shows that only about $1 out of every $2,500 of income (0.04 percent) received in the western states is directly associated with grazing on federal lands. In employment terms, one out of every 1,400 jobs (0.07 percent) is directly tied to federal lands grazing. For none of the eleven western states does the direct impact of federal grazing even approach 1 percent. Not all grazing on federal lands, of course, is threatened by proposed changes in federal grazing policy. Only some fraction is. Thus the impact of any of the proposed changes in federal grazing policy will be only a fraction of these estimated *maximum* direct impacts.

These potential job and income losses associated with federal grazing reform can be expressed in terms of the time it would take for the local western economies to replace these jobs and income through the normal expansion of the economy. If the average annual economic growth between 1990 and 1997 is taken as a reference point, the loss of *all* federal grazing in 1997 would have caused income growth in the eleven western states to pause for *eight days*. To make up for the lost jobs, economic growth would have had to pause for about two weeks. (See Table 1.)

Thus, even under the most extreme scenario, in which all grazing on federal lands would be eliminated, the direct income and job losses could be made up in a matter of a few days by the normal expansion of the economy. Obviously, some communities would be harder hit than others, given the uneven distribution of public grazing lands across states and the fact that some areas are more specialized in cattle raising. This potential will be analyzed in the following section. However, although the potential for such localized disruption should be analyzed and mitigated where necessary, it still is the case that economies of broad parts of the West will *not* be significantly harmed by a downsizing of the herds of commercial animals grazing on public lands.

Taking a Closer Look at Individual Counties That Are Dependent on Federal Lands

The preceding discussion was carried out in terms of whole states. One objection to this broad geographic focus is that it allows the metropolitan areas that dominate most states' economies to obscure the important role being played by agriculture and public lands livestock grazing. As shown earlier, however, the nonmetropolitan West is also not dominated by agriculture. Only 3 percent of total income in the nonmetro West originates in agriculture. There are counties, however, where agriculture is much more important, where ranching is the dominant type of agricultural activity, and where federal lands are a major source of cattle feed. In such counties, the role of federal grazing on the local economies could be much larger than the statewide averages reported earlier.

Before analyzing this possibility, however, it is important to understand that discovering such local dependency does not necessarily tell us what the appropriate public policy response should be. It is also valuable to know that for the vast majority of the population of the West and for the vast majority of economic activity in the West, federal grazing is not in any sense crucial; it is clearly quite peripheral. It is not obvious that public policy affecting millions of acres of public land—public lands with many other values in addition to their commercial forage value—should be dictated by the interests of a tiny fraction of the West's population.

The Bureau of Land Management (BLM) studied the quantitative role of federal grazing in the economies of over a hundred contiguous counties in seven western states as part of the Interior Columbia Basin Ecosystem Management Project (ICBEMP).[6] The area studied is that east of the Cascade Mountains drained by tributaries to the Columbia River. It includes all of Idaho, western Montana, eastern Oregon and Washington, and small parts of Utah, Nevada, and Wyoming. Federal grazing leases in this region support about 5.3 million AUMs (animal unit months—one AUM being the amount of forage required by a cow-calf pair, or five ewes with lambs, for a month). ICBEMP federal forage represents about a third of the total federal grazing supply in the eleven western states.

Table 3. Interior Columbia River Basin Counties Dependent on Federal Grazing, 1996

County	State	Economic Activity Due to Grazing Federal Lands	
		% of Income	% of Jobs
Adams	ID	0.7%	3.5%
Camas	ID	1.9%	2.4%
Clark	ID	7.5%	5.9%
Custer	ID	0.3%	1.3%
Owyhee	ID	1.4%	2.1%
Harney	OR	0.1%	3.1%
Malheur	OR	1.1%	2.0%
Wallowa	OR	0.2%	1.5%
Ferry	WA	1.4%	0.9%
Sublette	WY	1.0%	3.0%
Humboldt	NV	1.6%	1.4%

Source: Based on L. Frewing-Runyon, "Importance and Dependency of the Livestock Industry on Federal Lands in the Columbia River Basin" (unpublished report prepared at the Oregon State Office of the Bureau of Land Management for the Interior Columbia Basin Ecosystem Management Project, Walla Walla, Wash., 10 April 1995).

The economic importance of public lands grazing was analyzed in this BLM report as outlined earlier. The relative importance of agriculture in the economy, the relative importance of livestock activity in agriculture, and the relative importance of federal forage as a source of feed were combined to estimate the contribution of that federal forage to the county economies.[7] Of the 102 counties, only 11 were found to have more than 1 percent of total income or employment associated with public lands grazing (see Table 3).

Clearly there are some counties in the West where federal grazing plays a greater role than the tiny fraction of 1 percent found at the state level. It is important to put these more dependent counties in perspective. The five Idaho counties (out of forty-four) represent a little over 1 percent of the total Idaho economy. The three Oregon counties (out of eighteen) represent about 11 percent of the eastern Oregon economy. The one Washington county (out of twenty-one) represents about 2 percent of the eastern Washington economy. The parts of Wyoming and Nevada included in the study were too small to characterize in this way. None of the twelve western Montana counties had a dependence on federal grazing that exceeded 1 percent of income or employment.

These counties had a higher dependence on federal grazing because some, like Clark County, Idaho, derived a very large percentage of their income (58 percent) from agriculture. Others, such as Harney County, Oregon, derived almost all (88 percent) of their agricultural income from livestock. Others, such as Camas County, Idaho, and Humboldt County, Nevada, relied on federal grazing for as much as 40 percent of their livestock feed. These higher dependencies were clearly the exception, not the rule, in the broad seven-state region of the West that the ICBEMP grazing analysis studied.

Another way to investigate the potential local dependence on federal grazing is to pick a particular local region in the West where federal grazing policy has been very controversial and analyze the relative importance of federal forage there. Southwest New Mexico, including Catron County, makes an interesting case study. The "county ordinance" movement, which asserts that county governments can exercise control over federal lands, began in Catron County. It was conflict over federal land management policies, including grazing policies, that ignited this part of the "Sagebrush Rebellion." Table 4 reports on an analysis of the five counties in the southwest corner of New Mexico.

In that isolated rural area, about 3 percent of income and 6.5 percent of jobs are associated directly with agriculture. Because agricultural prices, especially cattle prices, were low in the mid-1990s, the income figure probably understates the relative importance of agriculture in more normal times. On the other hand, since 30 to 40 percent of income is derived from nonlabor sources (for example, investment and retirement income), the jobs figure overstates the relative importance of agriculture.

The relative importance of livestock as a percentage of all agricultural activity varies considerably in these five counties. In Catron County, it is nearly 100 percent. In Luna County, less than a quarter of agricultural sales are from livestock. Reliance on federal lands for forage also varies from 34 to 43 percent in Luna and Catron Counties, to only 10 to15 percent in Grant and Sierra Counties.[8] As a result, for four of the five counties, less than 1 percent of income is tied to federal grazing. Only in Catron County does the economic importance of federal grazing rise much above 1 percent. For Catron County, about 2 percent of income and 9 percent of jobs appear to be tied to public lands grazing. For this group of southwest New Mexico counties as a whole, less than half to three-quarters of 1 percent of the economy is tied to federal grazing.

Taking a More Dynamic View of the Economic Impact of Federal Grazing

The discussion thus far has assumed that if the forage available from federal lands is reduced, cattle production will be reduced proportionately. This is a rather simplified view of the local economy that can be criticized as both understating and overstating the impacts of changes in federal grazing policy.

Grazing interests are likely to make two points. First, the privately owned "base" ranch may depend on adjacent federal lands for grazing while the private lands are used to raise livestock feed for winter or dry-season use. Those ranches depend on federal grazing leases on surrounding land to be viable. Without access to that federal land, it is argued, the ranch operation ceases to be viable, and the decline in cattle production will be more than just propor-

tional to the lost federal forage. The ranches will cease to function, and the entire output will be lost. Second, it is usually argued that declines in agricultural operations will have amplified impacts on the rest of the economy because ranching is the central part of the local economic base, and as it declines, the locally oriented businesses that depend upon it will also decline. In short, besides the direct impact, there will be an indirect "multiplier" effect.

Both of these criticisms of the proportional approach I have taken make the same assumption the proportional approach made, which is that there will be no dynamic, business-like adjustments to changes in the availability of forage. This is a static view of the economy: When there is a reduction in the availability of an input, production just passively adjusts downward. Less is produced and previously productively employed resources now sit idle or underemployed.

That is not how an entrepreneurial market economy responds to change. Productive resources are almost never left unemployed for a substantial period of time. Changes in the availability and cost of inputs do not lead to permanent shutdowns. Profit-seeking or loss-minimizing businesses immediately begin adjusting what they produce and how they produce it to accommodate the changes in economic circumstances.

Ranchers will respond to reductions in the availability of federal forage or increases in the cost of federal forage just as they have responded to the constantly falling real price of beef, lamb, and wool or the rise in cost of other inputs relative to commodity values. When fuel or feed costs rise, ranchers do not just reduce cattle production proportionately and permanently. They find ways of cutting other costs and improving the efficiency of their operations. Federal forage is just one type of feed and just one cost of ranching operations. As the availability and cost of federal forage changes, ranching will change to accommodate it. There may be some reduction in production, especially in the short run, but the more likely response will be a reorganization of western ranch operations to adjust to the new circumstances. Land, capital, equipment, water, and buildings will be incrementally redeployed to accommodate the new economic circumstances. This is not a new phenomenon. It is how farms and ranches have survived for a century or more.

Ranchers are not tied to a single way in which to use their resources to raise cattle. A variety of livestock systems are available and in use. Some operations plan for calves in the fall, either from purchase or from the operation's

**Table 4. Economic Dependence on Federal Grazing in Southwest New Mexico:
Catron, Grant, Hidalgo, Luna, and Sierra Counties (Average, 1992–1996)**

County	% of Income from Agriculture	% of Jobs from Agriculture	% of Agriculture in Livestock	% of Livestock Feed from Federal Grazing	% of Total Income from Federal Grazing	% of Total Jobs from Federal Grazing
Catron	4.2%	20.8%	99.6%	42.8%	1.80%	8.88%
Grant	0.4%	2.9%	93.7%	10.0%`	0.03%	0.27%
Hidalgo	8.3%	10.5%	35.6%	29.3%	0.89%	1.09%
Luna	5.7%	7.5%	23.9%	33.6%	0.49%	0.61%
Sierra	3.5%	8.2%	64.2%	15.0%	0.34%	0.79%
5 counties	3.3%	6.5%	45.5%	25.5%	0.39%	0.76%

Source: K. Moskowitz, personal communication, 21 January 1999.

own breeding cows, then put the calves on rations of alfalfa hay and small amounts of grain or other concentrates until spring. At that time they can be sold or put on pasture and range. Other operations purchase heavier animals to start with, put them on spring and summer pasture, and then sell them in the fall. That way, no winter feeding is necessary, and the short ownership period protects against price fluctuations. A third system, of course, is the cow-calf operation that involves calving in the late winter or early spring, when feeding is usually necessary; grazing during the summer and early fall; then sale of the calves and feeding of the cows over the winter. These are just three examples. The point is that there is considerable flexibility in how cattle growers can deploy their resources.

Federal agricultural economists have modeled the likely impact of a significant reduction in the availability of forage from public lands, specifically a 33 percent reduction. The rising cost of irrigation water and high value of irrigated land for crop production were assumed to block any expansion of irrigated pasture. Problems associated with farm organization were assumed to block any expansion in the use of crop residue for forage. This left only private dryland pasture and rangeland available to fill in the gap caused by the reduction in grazing on federal lands. The federal researchers estimated that not only could this gap be filled, but a significant expansion in total cattle production could take place.

The basis for this assumed ability of livestock producers to adapt to the loss of access to some federal grazing lands was the fact that in the recent past, private lands throughout the nation supported significantly larger cattle herds than they do today. Cattle numbers peaked in the mid-1970s and are 25 percent lower today. This indicates that private grazing lands can support a significantly larger number of livestock than they do today. The federal researchers estimated that a 40 to 50 percent increase on private grazing lands was possible.[9]

A market economy is not a static one in which workers, business managers, capital, land, water, equipment, buildings, and so forth get permanently wasted every time there is a significant change in economic circumstances. Quite the contrary. The genius of a market system is its ability to adapt to change and keep valuable resources in productive use. Projections based on the static assumption of valuable economic resources being permanently unemployed are simply wrong.

There is an important corollary to this thesis of a dynamic economy. The demand for agricultural products is limited. Consumption does not surge when prices are low or when incomes increase. There is only so much food we are likely to consume, and this quantity increases only as the population grows. That is one reason why farms and ranches are perennially in trouble. Each agricultural operation seeks to increase its income by expanding production. The collective impact of this is to drive prices downward. Demand does not respond much to the lower prices, and farmers and ranchers are simply worse off.

This relatively fixed demand for agricultural products also means that when the government supports expanded production in one geographical area, the inevitable result is hardship in other geographic areas where farmers and ranchers are not getting similar support. In that sense, for the nation as a whole, government support for cattle production in the West is not a boon to cattle producers as a whole. In fact, cattle producers in a state like Montana, where federal forage plays only a limited role, regularly express hostility toward other western ranchers, whom they see as being on the federal "dole" at Montanans' expense.

Who Depends on Whom:
The Increasing Reliance of Farm Families on the Nonfarm Economy

The last two decades have not been good ones for farmers or ranchers in the West. In both the mid-1980s and in the mid-1990s, net farm and ranch incomes in many areas approached zero or actually became negative. Despite these serious problems in agriculture in the West, since the late 1980s the western states have led the nation in job, income, and population growth. This western growth has not just been a metropolitan phenomenon. During the 1990s, almost every single nonmetropolitan county in the inland West saw significant population growth.

This extensive economic vitality despite difficulties in the agricultural sector speaks to the diversity and resilience of the economies of the West. This region is no longer primarily a ranching or farming or mining or timber area. Despite relative or absolute declines in all of these sectors, almost the entire region has shown impressive economic vitality. This tells us something about what is and is not energizing the economies of the West.

Farm and ranch operations are increasingly made financially feasible only because farm families have access to nonfarm employment and income. Nationally and within the Rocky Mountain region, almost 90 percent of the income received by farm and ranch operator families comes from nonfarm sources.[10] Of course, many of these farms and ranches are small and not the main income-producing economic activity of the operators. If we look only at farms and ranches with sales of $100,000 or more, about half of household income still comes from nonfarm sources.

Beef cattle ranchers in the western states also depend significantly on off-farm work to support their families, with 40 to 60 percent reporting their main occupation to be something other than rancher or farmer. In addition, 60 to 70 percent of western beef cattle ranchers report that they do some paid work

Table 5. Off-Farm Employment of Beef Cattle Ranchers, 1997

State	Worked Off-Ranch Some of Time	Worked 20 or More Weeks Off-Ranch	Main Occupation Not Rancher or Farmer
Arizona	59%	50%	47%
California	57%	52%	47%
Colorado	61%	49%	46%
Idaho	60%	50%	46%
Montana	53%	58%	35%
Nevada	60%	50%	45%
New Mexico	59%	51%	49%
Oregon	63%	53%	54%
Utah	66%	57%	58%
Washington	57%	51%	47%
Wyoming	57%	53%	40%

Source: U.S. Department of Agriculture, National Agricultural Statistics Service, *1997 Census of Agriculture*, vol. 1, *Geographic Area Series*, Table 16, www.nass.usda.gov/census.

off the ranch. Over half of beef cattle ranch operators worked twenty or more weeks off the ranch. (See Table 5.)

These figures suggest that assertions about the dependence of urban economies on agricultural activity taking place on the rural landscape have the relationship reversed. It is not that towns depend on agriculture but that agriculture increasingly depends on the vitality of urban and nonagricultural rural economies to provide the nonfarm income that keeps farm operations alive. Agriculture is a subsidiary activity supported by the vitality of the nonagricultural economy. It is the growth in locally oriented jobs and income during the past two decades that has kept the agricultural sector from shrinking significantly more than it has, not the other way around. This suggests that those concerned with the financial viability of agriculture need to be focused on enhancing the nonfarm economy—the economic base supporting agriculture.

Looking at the Whole Economy, Not Just One Sector

The political pressure to reduce the level of grazing on public lands is not motivated by some irrational desire to harm western ranchers or some irrational dislike of cattle and sheep. Rather, livestock grazing increasingly is competing with other valuable uses of public lands. Our federal lands are capable of contributing to many different commonly held objectives: food, minerals, recreation, open space, wildlife habitat, clean water, biological diversity, and other environmental services.

Not all of these values can be pursued simultaneously on each piece of public land. In that sense we face a familiar economic choice: What is the highest-valued use or mix of uses for any particular area? Unavoidably, the choice of one set of uses requires us to forgo other uses. This dilemma does not mean that we suffer net losses no matter what we do. Quite the contrary, when we choose a set of uses on which we place higher value—whether that value is monetary in nature or not—we improve our well-being. Making such choices among competing alternatives is something we do every day. When we pay for a meal at a restaurant or purchase an automobile, most of us do not feel that that economic act made us worse off. The value of what we gained, if we made the right choice, justifies the cost in terms of what we gave up. On net, we have improved our well-being.

This is a point of fundamental importance. In the analysis thus far we have focused only on what we might have to give up if livestock grazing on public lands were reduced. Since, however, we would be reducing public lands grazing in the pursuit of other valued objectives, against that loss has to be set the expected gain: a healthier landscape, cleaner water, more diverse wildlife, highly valued recreation experiences, and so on. If these nongrazing uses are in fact the more highly valued uses, we as a nation will experience a net gain in well-being, not a loss because of the changes in livestock production.

This same focus on net gain applies to the local economy as well. The ongoing growth in the western states' economies, despite relative or absolute declines in their natural resource sectors, is tied to residential and business location decisions motivated by the pursuit of higher-quality living environments. There is no other plausible explanation.[11] Protecting natural landscapes, for better or worse, draws economic activity toward an area. Therefore, efforts to protect the character and quality of the public lands and waters of the West can contribute directly to the ongoing economic vitality of the region.

Offsetting the small decline in economic activity that may accompany reduced public lands grazing will be increased economic activity supported by the public lands amenities that are being protected. Again, from an economic point of view, it would be an error to focus only on the cost of protecting those landscapes (the reduction in livestock activity) while ignoring the gain (improved environmental quality and the well-being and economic vitality it supports). In that basic sense, if the right choices are made in pursuing more highly valued uses of our public lands, there will be no economic loss, just a net economic gain.

Being Clear About What We Seek

Changes in federal policy on grazing on public lands will not lead to a catastrophic collapse of the economies of the West. Only a tiny sliver of those economies rely on federal grazing. A much larger part of those economies rely on the region's higher-quality living environments. The economies can certainly adjust productively to almost any change in the price or quantity of federal forage. Regional economic impacts should not be the issue. Some ranches and individuals will be significantly impacted in a negative way. Some certainly will be hurt by changes in policy. Depending on the equities involved, we as a people may want to assist those negatively impacted. It is grossly inefficient, however, to make blanket public policies to deal with the problems of a few. It is far more efficient to deal directly with those who suffer significant impacts and to assist them in making the necessary adjustments.

Most Americans, however, have an interest in the well-being of western ranchers that goes beyond whether they are or are not the economic base of the rural West. Western ranchers are an important cultural icon. In addition, the wide open spaces of the West that ranchers work are part of the landscape we deeply value. We would like to keep both that ranching way of life and that landscape. We can do that.

Unlimited grazing access to federal lands is not central to either of these objectives. As discussed above, ranching in the West is not going to disappear if it loses access to some or all federal forage. It will simply reorganize and work the private, state, and tribal lands of the West.

In addition, despite the almost explosive population growth in the West, most ranchland is not about to be subdivided and settled. Although subdivisions are occurring on the periphery of our urban areas, the West will continue to be primarily an urban area with huge expanses of open space between settlements. The existence of extensive public lands assures that a considerable amount of that open space will remain so indefinitely. The more remote nonfederal lands will continue to be devoted to agricultural uses. If we want to ensure that certain private lands remain as open space and/or in agricultural use, we can work to establish conservation easements and purchase development rights. Identifying key private lands, well before they are under imminent threat of residential development, would enable conservation-minded interests to protect such lands while costs are still relatively low.

Retaining public lands livestock grazing, which compromises other environmental values, is not an efficient or effective way to protect open space or a ranching way of life. If those are our objectives, we have to choose far more direct and focused tools, rather than continuing to sacrifice huge areas of the West to a less valued use—livestock grazing.

MORTGAGING PUBLIC ASSETS
How Ranchers Use Grazing Permits as Collateral

Mark Salvo

The practice of banks lending to ranchers on the value of their Forest Service (and likely Bureau of Land Management) grazing permits is not well-known or understood by conservationists and the public at large. However, this arcane and questionable system has serious ramifications for range management on public lands, and for potential grazing policy reform.

Mark Salvo *serves as grasslands and deserts advocate for the American Lands Alliance in Portland, Oregon, where he coordinates American Lands' campaign to protect the sage grouse, the "spotted owl of the desert."*

Bank in New Mexico. Lending institutions play a powerful role in shaping grazing management policy on many public lands, but their influence has not been widely recognized.

Since the 1930s, the U.S. Forest Service has engaged in a dubious, albeit public, practice of aiding public lands grazers in offering their grazing permits to commercial lenders as collateral for loans. Operating under vague legal authority, the Forest Service facilitates this process by serving as an escrow agent for the rancher seeking the loan and the bank willing to lend on the value of a grazing permit. Without the borrowed funds, many marginal ranches in the arid West would cease to exist. With the extra money, ranching persists at heightened stocking rates, damaging the water, vegetation, and soil on public lands and usually resulting in economic harm to individual ranches and communities.

What Are Escrow Waivers?

Grazing permits issued under the Taylor Grazing Act of 1934 convey to the permit holder the privilege to use publicly owned forage. The permits do not instill a right in permittees to graze federal lands. This distinction was intended by Congress in the act,[1] articulated in Forest Service regulations,[2] restated in federal grazing studies,[3] confirmed by scholars,[4] and upheld by the Supreme Court as recently as 2000.[5] Taylor Grazing Act permits are revocable, amendable, nonassignable ten-year licenses to graze that do not convey property rights.

Despite the indefinite nature of grazing permits, they are recognized as having a monetary value by the real estate market,[6] the Internal Revenue Service,[7] and some economists.[8] The value of grazing permits is sustained by a preference system that advises federal agencies to reissue grazing permits every ten years to the same permittees if they are in good standing. The expectation that public lands grazers will retain their grazing permits for as long as they desire has allowed ranchers and banks to treat them as private property when they use them as collateral for loans. In some cases banks will even make loans with payback periods of more than ten years on the certainty that the permit will be continually renewed to the borrower.

The loan process is convoluted: The permittee seeks operating capital for his or her ranch. Because base properties, ranch buildings, and livestock are often insufficient security for a loan, the permittee offers his or her grazing permit as additional collateral. (This is especially true in the Southwest, where base properties may be 40 acres or less.) However problematic a revocable, amendable, term grazing permit may be as collateral, the bank is primarily concerned about obtaining the permit in the event of foreclosure. To resolve any uncertainty that the bank would receive the grazing permit, the rancher waives all rights to it in advance, in case there is ever a loan default.

Since the Forest Service must authorize transfers of grazing permits, the bank requires the agency to commit to reassign the permit to the bank in the case of foreclosure. The Forest Service thus enters into a multiyear agreement on the grazing permit with the permittee and the bank called an escrow waiver. Under the waiver, the rancher agrees to relinquish the grazing permit back to the Forest Service, which agrees to hold it in escrow until the loan is paid back to the bank. During the period the Forest Service holds the permit in escrow, the permittee retains only the privilege to graze.

Since escrow waivers are public mortgages, copies are filed in county courthouses across the West. A brief review of these records indicates that millions of dollars may be loaned on permits in any individual national forest, and billions have been loaned throughout the West.[9] There is evidence that the Bureau of Land Management also maintains an escrow waiver program.

Are Escrow Waivers Legal?

Congress may have authorized federal agencies to issue escrow waivers in the Taylor Grazing Act. The act sanctions the use of a "grazing unit" (that is, ranch buildings, private base property, public grazing allotment) as security for a loan and recognizes that the permit contributes to the value of the grazing unit.[10] However, it is unclear whether Congress contemplated permittees' using grazing permits as collateral for private loans. Indeed, Forest Service regulations[11] and grazing management guidance[12] prohibit the private transfer of grazing permits; only the agency can transfer or assign permits and only to qualified ranchers. These rules would appear to prohibit transfer of a grazing permit to a bank (in the case of foreclosure) that does not qualify as a public lands rancher because it would not necessarily own base property nor be in the grazing business.

In 1938, the Forest Service signed a memorandum of understanding (MOU) with the (then publicly financed) Farm Credit Association to document the developing escrow waiver policy. Although the MOU stipulated that grazing privileges are not a property right and that the Forest Service still reserves the right to revoke or amend grazing permits held in escrow, it stated that in case of foreclosure the agency would "recognize the loan agency as the logical successor to the [grazing privilege]."[13] Forest Service guidelines developed in conjunction with the MOU also stated that, as a courtesy to lenders, forest supervisors would notify a bank "when it becomes necessary to reissue, discontinue, modify, cancel, or suspend in whole or in part, a term grazing permit for which a permittee has executed an escrow waiver" (to alert banks to fluctuation in the value of their collateral).[14]

The 1938 MOU was canceled by mutual agreement of the Farm Credit Association and the Forest Service in 1990 because the association wanted to separate itself from the banks it oversees.[15] The MOU was replaced by a similar memorandum signed by the Forest Service and individual lenders that committed the Forest Service to give ranchers with escrow waivers one year's advance notice of permit modification due to declining resource conditions. It also pledged not to reduce livestock numbers by more than 20 percent per year because of unsatisfactory resource conditions or failure to comply with forest plans, federal laws, regulations, or policy.[16] The 1990 MOU also expanded escrow waivers to additional private lenders. In addition, the Forest Service promised to promulgate regulations to codify the escrow waiver practice in federal law, but these were never finalized.[17] Thus, the 1990 MOU and associated guidance in a Forest Service handbook remain the only "legal" basis for escrow waivers.

The Problems with Escrow Waivers

Escrow waivers help prolong an antiquated public lands grazing program by overlaying a financial system that requires high stocking rates and low grazing fees. Since banks have loaned billions of dollars on grazing permits, they use their considerable clout in Washington, D.C., to oppose grazing reforms that threaten their investment. The banks also become involved in agency decision making on individual allotments where the value of an escrowed grazing permit is in jeopardy.[18]

Although the Forest Service insists that it has authority to amend or dispose of escrowed permits as circumstances justify, in practice the agency's hands are tied. Although the value of grazing permits is based on a number of factors, chief among them is the stocking rate. Banks make loans on permits largely on the basis of the number of cattle the rancher is authorized to graze on the allotment. Once an escrow waiver is issued, the Forest Service is pressured by the bank and rancher to maintain a high stocking rate even in times of drought or degraded resource conditions so the permit does not lose value. Whenever the Forest Service is forced by litigation to reduce the stocking rate or adjust the season of use on an escrowed allotment, the bank is the first to complain because its collateral has been devalued (and the rancher has more difficulty repaying the loan).

The banks' resistance to reductions in livestock numbers or other restrictions on a grazing permit causes headaches for the Forest Service, which is mandated to protect and restore land, water, flora, and fauna by the National Forest Management Act, Endangered Species Act, and the National Environmental Protection Act. If the agency acts to save an endangered species from meandering livestock, the permittee, his or her grazing association, his or her bank, and the local congressional member vigorously oppose the decision. If the Forest Service fails to protect the species, conservation groups sue the agency for breaking the law. ·

The banking industry has also organized against attempts by the federal government to increase grazing fees. Banks condemned the Clinton administration when it proposed to raise the grazing fee a few dollars in Rangeland Reform '94, claiming that "banks . . . are already cutting back on lending to public lands ranchers due to uncertainty created by Rangeland Reform '94" and that the proposed fee increase might cause "financial institutions with large portfolios of loans to public lands ranchers to go out of business."[19] These doomsday forecasts, as well as those by national cattlemen's associations, have acted to preserve very low grazing fees for decades; the resulting shortfall in administrating the federal grazing program is subsidized by taxpayers.

No matter how attractive the banking industry makes borrowing money, public lands ranchers should avoid collateralizing their operations. Many public lands ranches have substantial long-term debt, high debt/equity ratios, and limited income. When public lands ranchers borrow on their grazing units, they are usually forced to graze more animals than the land can support to pay the loan, resulting in ongoing resource damage that can reduce ranch viability and future profits. The easy money also tends to float incompetent ranchers or bankrupt operations the free market would otherwise eliminate from production.[20]

Potential Challenges to Escrow Waivers

There are a few potential challenges to the Forest Service escrow waiver program available to citizens. None of these has been attempted in court; all of them require further research. However, should escrow waivers be declared illegal, the environment, taxpayers, and even public lands ranchers would benefit.

The first challenge is found in environmental law. When the Forest Service neglects to restrict grazing numbers commensurately with resource conditions or to protect an endangered species on an escrowed allotment, it violates any number of environmental statutes. An argument could be made that these federal laws supersede (and invalidate) the 1990 MOU that provides ranchers one year's notice of livestock reductions in case of resource protection and promises not to reduce livestock numbers on escrowed allotments by more than 20 percent per year.

The second challenge is based on process. Although the escrow waiver program has persisted for decades, it is relatively unfamiliar to the conservation community. This may be due to the Forest Service's failure to disclose the program, its extent, and its management implications in public documents, including scoping and planning documents prepared under the National Environmental Policy Act. An argument could be made that escrow waivers significantly impact the Forest Service's ability to manage public lands properly and that the agency's failure to discuss their potential impact violates this act.

Finally, those familiar with federal banking law might contend that banks that loan on grazing permits are risking their depositors' assets on a legally unsound, politically charged federal program that is subject to change. And one might argue that investing in small, marginal public lands ranches is a violation of a bank's fiduciary duties.

The West is systematically looted.
—Bernard DeVoto,
The Easy Chair, 1955

BAD ECONOMICS

Many lesser-known subsidies support the western livestock industry. For instance, cattle guards are placed on roads where they intersect fencelines (they save drivers from having to get out and open and shut gates). There are thousands of cattle guards across the West, costing roughly $10,000 apiece (some cost much more). Taxpayers foot most of the bill. Fences themselves cost about $5,000 per mile to install and also require routine maintenance and repair. Taxpayers shoulder most of the expense for the many thousands of miles of fence lining western highways and crisscrossing the public lands.

Although recreation and tourism are much more important to the economies of western states than public lands ranching—in terms of number of jobs as well as income—land managing agencies allot far more of their time and budgets to administering livestock grazing programs than to enhancing scenery, wildlife viewing opportunities, fishing, or the like. Occasionally, minor accommodations are made to the general public's interests, such as fencing a campground. Yet recreation funds, not range funds, are typically tapped. Similarly, livestock damage to a creek or riparian area may be addressed with wildlife funds.

Opposite: Top, cattle guard, New Mexico. *Bottom,* fence along highway, Wyoming.

Upper right, campground full of cow manure, Nevada.
Middle right, rocks piled across gully to slow erosion, Idaho.
Lower right, cottonwood sapling—planted, fenced, and irrigated—along Gila River, Bureau of Land Management lands, Arizona: livestock necessitated the planting; wildlife funds, not grazing fees, paid for it.

WHERE THE MONEY GOES

Half of the meager fees paid by grazing permittees are returned to the Forest Service and Bureau of Land Management for the "Range Betterment Fund." These monies go to building structures such as fences, stock tanks, water pipelines, and other developments that essentially make public lands into better feedlots. These funds also go to forage "treatment" and various types of projects that manipulate the present vegetation in hope of getting something better—that is, better for livestock.

At times, alteration of habitat is justified on the basis of improving conditions for wildlife. However, it's usually because of livestock grazing that the wildlife are in trouble! And removing livestock is typically not part of the overall plan.

The public could still complain if ranchers were paying for all this manipulation and mitigation on the public lands with their own dollars, but the truth is that the vast majority of this is paid for with taxpayer money. And none of it would be necessary were it not for livestock.

Opposite: Top, helicopter spraying herbicide on sagebrush. Livestock drive out native grasses and make conditions more favorable for drought-tolerant woody species, such as sagebrush, or exotic annuals, such as cheatgrass. Then, ranchers and range managers want to spray the offending plants, so they can get back more good forage—for their cows. *Bottom*, stock tanks are water bowls for domestic livestock, not enhanced wildlife habitat, as some ranchers and range managers purport.

Upper right, grass seeding. When there are no plants left, or none that are desired, artificially seeding an area is an option. Many acres of nonnative crested wheatgrass now carpet parts of the West. *Middle right*, juniper trees cut to increase grasses. *Lower right*, fence to protect spring and recreation site from cows—paid out of recreation fees, not grazing funds. Cows knocked down the fence and trampled the spring anyway.

ENJOY BEEF.
REAL FOOD
FOR REAL PEOPLE

IT'S WHAT'S FOR DINNER
The Health Costs of Meat

Virginia Kisch Messina

Reducing or eliminating meat in the diet promotes health. Meat consumption raises the risk
of various health problems, such as heart disease, some types of cancer, obesity, and hypertension.
Plant-based diets can meet all nutritional needs and in addition provide plentiful amounts of
protective substances, such as phytochemicals and fiber.

Virginia Kisch Messina *is a registered dietitian with a master's degree in
public health nutrition from the University of Michigan. She has been
director of nutrition services for George Washington University Medical
Center and editor of the American Dietetics Association's* Issues in Vegetarian
Dietetics. *She is currently senior editor of the* Vegetarian Nutrition and
Health Letter *from Loma Linda University. Ms. Messina is the author of
several books for consumers, including* The Vegetarian Way *(1996), and has
written a textbook on vegetarian nutrition for dietitians.*

Heavy beef consumption is promoted by the meat and fast-food
industries. Scientific research by nutrition and medical experts is
building a substantial case that not only is high meat intake not
necessary for human health, it can be detrimental to health.

A wealth of evidence suggests that reducing meat intake goes a long way
toward protecting health. Eliminating it goes even further. Diets high in meat
have been linked to risk for some types of cancer, heart disease, diabetes,
obesity, hypertension, gallstones, and kidney stones.

How Meat Affects Health

Meat affects health in two ways. First, it has some direct effects that may
raise disease risk:

- Compounds in meat may increase cancer risk. A growth promoter given to
  beef cattle has been shown to increase DNA synthesis in both normal and
  cancerous breast cells. Also, red meat has been shown to cause oxidative
  damage to DNA, possibly because of meat's high iron content.[1]
- Meat that is cooked to the well-done stage produces compounds that
  are mutagens and raise risk for cancer. Some people are more genetically
  susceptible, and consuming well-done meat can raise their risk for
  cancer sixfold.[2]
- Red meat causes the formation of toxic compounds in the colon that
  can cause mutations. The levels of these compounds in the colon go up
  dramatically when people increase their meat consumption.[3]
- Diets based on meat can be too high in protein. Protein from meat in particular may play a role in raising blood cholesterol levels, may increase risk
  of kidney damage in people who are prone to kidney problems, and may
  have a negative impact on bone health.[4] Some evidence suggests that
  people who consume diets high in meat have higher calcium needs. Although high-protein diets have recently been in vogue, the scientific
  evidence doesn't support any benefit for these diets.[5]
- Diets that include meat are higher in fat, particularly saturated fat and
  cholesterol. Excess fat may contribute to obesity and raise risk for cancer.
  Saturated fat and cholesterol raise blood cholesterol levels and greatly
  contribute to risk for heart disease.

In addition to its direct effects on health, meat displaces plant foods, reducing
intake of a number of healthful components:

- Plant foods provide fiber, which protects against colon disease.[6] Fiber also may help to protect against heart disease and diabetes.[7] Meat contains no fiber, and those who eat meat-based diets typically don't consume recommended levels of fiber. Vegetarian diets are two to four times higher in fiber than meat-based diets.
- Plant foods are rich in antioxidants. These are compounds that protect cells from the damage caused by free radicals—normal byproducts of metabolism that play a critical role in cancer and heart disease and perhaps in other diseases like arthritis.[8]
- Plant-based diets are rich in phytochemicals, which have important biological properties that can promote health. There are thousands of these in plant foods—vegetables, fruits, grains, nuts, seeds, legumes.[9]
- Plant foods are rich sources of a number of nutrients, such as folate, potassium, and vitamins C and E, all of which play important roles in preventing chronic disease.

Meatless Diets Protect Against Disease

ATHEROSCLEROSIS AND HEART DISEASE. Atherosclerosis is the buildup of fatty deposits in the arteries. It can impede blood flow to organs. If arteries leading to the heart are involved, the result can be a heart attack. Other organs can be affected as well. High blood cholesterol raises risk for atherosclerosis, and risk is raised further when the cholesterol is oxidized. Diets low in saturated fat and cholesterol and high in antioxidants—that is, diets based on plant foods—lower risk for atherosclerosis. Fiber may be protective as well. There is some evidence that high animal-protein intake, too much iron, and low levels of the B vitamin folate also raise risk. Those who avoid meat and other animal foods have lower blood cholesterol levels because they eat less saturated fat and no cholesterol. They also eat a diet rich in antioxidants, fiber, and folate and lower in iron and protein, all of which may protect against heart disease. Although not all studies show that vegetarians have less heart disease, risk does seem lower in vegetarian men compared with men who eat meat.[10] In studies at Loma Linda University in California, even occasional meat eaters were at significantly higher risk for fatal heart disease compared with people who never ate meat.[11]

CANCER. The evidence from populations throughout the world is that plant-based diets reduce risk for certain types of cancer. Colon cancer seems especially linked to meat diets. Studies in both the United States and Italy have linked red meat consumption to increased risk for colon, stomach, and pancreatic cancer.[12]

There are many ways that diet may affect cancer risk. For example:

- Dietary fiber can aid the body in excreting some carcinogens.
- Many enzymes in the body that deactivate carcinogens are themselves increased by phytochemicals.
- Promoters, like dietary fat, can prompt cancer cells to divide.
- Compounds in meat can directly act as mutagens, causing changes in DNA.

HYPERTENSION. Researchers have known for nearly a century that vegetarians have lower blood pressure and less hypertension than people who eat meat. In fact, a study done in 1926 showed that simply adding meat to the diets of vegetarian college students caused their blood pressures to rise significantly.[13] Likewise, a 1983 study showed that simply removing meat from the diet caused blood pressure to drop, even when other lifestyle factors were controlled for.[14] The reasons seem unrelated to sodium intake or body weight, both of which are closely related to blood pressure. Some combination of factors in plant-based diets appears responsible for lower blood pressures.

KIDNEY DISEASE. Each day, the kidneys filter the entire volume of blood in the body about sixty times. Diets high in protein cause the kidneys to work harder by increasing the rate at which they filter blood. This could raise risk for kidney problems in older people or in people already at risk for kidney disease. Not surprisingly, the filtration rate of the kidneys is lower in people who don't eat meat—presumably because their diets are lower in protein.[15] High cholesterol levels—associated with meat diets—can also contribute to kidney disease.

KIDNEY STONES. People who don't eat meat have a lower risk for kidney stones than those with high-protein diets.[16]

OBESITY. A number of studies indicate that people who don't eat meat have less body fat.[17] Most likely, this is due to the lower fat content and higher fiber content of plant-based diets, but exercise may be a factor, too.

Meeting Nutrient Needs Without Meat

In the past, nutritionists worried that those who reduced or eliminated meat from their diets would fall short of some nutrients like protein, iron, and vitamin B12. However, research over the years has shown concern to be unfounded. Today, the leading nutrition and health authorities in the country, such as the American Dietetic Association and the American Medical Association, recognize that diets without meat can provide for the nutritional needs of people at all stages in the life cycle, including pregnant women, young children, and young athletes. In fact, those who eat no meat actually have higher intakes of many nutrients, such as vitamins A, C, and E; the B vitamins biotin and folate; and iron, magnesium, copper, potassium, and manganese. In addition, better understanding about food sources and the body's utilization of certain nutrients has resulted in an increased appreciation for how easy it is to meet nutrient needs without meat.

Protein

There is little risk of protein deficiency on meatless diets since grains, legumes, nuts, and seeds are good protein sources. In addition, soy foods and dairy foods provide very high quality proteins. Vegetables provide protein as well, although their contribution is smaller.

The issue of protein quality received a great deal of attention in the past. Specifically, there was concern that plant foods were "inferior" sources of protein to meat, and that nonmeat diets required conscious planning to assure plant foods would combine at meals to provide adequate protein. However, a newer understanding of protein nutrition indicates that protein needs are met as long as people meet calorie needs and consume a variety of plant foods throughout the day.[18] Specific food combining at meals is no longer recommended by authorities like the American Dietetic Association.[19]

Vitamin B12

All of the vitamin B12 in the world comes from bacteria. These bacteria live in soil and the intestines of animals. The B12 they produce gets incorporated into animal tissue and products like milk and eggs. Bacteria on the outside of plant foods also produce B12, which we would, theoretically, ingest when we eat these foods. But normal cleaning of plant foods eliminates most of the B12, so they have become an unreliable source. Thus, those who do not consume any animal foods need to use fortified foods or supplements. People who do not eat meat and who have limited intake of dairy foods may also require vitamin B12 supplements. (And because of absorption problems that commonly occur with aging, all older people should supplement their diet.)

Although many argue that diets low in B12 can't be "natural," the evidence is that we evolved to do very well in a B12-poor environment—that is, on a plant-based diet. Our requirement for this nutrient is infinitesimal, and when we eat large amounts—like in a vitamin B12 pill—our body refuses to absorb more than a fraction of it. We also recycle and reuse the B12 in our body. Our ancestors may have been able to get all of the B12 they needed by picking it up here and there through dirt-contaminated water and food. In today's more hygienic (and healthful) conditions, we've had to rely on other sources.

Iron

Contrary to common beliefs, people who don't eat meat have diets that are higher in iron than those of meat eaters, since plant foods are very iron-rich.

Those who eat no animal products at all have the highest iron intakes since dairy foods contain hardly any iron. However, iron from plant foods is not absorbed as efficiently as that from animal foods. A number of factors also affect absorption, and plant iron is more sensitive to these factors than meat iron. Some nutritionists believe this sensitivity is protective, since high stores of iron have been linked to increased risk for heart disease. Meatless diets may actually represent the ideal situation. People who don't eat meat have lower stores of iron, which lowers their risk for heart disease, but they are no more likely to develop iron deficiency anemia than meat eaters.

However, iron is the nutrient most likely to be deficient in a number of populations, so it is important for everyone to make sure they get enough. Those who don't eat meat have higher iron needs than meat eaters. People who don't eat meat can enhance their iron absorption by consuming sources of vitamin C at meals and by avoiding tea, coffee, and milk with their meals, since these beverages decrease iron absorption. (It's fine to have these beverages between meals, however.)

Good Health Without Meat

Although the decision to give up meat is a personal one, there is no health reason to include meat in one's diet. Even small amounts of meat have been linked to increased risk for chronic diseases. Diets that include large amounts of meat clearly raise risk. Meat directly compromises health in a number of ways. It also displaces plant foods and the health-promoting compounds that are unique to those foods.

The beef industry has contributed to more American deaths than all the wars of this century, all natural disasters, and all automobile accidents combined. If beef is your idea of "real food for real people," you'd better live real close to a real good hospital.

— Neal Barnard, M.D., president of Physicians Committee for Responsible Medicine

EATING IS AN AGRICULTURAL ACT
Modern Livestock Agriculture from a Global Perspective

Richard Schwartz, Ph.D., and Mollie Matteson

Public lands ranching in the American West occurs within a global context of industrialized animal agriculture. A majority of the world's land base is now appropriated either for grazing livestock or for growing crops to feed to livestock. Many globally important issues are linked to the livestock industry and animal-centered diets—namely, food security, loss of arable land, desertification, tropical deforestation, and water scarcity.

Richard Schwartz *is professor emeritus, Mathematics College of Staten Island, New York. He is the author of* Judaism and Vegetarianism *(2001),* Judaism and Global Survival *(2001), and* Mathematics and Global Survival *(1991).*

Mollie Matteson *is a writer, editor, and environmental activist in Eugene, Oregon. She obtained a master's of science degree in wildlife biology from the University of Montana, studying a recovering population of gray wolves. Her work has included advocacy for wolf recovery, for wildlands protection and restoration, and against predator control.*

This cornfield in West Albany, Georgia, is part of the livestock industry. Most corn and soybeans grown in this country, along with several other crops, go to feeding livestock. Far more farmland is devoted to livestock feed than to food people eat directly. Meat consumption, and the amount of land devoted to growing livestock feed, is on the rise in many parts of the world, but this phenomenon is ecologically unsustainable and at odds with long-term human well-being.

Modern livestock agriculture and animal-centered diets not only contribute to the cruel treatment of billions of animals annually; they have devastating consequences for people, for the environment, and for scarce resources. One-half of the planet's total land area is grazed by cattle and other domestic grazing animals, and at least one-fourth of all cropland is now devoted to raising feed for livestock.[1] While more than 70 percent of the grain grown in the United States and close to 40 percent of the grain grown worldwide is fed to animals destined for the plates of the world's more affluent peoples,[2] hundreds of millions of people are chronically hungry, and several million die annually because of lack of adequate nutrition.[3]

This essay is an introduction to the international context in which livestock production occurs. Public lands ranching in the American West is one aspect of a global system of animal agriculture that is straining the earth's resources. There are parallels around the world to the problems caused by livestock grazing in the arid western United States. However, beyond this, the majority of public lands cattle eventually become, like the millions of other cattle in this country, grain-fed animals. As such, they become unwitting participants in many of the globally important matters discussed below, such as world food security, energy use, and loss of arable land.

So much of the planet and its resources are now claimed by livestock production that one biologist suggested, "An alien ecologist observing . . . earth might conclude that cattle is the dominant animal species in our biosphere."[4] Although other species face ever more assaults from a myriad of human activities, livestock agriculture is linked to many of the leading causes of species endangerment and extinction: desertification, deforestation, and water degradation, to name just a few.

Overgrazing by domesticated herd animals is an ancient phenomenon, replicated throughout history in many parts of the world. It is no less a problem today. Over 3 billion acres (or 11 percent of all vegetated land on the planet) have been seriously degraded since 1945, and overgrazing alone accounts for more than a third of this loss.[5] Yet cattle production, not simply cattle grazing, contributes to all the causes of desertification, which is the process by which seasonally dry landscapes are degraded into artificial deserts.[6] These causes, in addition to overgrazing, are overcultivation of the land, improper irrigation techniques, and deforestation.[7]

In the United States, "at least moderate desertification . . . [affects] 98 percent of the arid lands,"[8] and around the world, desertification has hit hardest in cattle-producing regions, such as the western United States, Central and South America, Australia, and sub-Saharan Africa.[9] The grazing pressure exerted on the world's grasslands by livestock is about ten times higher than what such landscapes would naturally support in the way of native herbivores.[10]

Not surprisingly, many wild species are losing ground. In the United States, livestock grazing alone is the second major cause of plant species endangerment. Of all 1,208 plant and animal species federally listed as endangered or threatened, or proposed for listing, mining affects 11 percent, logging affects 12 percent, and livestock grazing affects 22 percent.[11] Yet, because livestock production involves much more than mere grazing, even these figures do not capture the full extent to which livestock agriculture affects biodiversity.

In the tropics, cattle production is responsible for the degradation and clearing of forest on a vast scale, particularly in Latin America. Since 1970, over 50 million acres of moist forest in South and Central America have been converted to pasture. More than a third of Central America's forests have been cleared since the early 1960s. Nearly 70 percent of the deforested land in Panama and Costa Rica is now cattle pasture.[12] Tropical forest is cleared for cultivation of crops for livestock, as well. In northeast Thailand, a quarter of the deforestation can be attributed to farmers growing cassava, an export sold as cattle feed.[13]

The implications for global biodiversity are ominous. Tropical forests may contain 50 percent of the world's species, though they occupy only 7 percent of the earth's total land area.[14] Most of these plants and animals are still unknown to science, and many will become extinct before they are studied or even identified. Finally, millions of tons of carbon dioxide are released into the atmosphere each year by the cutting and burning of tropical forests. This is a major factor in global climate change. In essence, for one cheap, imported quarter-pound hamburger, about 55 square feet of tropical forest must be leveled.[15] It is a trade-off that becomes costlier every year.

The amount of grain in the diet of cattle and other grazing animals has soared over the past several decades. Furthermore, the numbers of all types of livestock are steadily increasing, with global meat production quadrupling since 1950.[16] Meanwhile, productivity of cropland in many parts of the world appears to have peaked, with declining yields either imminent or already beginning.[17] Future world food security may depend on the willingness of nations and individuals to reduce consumption of grain-fed animals.

Americans eat far more meat than most other people of the world. In 1990, the typical diet in the United States included 246 pounds of meat (beef, pork, mutton, lamb, and poultry, based on carcass weights). In the United Kingdom, meat consumption was 156 pounds per capita, and in China, 53 pounds.[18] Americans consume five times more beef than the average human on earth—about 65 pounds per year.[19] The high meat intake of Americans and other relatively wealthy people of the world equates to a disproportionate consumption of grain, with the typical diet in the United States requiring the equivalent of

one ton—2,000 pounds—of grain each year. In contrast, the traditional Asian diet depended on about 300 to 400 pounds of grain annually.[20] Asians still consume a majority of their grain directly, but Americans "eat" well over 600 pounds of grain in the form of beef alone, and close to another 500 pounds in the form of pork and chicken.[21]

Asian diets are changing, however, and after population growth, the greatest source of strain on global grain stocks is rising affluence in Asia and elsewhere, with its concomitant increase in meat consumption.[22] In nearly all Asian nations, the demand for grain for feed is rising rapidly. For example, China in 1995 changed from a grain-exporting country to a major grain-importing country.[23] Asian appetites appear to favor pork, poultry, and fish over beef, yet the net effect is still an acceleration of a basically inefficient system: processing grains through domestic animals, rather than feeding plant crops directly to humans.

The rise of global meat consumption and production is paralleled by changes in agriculture around the world, and thus in human societies. In poorer, less-developed countries, rural subsistence economies are giving way to commercial, especially export-oriented, agriculture. In Mexico, for example, land area devoted to raising the staple foods of corn, rice, wheat, and beans has declined since 1965, while the share of cropland growing feed and fodder for livestock more than quadrupled between 1960 and 1980. While 22 percent of Mexico's population is malnourished, 30 percent of its grain goes to livestock.[24]

Industrial agriculture, especially cattle ranching, employs many fewer people than traditional farming economies.[25] In Central and South America, as well as other parts of the globe, growing rural impoverishment and unemployment is driving many to seek opportunity in urban areas. Unfortunately, what migrants often find instead is continued joblessness and poverty, in overcrowded and unhealthy settings.[26]

The tremendous quantity of grains grown to feed animals requires extensive use of chemical fertilizers and pesticides. The amount of synthetic pesticides produced annually has increased by 400 percent since 1962, when Rachel Carson wrote *Silent Spring*, the book that so eloquently sounded the alarm about the dangers of pesticides to our health, rivers, and wildlife.[27] Over half of U.S. water pollution can be traced to these chemicals.[28] Nitrogen from artificial fertilizers, along with manure nitrogen, moves into groundwater and causes nitrate contamination—a pervasive problem in many agricultural areas of Europe and the United States. Nitrates in water have been linked to cancer and nervous system disorders, among other maladies.[29] Also, in a "circle of poison," pesticides banned or heavily restricted in the United States are legally exported to poor countries, where they are sometimes used on foods imported into the United States.[30] Because of the biological accumulation of pesticides in the body fat of animals, people eating meat and other animal products ingest large amounts of pesticides.

Soil depletion and erosion on cultivated lands is a worldwide phenomenon.[31] The United States is not exempt. According to mathematician Robin Hur, nearly 6 billion of the 7 billion tons of eroded soil in the United States is directly attributable to cattle and feedlot production.[32] Agronomist David Pimentel indicated that about 90 percent of U.S. cropland is losing soil at a rate at least thirteen times faster than the sustainable rate. Nearly 2.5

million acres of cropland, an area larger than Yellowstone National Park, is being abandoned annually in the United States because of erosion and other agriculture-induced degradation.[33]

Manure, once a useful material for farmers, has become another ecological threat stemming from large-scale, intensive livestock agriculture.[34] Cattle concentrated on feedlots produce mountains of manure that run, untreated, into surface waters and aquifers. U.S. livestock produce an astounding 2 billion tons of manure per year, or about 185 times that of the U.S. human population.[35] Food geographer Georg Borgstrom has estimated that American livestock contribute five times more organic waste to water pollution than do people, and twice as much as does industry.[36] Globally, animal wastes threaten water bodies with eutrophication (a process by which increasing levels of nitrogen and phosphorous promote growth of oxygen-depleting algae), from lakes in Italy to northern Europe's Baltic Sea.[37]

Current livestock agriculture and the consumption of meat contribute greatly to three of the four major gases associated with global warming: carbon dioxide, nitrous oxides, and methane. The burning of tropical forest to create cattle pasture and land for growing feed crops not only releases tons of carbon dioxide; it eliminates trees that formerly were absorbing carbon dioxide. The highly mechanized agricultural sector uses a significant amount of fossil fuel energy, which emits carbon dioxide gas.[38] Manufacture of the many products used in agriculture, such as pesticides and chemical fertilizer, further contributes to carbon dioxide emissions.

Cattle emit methane as part of their digestive and excretory processes (as do termites feasting and proliferating on the charred remains of tropical forests). The world's 1.3 billion cattle annually emit about 100 million tons of methane, about 29 percent of the total amount of this gas placed in the atmosphere.[39]

The large amounts of petrochemical fertilizers used to produce feed crops create significant amounts of nitrous oxides.[40] U.S. cornfields, 80 percent of which produce feed for livestock, use about 40 percent of the country's nitrogen fertilizer.[41]

Finally, the refrigeration necessary to prevent animal products from spoiling, including during transport, adds chlorofluorocarbons to the atmosphere. These gases are responsible for thinning of the ozone layer.

A meat- and dairy-centered diet requires about fourteen times as much water as a completely plant-based diet. Globally, agriculture claims more than two-thirds of all water used by humans.[42] Nearly half the water consumed in the United States is used to raise livestock, primarily to irrigate land growing livestock feed.[43] Whereas a typical meat eater's diet requires 4,200 gallons of water daily (with an estimated 80 percent going to produce animal products), a pure vegetarian's diet only uses 300 gallons.[44] Freshwater shortages are becoming increasingly commonplace around the world.[45]

Livestock agriculture requires far more energy than does the production of plant foods for direct human consumption. Whereas 78 calories of fossil fuel are necessary to produce each calorie of protein from feedlot beef, only 2 calories of fossil fuel go into each calorie of protein from soybeans.[46] The annual beef consumption of a typical American family of four requires more than 260 gallons of fuel, as much as the average car uses in six months.[47] Nearly 50 percent of the energy used by agriculture in the U.S. supports livestock production.[48]

When one considers the above facts, as well as the global epidemic of diseases such as heart disease, strokes, and many forms of cancer that can be linked, in large part, to high-meat, high-fat diets,[49] it becomes increasingly clear that a shift away from animal foods is not only an important individual choice, but also imperative for the well-being of humanity, and the ecological systems of the earth.

How we eat determines to a considerable extent how the world is used. . . . Eating is an agricultural act.

—Wendell Berry

FALSE
COUNTER

OPEN
RANGE
LOOSE
STOCK

HOPES AND ARGUMENTS

Ways to Stay Blind to the Critical Plight of Western Ecosystems

PART VI

There are few who would baldly deny that the West has been damaged by livestock grazing. Even most ranchers would probably acknowledge that abusive grazing practices were characteristic of the past and still occur in some places today. In response to mounting criticism, supporters of western livestock production have developed more sophisticated counterarguments. This section explains the current leading arguments in favor of maintaining livestock grazing on public lands and methodically rebuts each one.

"Holistic management"—an approach to problem solving as well as a particular set of livestock husbandry practices, popularized by Allan Savory—holds out hope that ecosystem protection and livestock production can occur on the same piece of ground. Among more "progressive" ranchers, holistic management has a strong following. Yet scientists have documented a great deal of damage occurring on rangeland where holistic management is applied. George Wuerthner explains why even "better managed" livestock grazing is destructive in the arid West.

Among many conservationists, there is hesitancy to advocate the removal of livestock from public lands because of an overriding fear that sprawl and urban development will inevitably follow on private lands. In fact, some environmental organizations strongly defend livestock grazing, believing it to be the last bastion against a perceived condo-ization of the West. But this belief is deeply flawed in its assumptions and logic. Read "Cows or Condos" if you are one of those people still of the mindset that ranching is the lesser of two evils.

The "cattle as bison" argument is basically an assertion that cattle fill the same ecological role as bison. Yet, the behavior and physiology of cattle and wild bison differ greatly, as do plants found in historic bison range versus those that evolved in the absence of bison. And most public lands occupied by cattle today fall outside the historic range of bison. Cattle, and their impacts on the land, are unprecedented and unnatural.

Finally, some livestock proponents claim that cattle grazing can be an important "tool" for achieving specific management goals, such as weed reduction. This view is not so much wrongheaded as it is extremely limited. For example, fire can also be used as a tool to control certain plants. Also, in focusing on one goal, range managers may lose sight of the numerous other, negative impacts of livestock.

These essays may seem to be largely about opposition to something—namely, public lands livestock grazing. However, this opposition arises because of things we are fundamentally *in favor* of: effective, long-lasting land protection, wise and conscientious use of the taxpayer's money, and whole, healed natural communities of native plants and animals.

286–287: The Nature Conservancy's Red Canyon Ranch, near Lander, Wyoming.

Opposite: Cattle graze on public lands in the Animas Valley, Malpai Borderlands region, New Mexico.

THE DONUT DIET
The Too-Good-to-Be-True Claims of Holistic Management

George Wuerthner

One component of holistic management is a grazing system based on the theory that western arid lands require periodic heavy disturbance for ecological health. Stocking the land with larger cattle herds but with more frequent moves is a key component of this system. Ranchers and some government agencies have found holistic management ideas appealing, but the benefits promised often have not materialized in practice, particularly with regard to improvements in land condition.

No treatise on western ranching and its effects on the environment would be complete without a discussion of holistic management. Holistic management (HM) is billed as a plan that will "improve the quality of life . . . while restoring the environment that sustains us all."[1] Although there is nothing in this statement about livestock production, the best-known application of HM occurs in livestock husbandry. HM doctrine defines the major problem facing rangelands as "overrest," not overgrazing. HM founder Allan Savory maintains that "rest is probably the most destructive tool known to science."[2] More cows, not less, say HM supporters, is the solution to a host of rangeland problems. It's not surprising that this strikes a responsive chord in most ranchers.

Most HM doctrine has nothing to do explicitly with livestock production but instead focuses on goal setting and operating a business in accordance with widely accepted practices. The major area of contention and the focus of the remainder of this critique revolves around HM's assertions that livestock are necessary to maintain healthy ecosystems and can restore biologically impoverished western ecosystems.[3]

Although HM advocates would claim otherwise, their solution to nearly every woe on western rangelands requires the use of livestock management to correct the perceived problem. They believe that without livestock (managed according to HM prescriptions, of course), rangelands would suffer desertification, declining productivity, and diminished biodiversity.[4] Managed properly under HM guidelines, proponents assert, livestock can be used to reduce weeds and soil erosion, increase productivity of rangelands, improve water quality and wildlife habitat, increase biodiversity and water infiltration, and restore riparian areas, all while simultaneously enriching the rancher's bottom line.[5]

If you think this sounds a bit like the magic elixir that snake oil salesmen once purveyed, you're not the only one. Many activists and scientists question HM's basic ecological assumptions.[6]

Many HM supporters assiduously deny they like livestock or even support the livestock industry; rather, they assert that they are only interested in ecosystem health.[7] (Taking a cue from HM, most timber companies today advocate more logging, not to further their profits but out of their heartfelt concern for healthy forest ecosystems.) The need to restore and repair degraded landscapes through controlled livestock grazing, is, of course, a very happy coincidence for the livestock industry.

Some aspects of HM livestock management techniques are not in and of themselves flawed, and indeed have an ecological basis that is fundamentally sound—assuming that you want to graze livestock at all. HM doctrine requires confining large numbers of animals (that is, livestock) into relatively small areas, under tightly controlled conditions. Although the stocking rate is high, the duration of grazing in any one pasture is short. Ranchers monitor plant utilization and, at the time deemed proper, move their cattle to the next grazing site, allowing ample time for plant recovery.[8] If followed meticulously—and that is the big *if*—such a grazing scheme has some merit from a livestock management perspective.

It is when HM doctrine strays beyond basic livestock husbandry and gets into ecological theory that it begins to elicit the ire of critics. For instance, HM proponents flatly declare that rest from livestock grazing is destructive; they claim that arid lands need more livestock grazing, not less.[9] Related to these beliefs is the notion that livestock grazing promotes higher productivity of plant communities. In addition, HM advocates like to say that "hoof action" of livestock is necessary to incorporate organic matter into the soil, to push seeds into the

Ranch hand points to two sides of a fence. The area on the left is the Appleton-Whittell Research Ranch in Arizona, owned by the Audubon Society. It has not been grazed in over three decades. The ranchland on the right employs holistic management techniques. Many holistic management proponents point derisively at the Audubon land as an example of the disaster that befalls an area when cows are removed. Holistic management originator Allan Savory has said that the grasses on the Appleton-Whittell Ranch are "moribund," with "bare spots opening up." However, scientists conducting research on the Audubon land found that species richness increased from 22 species in 1969, one year after livestock grazing ceased, to 49 species in 1984. Plant cover increased from 29 percent in 1968 to 85 percent in 1984. In another study on the Appleton-Whittell Research Ranch, total grass cover was significantly higher on ungrazed than grazed sites.

ground for germination, and to improve water infiltration into the soil.[10] All of these assumptions will be challenged below.

Before taking up each of these claims in turn, it is important to discuss a key operating principle of HM, something that allows HM proponents readily to adopt a livestock management strategy that on the face of it, seems too good to be true. We might call this principle the "Donut Diet" phenomenon. That is, the Donut Diet, or HM, as the case may be, offers a counterintuitive, even shocking, but ultimately tantalizing solution to a perennial problem. The conventional wisdom about how to solve the problem is not very appealing—for example, you're overweight, so eat fewer calories; your range productivity is diminished, so reduce the number of cattle on the range. Then some person or concept comes along that offers a way to solve the problem without requiring any sacrifice. In fact, you can have what you want—only more of it! Some people immediately scoff and will hear no more about this "revolutionary" approach. Others, however, are intrigued. Eat nothing but donuts, and lose weight! Put more cows on the range, and get more forage! Heck, why not?

The devil is in the details, of course, which is where HM and a "Donut Diet" start to break down. To implement HM properly, one must monitor range condition very closely. This requires a great deal of self-discipline, is labor intensive, and is often expensive. Success, of a kind, is possible in theory but often is very difficult to realize in practice. The same principles hold for a diet that would allow one to dine on donuts and other junk food; one *can* lose weight, but only with greatly restricted caloric intake.

Sustaining programs such as these is extremely taxing, and the temptation to slack off or cut corners is extremely high. Nonetheless, many people keep on trying, despite their own setbacks and despite outside evidence that what they are doing will not work. They want the program to succeed very badly. They do not blame the method, but their own failings. As an HM practitioner is quoted as saying, "After 13 years I can say it is still the hardest thing I have ever tried to do. The lack of success we have had in some areas has not been because holistic management doesn't work; it is because we haven't practiced it properly."[11]

Yet there *are* ranchers who testify that they have measured improvement in range condition and/or livestock production under HM, just as some people may indeed lose weight eating nothing but donuts. How can this be? One answer is fortuitous timing. In some instances, the positive results observed by ranchers occurred during periods of above-average precipitation, when grass production was naturally higher.[12] However, the main reason that some livestock operators see a change for the better after switching to HM is that they begin to pay close attention to something—livestock husbandry—to which they formerly gave little thought.

Under traditional grazing schemes, most ranchers dump their cattle out on rangelands to fend for themselves. Both the cattle and the rangelands are left unmonitored for weeks or even months at a time. HM, on the other hand, requires intense and frequent monitoring, and regardless of its other aspects, this is a good thing. (It is worth noting that researchers comparing HM techniques with other grazing strategies have found no inherent superiority to HM techniques. Indeed in some cases, greater improvement in range condition, at lower cost, is realized under other traditional grazing schemes, if live-

stock operators give the same strict attention to stocking rate and monitoring range condition.[13]) Ranchers, with greater awareness, can become more responsive to the condition of the land as well as that of their animals. Intensive grazing can also force livestock to use more efficiently the forage in an area. It is not so different from any weight loss diet that gets the dieter to become more conscious of the act of eating and the food's caloric value. No matter what's on the menu, if one carefully observes what is being eaten, chances are that sensations of satiety will be felt sooner, and, correspondingly, fewer calories will be consumed. It's not the donuts that help one lose weight but the discipline and restrictions of the diet.

Just as nutritionists would argue with anyone who asserted that donuts were necessary for a healthy diet just because someone managed to lose some weight consuming them, ecologists and livestock activists object to HM's assertions that livestock grazing is necessary for arid land health.[14] Numerous studies of both livestock-grazed and livestock-free lands provide scientific evidence supporting opposition to HM.

One major assumption of HM is that plants *need* to be cropped. This assumption is based on the observation that plants regrow new leaf material to replace that removed by herbivores. Yet plant responses to the loss of aboveground biomass can more properly be considered a coping mechanism to plant material losses, rather than a positive response to a beneficial event.[15] This is not unlike the documented ability of coyotes to breed at a younger age and produce more pups in the face of predator control. One would be remiss to conclude that coyote populations' tolerance of exploitation translates into coyotes' "need" to be shot, poisoned, and trapped for health.

Areas protected from livestock grazing offer the most telling evidence that munching cattle are not a prerequisite to ecosystem health. Forest Service researchers recently published a study of Dutchwoman Butte in Arizona. This isolated mesa top had never been grazed by livestock yet was "striking in the diversity, density, and vigor of the grasses" and remarkably free of plants such as curly mesquite and snakeweed, which are undesirable forage plants and quite common on sites grazed by livestock. The amount of forage on the butte was four times that found in similar livestock-grazed areas despite the occurrence of a severe drought at the time of the study.[16] There are other livestock-free places throughout the West—though rare due to the ubiquity of livestock—that further make the case that plant communities thrive in the absence of grazing domestic animals.[17]

Another basic ecological problem with the HM livestock bias is that it ignores the evolutionary history of entire biotic regions. Although some parts of the Great Plains were grazed by mobile herds of large herbivores, most plants west of the Continental Divide evolved in the absence of large herding animals such as bison—the native species that HM advocates suggest their cows mimic. Except for small areas along the western fringe of their natural range, bison were not found during historic times in the Southwest, the Great Basin, California, the Pacific Northwest, or in the higher subalpine and alpine mountains of the Rockies. Plants across this vast region lack mechanisms to cope with significant grazing pressure from large herbivores.[18] Yet HM proponents argue that these very dry regions would benefit the most from livestock grazing and trampling effects, even though there was no native analogue to domestic cattle.[19] Some also question the claim that trampling can increase herbage production.[20]

"Overrest" is another term HM proponents use frequently to describe areas not sufficiently grazed by livestock.[21] They warn that with too little grazing, or too much rest, plants become "overmature" and "decadent," and areas of bare, eroding soil increase in size over time.[22] These words may be familiar to conservationists since they reflect the same attitude that foresters have held toward old-growth forests. Today, we appreciate that so-called "overmature" and "decadent" trees are essential to the ecological health of forests.

It's worth noting that almost no plant communities are really "overrested," since all rangelands are grazed whether a cow steps foot on them or not. A host of native herbivores, from grasshoppers to jackrabbits to elk, consume plants even in the most isolated meadows and mesas. Even the focus on large mammals may be misguided. In livestock-free Yellowstone National Park, researchers have found that grasshopper biomass on the northern range exceeds that of all ungulates combined (bison, elk, pronghorn, moose, deer, and bighorn sheep) by three times and that grasshoppers are a major consumer of above-ground biomass.[23] Thus, what HM advocates really mean when they talk about "overrest" is not whether an area is grazed, but whether it's grazed by livestock.

HM advocates assert that livestock grazing increases plant productivity, often using "forage production" as a gauge of ecosystem status when it is really a reflection of the economic concern ranchers have for quantity of livestock forage. Scientists readily acknowledge that many plants compensate for injuries by producing new growth. This regrowth is often higher in nitrogen and other nutrients, and hence more palatable to herbivores. But regrowth of a plant is not evidence that the plant has benefited from being eaten. Indeed, grazing has a cost to plants. After losing its leaves to an herbivore, a plant must redirect energy from seed or root production toward production of above-ground photosynthetic material. In other words, those who claim that grazing "increases" forage production are correct in a limited sense, but such increased production interferes with other plant functions, such as root development, making plants far more vulnerable to drought and other stresses.[24]

Research has shown that grazing cannot increase overall plant biomass production, except under growth chamber or cultivated conditions.[25] Furthermore, regrowth is dependent on moisture, and in many parts of the West, if grasses are intensively grazed, they may not have access to sufficient moisture to regrow in the same season, or even in subsequent seasons.[26] Yet even using forage productivity as a measure, HM techniques are not inherently superior and often fail to produce as much forage per acre as other grazing techniques.[27]

HM advocates claim that the hooves of livestock are necessary to integrate organic matter into the soil and improve soil fertility. Yet research has shown that soil fertility is not the limiting factor in most western ecosystems—water is. And with regard to soil fertility, livestock actually interfere with nutrient cycling. Since livestock tend to reduce soil moisture—by removing shading vegetation and by compacting soil so water cannot penetrate as deeply— they limit microbial decomposition, which is moisture-dependent.[28] One study in Alberta found that short-duration grazing reduced soil organic matter and nitrogen when compared with ungrazed controls. Trampling by hooves played a limited role in this decomposition.[29] In fact, in a review of the literature, one range scientist stated, "In our search of the literature we could find no studies that substantiate Savory's claims on the benefits of hoof action on range soils."[30]

HM doctrine claims that hoof action will enhance water infiltration through trampling of the ground. This, according to HM proponents, breaks up the soil surface so that runoff is slowed and the rain is better able to soak into the ground. But research has shown that cattle hoof action actually impairs soil health in two ways. First, it compacts the soil's upper layers, which reduces water infiltration and increases runoff.[31] At the same time, the destruction of the living soil crusts, known variously as biological crusts, cryptogamic crusts, and so forth, further accelerates erosion by making the surface soil more easily washed away.[32] The loss of cryptogamic crusts is also considered one of the factors that favor the spread of weeds such as cheatgrass.[33]

Finally, the way that HM measures and defines success needs to be examined closely. For instance, HM purports to improve biodiversity.[34] Typically HM supporters consider any increase in species numbers an improvement in biodiversity. But conservation biologists use very different and more complex measures of biodiversity and improvement in biodiversity. To conservation biologists, biodiversity is not just about having a lot of different species on any particular site or even an increase in a few key species; rather the goal is to preserve or restore native species to something approaching their historic distribution and numbers as well as to preserve the important ecological processes that direct species' evolution.[35] Under such a definition, an increase in the number of species may actually signal a departure from the goal of biodiversity preservation, if many of those species are exotic or were historically rare or absent.

Livestock production is destructive to biodiversity. The resource pie is only so big. The majority of the West's water, forage, and space cannot be going toward domestic livestock production and *not* significantly reduce the biological potential of native species, from grasshoppers to trout to elk. HM, by its single-minded reliance on, and advocacy of, livestock as the cure for just about every woe on western rangelands, contributes to the destruction—not the enhancement—of biodiversity and wildlands ecosystems.

HRM [holistic resource management, now shortened to holistic management] promotes the dangerous philosophy that humans are capable of, and should be, managing a planet. It does not recognize the integrity of the natural environment, its right to free existence, or humans' place in it.

— Lynn Jacobs, *Waste of the West*, 1991

JUST A DOMESTIC BISON?
Cattle Are No Substitute for Buffalo

George Wuerthner

It is commonly presumed that cattle are the ecological equivalents of the West's extirpated native bison. This is a political justification for continued livestock grazing, and it is also a biological fallacy. Bison and cattle differ in many ways, including physical and behavioral adaptations to life in semiarid and rugged landscapes. Some western ecosystems now grazed by cattle never did support many bison, and plant communities in these areas have low tolerance for grazing pressure.

Bison once ranged across much of North America. The greatest numbers were found on the shortgrass plains east of the Rocky Mountains, in a zone stretching from Alberta to Texas.[1] Many plants in this region have flexible growth strategies enabling them to tolerate herbivory by bison, as well as by other ungulates, rodents such as prairie dogs, and even invertebrates.[2]

Due to hide hunting, sport hunting, and perhaps also the introduction of the horse to Native Americans, which increased their hunting efficiency,[3] bison numbers had plummeted nearly to the point of extinction by the late 1800s.[4] Domestic cattle have subsequently become the major large herbivore grazing most of the West.

As a consequence of gross similarities in shape, size, and foraging habits between bison and cattle, livestock proponents, range managers, and others have argued that cattle merely fill the vacant niche left by the virtual extinction of bison. Since bison herbivory was an important influence on many grassland ecosystems, many people assume that properly managed cattle have no negative impacts on western rangelands.[5] For example, in a 1991 article in *Rangelands* (published by the Society for Range Management), the author states:

> Long before the American pioneers laid eyes on the mountains and plains of North America, there were "cattle" on our western ranges. Bison roamed the hills, migrating from winter to summer grazing areas, making seasonal use of these lands much as our domestic livestock do today. The bison and the domestic cow belong to the same family (Bovidae) and are genetically similar. They are also very similar in their grazing habits and preferences.[6]

There are those who even assert that western rangelands, particularly the most arid landscapes, require the disturbance impacts of cattle to replace the supposed impact of bison, pronghorn antelope, and other wildlife, now greatly diminished in number or absent altogether. The same author quoted above

Bison herd, National Bison Range, Montana.

concludes that extirpation of the bison "would have put thousands of acres of rangeland into a stagnant and very unnatural situation except for one saving grace: we substituted domestic livestock for the buffalo."[7] Others suggest that herding animals are necessary to break up soil crusts, trample seeds, remove "decadent" plant material, and increase rainwater penetration of the soil.[8] The argument that cattle are ecological equivalents to bison is frequently used as justification for continuing domestic livestock grazing on public lands.[9] So ingrained is the idea that livestock and grazing are synonymous that some livestock supporters assume removal of domestic cattle and sheep equals "no grazing" and caution against the presumed deleterious effects on rangeland "health."[10]

However, the absence of livestock is hardly the same as no grazing. Most native rangelands experience a wide variety of herbivory pressures, from animals as diverse as nematodes, grasshoppers, prairie dogs, pronghorn antelope, elk, and bison.[11] Under natural conditions, populations of these native animals fluctuate seasonally and annually because of predation pressure, competitive interactions between and within species, and availability of forage. Near-constant numbers of cattle are questionable replacement for naturally fluctuating numbers of wildlife.[12] In addition, plant response to herbivory is merely a defensive mechanism that should not be interpreted as a benefit or as promoting rangeland "health."[13]

In any event, a wide variety of evidence—evolutionary, historical, behavioral, and physiological—strongly suggests that bison are very unlike cattle in both their biology and how they use the landscape. The ways in which livestock graze the West, and the ways in which they so dominate American rangelands, is unprecedented. The differences have serious implications for western ecosystems.

Evolutionary History

Bison and cattle evolved from a common ancestor in Asia. Early in their evolutionary history, bison resembled cattle in many features, including horns that pointed forward, a straight back, and a few secondary sexual characteristics. During the Pleistocene, bison gradually adapted to exploit the steppe-tundra ecosystem and spread into available habitat across Asia and Europe.[14] Bison colonized North America via the Bering land bridge.[15] Whereas North American bison became inhabitants of open landscapes such as plains and grassy savannas, most of the evolutionary precursors of domestic cattle were inhabitants of subtropical lowland regions, where they lived in swampy, humid forests.

During the Pleistocene, vegetation in much of North America changed substantially. Broadleaf evergreen species were replaced with a greater abundance of coniferous species, resulting in a deterioration in herbivore forage opportunities. However, at the same time, there was an expansion of steppe areas with fibrous and abrasive foods.[16] The adaptive response of bison was to increase body size and expand their ruminant digestive system to process large amounts of low-quality forage.[17] Bison also developed a high degree of social behavior, manifesting itself in strong herding characteristics, a higher reproductive potential than earlier forms, and rapid maturation rates.[18]

Additional savanna-steppe adaptations displayed by bison are the possession of nonlethal fighting apparatus, movement in large groups, class hierarchy, elaborate social organization, migratory-nomadic behavior, generalist diet, and the ability to digest coarse fiber. Cattle, on the other hand, despite centuries of domestication, display traits characteristic of woodland-dwelling animals: lethal fighting apparatus, small groups, linear or modified-linear hierarchy, territorial fidelity, selective feeding strategies, and reduced seasonal adaptations.

Morphological adaptations of the bison to facilitate existence in a grassland environment include the downward rotation of the head relative to the vertebral column, along with the lateral placement of the eye orbits, which permits maintenance of visual contact with the herd as well as predator detection while grazing. Short limbs also permit easier access to short grasses.[19] Cattle lack these features, having (usually) longer legs, plus straight backs, which may permit short bursts of speed but are not useful for long-distance movement. Of course, not only are cattle less mobile by nature, but selective breeding has resulted in animals that maximize weight gain, making them unfit for long-distance travel.

These morphological differences between cattle and bison also account for the disparity in their ability to sustain flight from predators.[20] Bison possess muscular front quarters and small hind quarters with lightweight leg bones. The bison's hump acts as a fulcrum on which the hind legs can be easily and efficiently swung forward, thereby allowing the maintenance of speed over long distances.[21] Cattle, however, tire rapidly from running. They have largely lost their ability to escape or ward off predation without human intervention.

Bison move frequently, shifting habitat use suddenly. Such sudden movements were noted by many early travelers on the plains. For instance, John Kirk Townsend, traveling along Wyoming's North Platte River in 1834, commented that "buffalo still continue immensely numerous in every direction, and our men kill great numbers." But the next day he wrote, "When we rose this morning, not a single buffalo, of the many thousands that yesterday strewed the plain, was to be seen. It seemed like magic. Where could they have gone? I asked myself this question again and again, but in vain."[22]

More recently, many studies have documented the tendency for bison to move extensive distances compared with cattle. A researcher in the Henry Mountains of Utah noted that bison seldom stayed in one location more than three days.[23] A scientist studying bison in Theodore Roosevelt National Park noted that the animals seldom stayed in the same location for more than forty-eight hours and characterized them as being "highly mobile, moving to new localities and habitats almost daily." The plants were "potentially grazed only once, if at all, in a 3–4 week period."[24]

Not only do bison move more frequently than cattle, their selection of habitat within the landscape is also different. In the Henry Mountains, an introduced herd of wild bison strayed farther from water sources, and used steeper terrain and higher elevations, than did cattle.[25] In Theodore Roosevelt National Park, bison stay at watering areas for a "short duration—one hour or less for even the largest herds."[26] Bison appear to prefer drier forage and spend less time in swales and depressions (where soil moisture is higher) than expected. They select rougher, less digestible forage than cattle, giving them a competitive advantage on native grasslands, where forage quality varies seasonally.[27] This behavior also results in better distribution of grazing pressure on rangelands grazed by bison than on rangelands grazed by cattle.

Cattle are less efficient water users than bison and display a marked preference for moister forage. Researchers in Wyoming reported that 77 percent of the observations of cattle grazing in foothill ranges were within 366 meters of water and noted that the majority of use was on wetlands or subirrigated, level sites.[28] Similarly, other investigators have found that cattle select a higher percentage of floodplain habitat and a lower percentage of upland habitat than these habitat types represent in the study areas.[29]

Because of their propensity to linger in riparian areas or wetlands, domestic cattle pose a far greater threat to arid land biodiversity than do native species such as bison. Preventing damage by livestock to riparian areas requires capital investments in upland water development, fencing, salting, and herding—all of which increase the costs per unit of production, quickly exceeding the financial return on investment in many arid western rangelands, unless costs are subsidized.[30]

Bison have the edge over cattle in areas of deep snow, being able to plow through and push aside snow to get to vegetation,[31] whereas cattle must have supplemental feed. Bison also have a hide of higher insulative value than that of cattle,[32] another adaptation to harsh winters and seasonal food limitations.

One more difference between domestic cattle and bison may be habitat utilization. Bison bulls make up a higher proportion of the herd than bulls in a typical cow-calf operation or other livestock operations. Bison expert Valerius Geist has pointed out that competition for forage is keen in a bison herd, resulting in differences in foraging behavior based on the sex and age of the animal. Bison bulls compromise security in favor of food, whereas cows and calves compromise food in favor of security. Male bison tend to wander farther to find isolated patches of high-quality forage than do females and calves.[33] As a consequence, a herd of wild bison likely grazes a landscape more uniformly than does a herd of domestic cattle.

Although there are historical accounts sometimes documenting heavy grazing by bison, it would be incorrect to assume that bison carpeted the plains as one great mowing machine. Many nineteenth-century travelers on the plains noted both the abundance and the absence of bison and other large ungulates due to seasonal movements and other factors. There was a shifting mosaic of grazing pressure, resulting from the near-constant movement of bison herds. There were also periodic declines in bison populations, due to variability in climate. Consequently, areas that had been heavily grazed would have a respite from grazing pressure, for months or even years. This is a rare situation under most management schemes for domestic livestock, however, in which the focus is on maintaining constant numbers with at most a rest of a year or two.

Historic Bison Distribution

Large portions of the West, including most of the public rangelands in the Intermountain West and Southwest, consist of ecosystems that historically did not support large herds of bison, if any at all. These include most of the sagebrush steppe, the Southwest deserts, and the Palouse grasslands of Washington and eastern Idaho.[34] Other large ungulates, such as elk, deer, pronghorn, and bighorn sheep, were also unevenly distributed over large areas of these semidesert and desert regions, with bighorn sheep and antelope being the most numerous large animals.[35]

There is evidence suggesting that even where bison were found west of the Great Plains, their numbers were small, and distribution was patchy. Periods of favorable climatic and forage conditions probably enabled intermittent recolonization by herds moving in from the plains. However, deformities among the skulls and teeth of bison remains from eastern Oregon suggest such periodic recolonizations were infrequent and that these animals were isolated, locally inbred populations.[36] Rexford Daubenmire argued that protein deficiencies of native bunchgrass rangelands, along with occasional deep snow, limited bison populations along their western margins[37]—although as seen earlier, bison are more efficient at extracting nutrients from forage than are other ruminants.[38]

Richard Mack and John Thompson suggested that grass phenology may have limited bison productivity in the Intermountain West, compared with the plains.[39] Although cool season grasses provide plenty of protein early in the spring, they enter dormancy in summer. This may have posed too great a nutritional stress to lactating female bison. On the plains, a mixture of cool-season and warm-season grasses extends the availability of nutritious forage.

Dirk Van Vuren has postulated that on rangelands in the Intermountain West, forage was inadequate to sustain large numbers of bison except in a few locations.[40] Frequent local extinctions due to weather, hunting by humans, and slow recolonization rates may have also contributed to keeping bison numbers exceedingly low over this region.

Further evidence for the absence or limited distribution of bison throughout the Intermountain West comes from the native vegetation itself, which is not adapted to persistent, heavy grazing and trampling.[41] Vegetation on the Great Plains was dominated by blue grama *(Bouteloua gracilis)* and buffalo grass *(Buchloe dactyloides)*, which seem to tolerate grazing. The native vegetation in the Intermountain West is dominated by species such as bluebunch wheatgrass *(Agropyron spicatum)*, Idaho fescue *(Festuca idahoensis)*, and Indian ricegrass *(Oryzopsis hymenoides)*, which are caespitose or bunchgrasses. These species tend to decline in the face of heavy grazing pressure.[42]

Cattle Not Ecological Stand-ins for Bison

Due to their respective evolutionary histories, a variety of behavioral, physiological, and ecological differences exist between bison and cattle. Cattle are poorly adapted for semiarid landscapes and for rugged terrain; cattle grazing leads to degraded rangelands. Wild, free-roaming bison, on the other hand, are better adapted to such environments and were sustained for thousands of years without degrading rangeland ecosystems.

Since substantial differences in behavior, habitat use, and selection exist between bison and cattle, the suggestion that cattle fill the ecological niche left by the extirpation of the bison is erroneous. Rather, cattle should be viewed as a new ecological force that differs significantly from the native species, in highly negative ways.

Furthermore, some western ecosystems apparently did not support bison in any great numbers or at all. In particular, much of the Great Basin, Palouse prairie, Southwest deserts, and California annual grasslands evolved without the presence of bison. These native rangeland ecosystems display limited tolerance to grazing pressure of any kind.[43]

The assumption that exotic animals are a replacement for native species should be scrutinized closely. The gross appearance of similarity often does not translate into similar use of the landscape.

COWS OR CONDOS
A False Choice Between Public Lands Ranching and Sprawl

George Wuerthner

Fear of sprawl and urbanization is a major obstacle to effecting change in public lands ranching
policy, but the perceived connection between loss of grazing privileges on public land and loss of private
ranchland to development has little basis in fact. The impact of livestock production is also minimized
by many people who do not appreciate the geographical scale at which it occurs in the West.
There are effective ways to protect open space and other values on private lands, but
maintaining livestock on public lands is not one of them.

I have been giving talks and slideshows about the negative effects of western livestock production for many years. I go through a litany of ecological, economic, and human health costs until members of the audience are awash in facts and statistics as well as dozens of images of cow-trashed landscapes. Often my audiences are very sympathetic to environmental causes and are troubled by what they hear. But inevitably, when I suggest that at least on the public lands, livestock grazing be eliminated, someone will raise an objection. It always goes something like this: "Well, I agree livestock do damage. But if you eliminate grazing on public lands, the ranchers will be forced to subdivide on the private lands. Then we'll get more houses, condos, and people. Isn't that far worse than what the cows do?"

The answer, in a word, is no.

First, condos and sprawl, bad as they are, are not worse than ranching. Primarily, this is because "sprawl" and all other urban/suburban and second-home development take up a relatively tiny area of the West, whereas livestock grazing and crop production to support livestock take up immense acreages. Although I do not dispute the damage done to natural systems by sprawl, livestock production also costs a great deal in terms of ecological health and taxpayer dollars.

Second, the notion that protecting ranchers will preserve open space is wrongly premised on the belief that without access to public lands forage, permittees will go out of business and sell their ranches for development. I believe there is compelling evidence to suggest that a very different dynamic from rancher hardship is driving development in the West.

Lawn chairs by pool, Arizona. Many people think subdivision is a greater threat to wildlife habitat than are cows—which on a per acre basis might be so. But the geographic scale of livestock production in the West absolutely dwarfs urbanization and sprawl, so it cannot be simply dismissed as the "lesser of two evils."

Finally, "cows versus condos" is not only a falsehood, it is an impediment to clear thinking and effective action regarding the problems of habitat conservation and preservation of open space on both public and private lands. So long as land protection advocates focus on a false choice between cows or condos, they ignore proven ways to protect ecological values on private property while they also allow livestock to degrade natural systems. Conservationists must move beyond "cows versus condos" if they are serious about long-term protection of western lands.

The Geography of Sprawl and Agriculture

The ecological costs of growing livestock are enumerated elsewhere in this book. Here, I focus on the scale of that activity for the simple reason that most people seem to have very little sense of comparison between the physical footprint of cities and subdivisions in the West and that of livestock production.

To help you think about the geography of the West, let's pretend you are going on an airplane flight. Your journey begins in Denver, say, or Phoenix or Salt Lake City. As your plane waits in the queue for takeoff, you are surrounded by asphalt. Not far off are city streets, buildings, and bustle. When you land, in Sacramento perhaps, or Portland or Los Angeles, it's the same thing. But if you look out the window while you are flying, that is not what you see hour after hour. If you are fortunate to have a clear day, you see this: mountains, valleys, plains, deserts. Occasional towns, if you happen to be peering out at the moment the jet rushes over them. Now and then, especially if your route is along the Pacific coast, you see the telltale gridwork of urban centers. But the dominant impression—if you judge it fairly, if you bother continually to watch that window between takeoff and landing—is open land, land without human residents, or at most very few. Indeed, once outside of the major urban centers and resort communities, open space is the dominant feature of the West.

Let's pretend again that you are flying. This time, you are wearing very special eyeglasses. They are designed to recognize and alter the hue of any land that is dedicated to livestock production, much like Landsat photos that shade areas differently according to dominant plant communities. I'll call these glasses "livestock lenses." Let's say the land looks red wherever it is utilized in some fashion for the raising of livestock. In the West, that's primarily cattle, a few sheep. So, when you fly over rangelands, public or private, you see red. Over the West, there's a whole lot of land used as livestock range, so you see lots of red—flying over mountains, over forests, over deserts. But there's also cropland that is dedicated to raising feed for cattle—hay and alfalfa, primarily. And thus you see valley after valley, extensive flatlands, all red, or nearly so.

And then, these very special livestock lenses have a mechanism for detecting the degree to which water is also used for livestock. Rivers that are partially diverted for irrigation, to grow cattle feed, are pink. From so high up in a jet, you probably cannot see all the tiny rivulets and streams threading, crimson, vermilion, across the landscape. But they are there—some impounded or diverted for irrigation, many more serving as watering troughs for grazing animals, and also as conduits for manure and soils eroded by pounding hooves.

By the time your plane descends and you pull off the livestock lenses, you have seen a landscape dominated by one color—and one use. For that is what the West—especially the arid West—is today: a geography dominated by livestock use.

Indeed, livestock production dominates the entire country, not just the West. The land area utilized for livestock production—including rangelands, pasture, and the production of forage crops (corn, soybeans, alfalfa, and so forth)—occupies 65 to 75 percent of the total U.S. acreage, excluding Alaska, according to U.S. Department of Agriculture statistics.[1] Four crops account for approximately 80 percent of all acreage planted per year in this country: hay, corn, soybeans, and wheat. All but wheat are grown primarily to feed livestock.[2]

In comparison (and again, not counting Alaska), the amount of land taken up by sprawl and development is slightly more than 4 percent.[3] In the West, urban and suburban landscapes, including fairly low density subdivisions, occupy an even smaller fraction of land than in the country as a whole. Sprawl, though a serious and usually permanent blight where it occurs, is not the major ecological threat to the natural systems of the West for the very reason that it is—despite the connotation of the term—confined to a limited area. (I readily acknowledge that cities are drawing resources from a huge area, and their *ecological* footprint is great—but that is a different debate than the matter of sprawl eating up the western landscape. Per capita resource use is an issue of lifestyle for *all* Americans, urban and rural.)

The latest Geographical Analysis Program reported that less than 4.5 percent of California—the most heavily populated western state—is urbanized, and that figure includes all highways, malls, subdivisions, and industrial parks.[4] Most of the human population is concentrated in a few large metropolitan centers, such as San Diego, Los Angeles, San Francisco, and Sacramento. Agriculture is far more pervasive, affecting about 70 percent of the state, by a conservative estimate. This includes croplands, as well as pasture and rangeland. The majority of this land is dedicated to livestock production. Very little grows crops directly consumed by people. For example, about 1.5 percent of California's land area is used to grow vegetables.[5] And from this relatively small amount of land comes about half of all the vegetables grown in the United States.[6]

In Montana, forty-five of fifty-six counties, or 87 percent of the state's land area, have a population density below six persons per square mile, which meets the current census definition of "frontier." Not only this, but twenty-four counties, or 43 percent of the state's land area, meet the 1890 census definition of "frontier," or less than two people per square mile.[7] Yet despite the fact that most of the state is nearly uninhabited by people, numerous native species in Montana are imperiled or significantly reduced in numbers, primarily because of agriculture—which in Montana usually means livestock production. These species include bison, wolf, grizzly bear, swift fox, black-footed ferret, Columbia sharptail grouse, and sage grouse. What is particularly disturbing about this list is that all these species were once widespread and abundant in Montana. None of the aforementioned animals have specialized habitat requirements. It is clear that "open space" is not the same as good-quality wildlife habitat.[8]

Thus, it is the pervasiveness of livestock impacts, and the huge geographical scale at which livestock production occurs, that makes them a far greater threat to the native plant and animal species of the West than sprawl. This is not to minimize the serious consequences of sprawl and development where they are occurring. Still, it should be recognized that this development is relatively concentrated and occupies a small proportion of the western landscape.

Demand Drives Development

Now, even if one is inclined to disagree with my assertion that livestock production is a disaster for the West's native species and ecosystems, that doesn't mean ranching can preserve open space. Even if you think livestock are ecologically benign, supporting ranchers does not safeguard ecosystem values. That's because ranching does not and cannot prevent subdivisions. The problem is complex, but one has only to realize that most western cities sit on land that was once ranched, farmed, or grazed to see that the mere presence of agricultural land did not stop urbanization in the past. And it is not stopping it now.

The growth of subdivisions and sprawl is driven by demand, not the mere availability of land. In fact, sheer population growth accounts to a significant degree for the expanding boundaries of most western cities. A study reviewing census data since 1970 shows that per capita land consumption, or the average area of land physically occupied by people, is actually declining in many western cities.[9] And at the regional level, sprawl in California, the Southwest, and the mountain West is overwhelmingly due to population growth; very little is due to increases in per capita land consumption.[10] Net in-migration, the major reason for population growth in the West as a whole, is fueled by a number of factors, including availability of employment and amenities. Most sprawl is occurring near existing large cities, where jobs, good schools, transportation centers, and diverse cultural offerings are located.[11]

Recreation-related development ("condos") is another type of sprawl occurring in the more rural areas of the West. It is a phenomenon of highly scenic areas with superlative opportunities for activities such as skiing, fishing, boating, and other outdoor pursuits.[12] Again, however, the growth of select recreation/resort/retirement sites in the West probably cannot be separated from population pressures overall and accompanying declines in urban quality of life. Whether one looks at spreading cities or burgeoning "hot spots," the fact is that without addressing the demand for land created by increasing numbers of people in general, any effort to prevent sprawl is ultimately doomed to failure.

It is easy to see why the simple availability of land is not the driving force behind sprawl when you look at places that are *not* experiencing population growth. You do not find much threat of subdivision in the middle of North Dakota or eastern Montana—places where tens of millions of acres are for sale. Why not? Because marginal agricultural economics plus mere availability of private land does not add up to sprawl. A landowner may greatly desire a sale to developers, but he or she will not get it unless there is already demand for land. Very few people want to live in North Dakota except the people already there. No demand, no sprawl.

Low demand has several effects. First, it keeps land prices low. Low land price means that another rancher or farmer can afford to purchase the land of a neighbor and pay off the mortgage running cows on it. When land prices rise—as they have done in some of the more scenic parts of the West—it becomes next to impossible to get into ranching, or to expand one's existing operations. The rising cost of getting into ranching is aggravated by declining profitability of livestock production.[13] Only wealthy "hobby" ranchers can afford to purchase ranches.[14] Indeed, many ranchers think of their ranches as retirement nest eggs and have every intention of eventually selling their properties for development. One study in Utah found that 43 percent of public lands ranchers approaching retirement age stated a desire to sell their land to developers.[15]

High land prices (that is, high demand for real estate) in an area can hurt the ability of ranchers to pass their land on to the next generation, even when that is their wish.[16] In addition, many children of ranching families are simply not interested in taking over the business.[17] There are many factors driving this trend, including better economic opportunities outside agriculture. The high price of land, where this is the case, not only makes selling to developers more attractive to present owners, it becomes one more reason children can't or won't continue to run the ranch. If there are several children in a family, deciding who gets to keep a ranch potentially worth millions of dollars becomes a thorny issue. For many, the easiest solution is to sell it and split the profits among all heirs.

In the past, low land prices permitted western producers to compete with more productive agricultural regions through an economy of scale. Western lands generally support fewer animals per acre than more equable climes, but ranchers could easily buy and own thousands of acres or acquire vast tracts of public lands, compensating somewhat for low productivity by maintaining large holdings. Rising land values have undercut the viability of this option. Ranchers can no longer expand their land holdings and pay off the mortgage with a low-value product, such as beef.[18] Yet the minimum herd size, hence land base, needed to operate an economically sustainable operation continues to rise, further undercutting the long-term stability of the western livestock industry.

An increasing problem for the livestock industry is simply the higher cost of doing business. For generations, ranchers have externalized many of their operational expenses—primarily to the environment and also to taxpayers, who subsidize ranching in a myriad of ways. Whether one is talking about below-market-value grazing fees on public lands, taxpayer-subsidized

irrigation projects, or the numerous environmental costs that the land and society must bear, ranchers have lowered production expenditures because the rest of us have carried the true debt for them. Now, as the American citizenry wakes up to the losses—ecological and economic—ranchers are being asked more and more to pay the full costs. Given the financially marginal nature of most western livestock operations, this can only hasten the demise of ranching in the West.

All of these difficulties are exacerbated by globalization of the market. Increasingly, the price ranchers get for their cows is determined by the world market, not regional or even national economic forces. Yet, production costs are local. Cheaper beef can be grown elsewhere—either because in other, moister, milder regions, the costs are inherently less, or because in other parts of the world, labor and land are less expensive. There is very little the rancher can do to alter these distant situations.

The False Dichotomy of Condos or Cows

The final problem is the false dichotomy of condos or cows. In truth, over much of the West the current economic choice is cows, or . . . well, there aren't a lot of other options. Some ranchers sell out to other ranchers—increasingly, the new owners are corporations, or distant millionaires.[19] Other ranchers turn to game farming or other pursuits that are dubious from both ecological and public interest perspectives. In some places, the unfolding reality is cows *and* condos: livestock grazing continues on rangelands, while the limited wildlife habitat that did exist on private lands shrinks still more.

Critics of eliminating livestock on public lands erroneously assume that the only way of forestalling private land subdivision is by keeping ranchers going, by whatever means possible. Yet, this is wrong-headed for two reasons. First, as I've explained above, the economic forces at play are both complex and powerful. For the most part, there is little ranchers or ranching proponents can do to influence beef prices, nor are they going to stop the public cry for cleaner water, restored species, intact ecosystems, and the like. And unless laws are passed to halt newcomers forcibly at state or county borders, it will be very difficult to put a lid on demand for real estate in places either picturesque—like Paradise Valley, Montana—or booming with opportunity—like Silicon Valley, California. Where land prices rise high enough—in other words, where the demand is great enough—most ranchers are tempted to cash in, if not this year or next, then a decade hence. Relying on the good will and endurance of ranchers is not a good strategy for ensuring long-term land protection.

Furthermore, despite the either/or dynamic implied by the "condos or cows" mantra, there is not a direct relationship between loss of public lands grazing privileges and subsequent sale of private ranchland. Surveys among livestock producers have shown that lifestyle and independence are the prime motivations for remaining in ranching.[20] If access to public lands forage is reduced, many ranchers will seek to stay in the business by modifying their operations: buying more private land, reducing herd size to fit existing private land holdings, and obtaining outside employment to bolster family income.[21]

Perhaps one of the most unfortunate consequences of the "condos or cows" mentality is the lack of initiative among a variety of conservation groups and open-space advocates in taking up truly effective private land conservation strategies. There are examples from around the country of approaches to open-space protection that don't depend on the acceptance of continued degradation of both public and private lands. I briefly describe a few below. However, there are probably many more creative solutions that could be imagined and implemented, if only we could get away from the paralyzing fear that without cows, our only option is houses and concrete.

ZONING AND PLANNING. Zoning and planning are fighting words in much of the West, but if you care about protecting both social and ecological values, zoning and planning really work. Oregon has a statewide zoning system that limits all new development within designated urban growth boundaries. This automatically protects open space outside of the urban regions. It also has the effect of keeping agricultural land prices low, since these are unavailable for residential development. In the Willamette Valley, home to 70 percent of Oregon's population, including the cities of Salem, Eugene, and Portland, 95 percent of the land area is in agricultural production (with plenty of ecological impacts as a result), timber production, or other rural land uses.

LAND ACQUISITION. Many ranchers don't like the land acquisition option too well, either. But the public can decide to make funds available for willing sellers of land that hold important wildlife, scenic, or recreational values. Or private organizations, like land trusts, may purchase significant properties and either donate them to the government or keep them as private preserves. Of course, if cows remain on the purchased lands, I would argue that much of the ecological benefit of the acquisition is lost.

DEVELOPMENT RIGHTS. Development rights can be purchased or traded. In the Pine Barrens of New Jersey, for instance, landowners can "sell" their development rights to developers in urban areas. The urban developers can then apply to city governments to build higher-density housing than is normally permitted. The law allows them to mitigate, in essence, for the high density in the city by preserving open space in the barrens. In either the case of land acquisition or acquisition of development rights, protection against sprawl is far more secure than with a policy of hoping ranchers will act against their economic self-interest, even as the market pressures on them increase. And remember, although outright purchase and acquisition of development rights can be expensive, development is not a threat over most of the West. We don't have to buy all the private ranchland to afford reasonable protection against condos or subdivisions. Many properties will remain open space, no matter what conservationists do or don't do.

Those who suggest we don't have the money to buy up critical lands forget that we currently bestow billions of dollars on the agricultural industry in the form of subsidies and direct payments. In the fall of 1999, for example, Congress granted an emergency $8.7 billion relief package on top of the $26 billion it was already doling out that year to agriculture. Of this, tobacco growers alone received $340 million to make up for a decline in tobacco sales—the result of antismoking campaigns (for which taxpayers have also paid, to a large extent). To give some perspective, $340 million is more than was spent in 1999 on *all* federal land acquisitions, in all fifty states. There is plenty of money in the federal budget, if the political will can be mustered to prioritize permanent protection of habitat and open space. Political will for such investments is undermined by those advocating ranching as a mechanism to protect and preserve open space and wildlife habitats.

Americans are clearly willing to fund land acquisition if they believe no other alternatives are viable. Florida—not known as a particularly liberal, or "green," bastion—has spent more than $450 million a year on land acquisition programs since 1991.[22] In a state that has seen more development pressure than most of the West will see for the next several centuries, Floridians realized that the only effective way to ensure open space would be preserved was to buy it. They have reiterated their commitment to this strategy by voting several times in favor of land protection bond measures.

We must get beyond the misleading and destructive belief in "condos or cows." While thousands of acres go under the bulldozer because of a misplaced faith in ranching as a land protection strategy, hundreds of millions of acres continue to be pounded under the hooves of cattle. While the search goes on for "win/win" solutions between stock growers and conservationists, what is more likely to happen is the "lose/lose" reality of unguided, uncontrolled development in the beauty spots and hot markets of the West, and unabated abuse of the lands and waters that belong to all the people—the public lands—and ultimately, to all the wild creatures that inhabit them.

What would a West without cows be like? Endless subdivisions and cities? Hardly. It would be just this: millions of acres, rich with newly invigorated native grasses; robust with sagebrush and other shrubs no longer bulldozed or chained to make way for cattle feed; swept by growing herds of elk, wild sheep, pronghorn antelope, and bison; vibrant with the energy of predators large and small—from wolves to black-footed ferrets, from grizzlies to swift fox, kestrel, and burrowing owl. The West, without cows, would be thousands of miles of clear streams running deep, filling up with fat native fishes, welcoming back along their margins flocks of raucous songbirds, and a slow, quiet tide of lesser-known beasts: reptiles, amphibians, and invertebrates of all kinds. Relieved of livestock, the West would see the reappearance of the great cottonwood galleries, the regreening of lowland meadows, the regained curvature and grace of flat valley rivers. This, and much more, would be the West without cows.

Next time you fly over it, imagine a West like that.

Farm fields stretch across the Gallatin Valley near Bozeman, Montana, one of the fastest-growing areas of the West. Urbanization and rural sprawl are certainly cause for concern. Yet, as can be seen in this aerial view, agriculture takes up vastly more acreage than sprawl and subdivisions. Consequently, agriculture has tremendous impacts on biodiversity and land health. Agriculture has converted habitat to monocultures of exotic species and has led to fragmentation of the valley's natural landscapes. Livestock production is the dominant type of agriculture in the West (rangelands and crops fed to livestock). By comparison, food consumed directly by people requires relatively little land and water.

USING A HAMMER TO SWAT MOSQUITOES
Livestock as Management "Tools"

George Wuerthner

Some livestock grazing proponents maintain that livestock are important management "tools,"
and therefore commercial grazing should not be banned on public lands. A four-step process is offered
to assess the "livestock as tool" argument in specific situations. More often than not, the cure is probably
worse than the disease. However, if the final conclusion is that livestock grazing is the best way
of dealing with a certain resource problem, a special permit can be granted
while a general ban on livestock is in place.

In response to the growing call for the elimination of livestock production on public lands, those opposed to an outright ban seek various justifications for the continued acceptance of ranching on public lands. One frequently employed argument is that we need livestock as "tools" to improve or restore rangeland health. In other words, livestock—utilized properly—would allow the achievement of management goals such as the control of exotic weeds, reduction of fuel loading, or the creation of better wildlife habitat.

However, before you accept this argument, make sure that livestock are the right tool for your specific problem. A hammer can be used to swat mosquitoes on your face, but the collateral damage your face would suffer makes a hammer a dubious instrument for avoiding bug bites.

Most often, those who advocate using livestock as "tools" are more interested in preserving the cowboy lifestyle or the ranchers' privileged use of public resources than in searching for ways to restore ecological health. Nevertheless, there *may* be a few instances in which livestock grazing furthers a specific management goal. The key issues are whether alternatives to using livestock exist, and what the comparative costs—including and especially ecological costs—of the various options are. If a "solution," like using a hammer to get rid of mosquitoes, causes more damage than it prevents, then obviously a different solution should be found.

In any event, livestock need not be lost as management tools under a general ban on commercial livestock grazing of public lands. Exceptions could be made where livestock are used for specific, legitimate management purposes. For instance, the Wilderness Act precludes the use of motorized vehicles in designated wilderness yet allows motorized access in emergency situations, such as the rescue of injured people. If, after investigating all other alternatives, livestock grazing is the only option that works—a situation akin to an emergency in a wilderness area—then Congress could allow for these isolated exemptions.

Clearly, there must be a process for evaluating proposals for maintaining or introducing livestock as management "tools." Managers, activists, and the concerned public should consider the four-step procedure outlined below as a way to determine whether livestock are indeed the best tool for the job. The goal is to avoid wielding "hammers" when less dangerous remedies are available.

Step One

Ask, "Is the 'problem' really a problem?" Claims that livestock can solve a perceived problem need to be scrutinized closely. Often there is nothing broken that needs fixing, though livestock proponents may see it that way. For instance, in some places, land managers argue that we need to graze public lands to increase shrubs to produce more deer. If greater deer production were the goal and need, there might be some logic to this. However, one can easily refute the need for more deer—deer are among the most common and pervasive large animals in the United States. We do not need livestock to create more deer habitat—there's plenty already.

If, after careful evaluation, there appears to be a legitimate resource problem, go to step two.

Step Two

Ask, "Are there negative 'side effects,' and what are they?" Even if livestock can be shown to achieve some specific management objective, livestock grazing does not occur in a vacuum. There are consequences for putting livestock on a piece of ground, other than just the ones managers are trying to effect. In essence, the cure may be worse than the disease.

For example, some livestock proponents argue that concentrated, early grazing of cheatgrass by cattle can reduce the vigor of this exotic annual and also

Wet meadow pulverized by cattle, Boulder Mountain,
Dixie National Forest, Utah.

eliminate it as a fuel source; this tends to reduce the competitive advantage of cheatgrass over other plants on a site. Controlling cheatgrass is desirable from an ecosystem restoration perspective because cheatgrass has taken over millions of acres in the Great Basin region, it is highly flammable, and the frequent fires that can result when cheatgrass moves in may eventually drive the native perennials out.

Yet what are the other effects of attempting to control cheatgrass with livestock? A holistic analysis does not focus on a single species. Livestock production has many impacts. For instance, since livestock hooves disturb soils, and the animals themselves are often vectors for the distribution of exotic plant seeds, the use of livestock to control one species like cheatgrass may increase the overall number of weedy species on the site. In addition, heavy grazing—if not done very carefully—can also damage native perennial grasses. Thus, while it may reduce the vigor of cheatgrass, livestock grazing can also hurt the very grasses that land managers are trying to conserve.

In drought years, cheatgrass, an annual, doesn't sprout. When cattle are turned out during such dry periods, the only thing left for them to consume are the native grasses, thus speeding the loss of native perennials on the site. (Again, this points out the problem of commercial use. Most ranchers will simply not volunteer to remove their cattle from an allotment, especially in a drought year, when other forage options are limited, and most managers are loathe to buck the desires of the politically well-connected livestock industry.)

Whether they are "tools" or not, cows still compact soils, reducing water infiltration. They still destroy biological crusts. Cows compete for forage with native species. Predators may still be persecuted and killed. Livestock can still transmit disease to wildlife. Range developments and fences may still be needed.

Another problem is that most research purporting to demonstrate how livestock can be useful for controlling weeds, and so forth, is conducted under very strict conditions—the kinds of conditions not typical of a commercial livestock operation, and perhaps not even achievable on most public rangelands. The fact that livestock grazing may successfully accomplish a particular end under highly controlled and experimental conditions does *not* lend support to the argument for the continuation of commercial livestock operations on public lands.

Step Three

Ask, "Are there alternative methods for achieving the same goal?" Unless a legitimate attempt has been made to seek alternatives to livestock, it is likely the "livestock as tool" argument is a convenient excuse for keeping the status quo. When other options are seriously explored, especially those that allow natural processes to be reinstated or revitalized, livestock grazing may become a much less attractive management tool.

For example, some livestock grazing advocates assert that livestock are needed to "control" smooth brome on the plains. But prescribed fire in the spring and early summer can achieve the same goal, and in most places, fire has fewer ecological liabilities than livestock. In some instances, as in the use of prescribed fire, the cost of these alternative tools may appear marginally higher than the use of livestock—but not if you consider all the negatives associated with livestock grazing.

Step Four

If livestock are the best tool for a particular situation, seek special, limited permission for their use. If, after considering all the alternatives for solving a problem and after considering all the ecological costs of various solutions, livestock grazing comes out as the only viable option, a special use permit can be granted for the particular sites and situations that have been evaluated. Meanwhile, a general ban on commercial livestock production would remain in place.

Small, wet swale chewed up by cows, Big Spring Creek drainage, Upper Ruby River Allotment, Beaverhead National Forest, Montana. Whether they are just plain old cows or "management tools," cattle still want to be near water, still want to eat moisture-loving vegetation, still consume plants that other animals need for their own food and homes, and still beat down the soil with their hooves.

FOR
ONS

Restoring the West and Wildlife

PART VII

What can be done to address the problems associated with public lands livestock grazing? There is a simple answer: end it. Get the cows and sheep off, let the wild creatures reclaim their native habitat, and send the ranchers a bill for the cost of restoration.

Of course, as a practical matter, this is easier said than done. The goal may be clear, but the way to it is not. A lot of time can be wasted debating which is the "right" answer to ending livestock grazing on public lands, but as believers in diversity—biological and otherwise—we encourage a multiplicity of approaches.

In this last set of contributed essays, we aim to expand ideas of what is feasible—to suggest that "political reality" is, to some degree, what we make of it. We offer the thoughts of two experienced and successful strategists who have worked for years to rescue public lands from the abuses of livestock grazing.

Bill Marlett acknowledges the impossibility of "killing the myth of the cowboy" but suggests that what real cowboys have done to western lands is no longer completely veiled by public ignorance and indifference. He discusses various approaches to the challenge of phasing out public lands ranching; further exploration of their individual merits and drawbacks awaits conservationists, the ranching industry, political leaders, and the public at large. Certainly the field is wide open for other creative solutions.

Attorney Stephanie Parent uses the law to seek protection for public lands. Litigation is often seen as environmentalists' "hard line," yet as Parent points out, it is based simply on the goal of getting government agencies to properly enforce laws and standards already on the books.

Although it is our desire to make the end of commercial production of livestock on public lands as painless as possible for the affected ranchers, we recognize that it won't be pain-free. Change, even positive change, can be stressful and disconcerting. Ultimately, however, it is the natural world that supports us all. And if we wish to behave compassionately toward future generations—human and nonhuman alike—we must not postpone or shirk the work to be done today.

308–309: Beaver Meadows, Rocky Mountain National Park, Colorado.

Opposite: Battered ground around stock tanks, eastern Oregon.

THE LAST ROUNDUP
Options for Ending Destructive Public Lands Ranching

Bill Marlett

Livestock grazing on public lands is not sustainable or desirable in terms of costs to the land and society. There are various means, not mutually exclusive, by which public lands grazing can be phased out. These include allowing nonranchers to hold or retire permits for conservation purposes; allowing ranchers to choose not to run livestock on their allotments; allowing the government to buy out permits voluntarily surrendered; granting payments to counties that have lost federal revenues because of permit retirement; and prohibiting the use of federal grazing permits as loan collateral.

Bill Marlett *has been executive director of the Oregon Natural Desert Association (ONDA) since 1993. Prior to joining ONDA, he founded and directed the Central Oregon Environmental Center in Bend, led a successful statewide ballot measure protecting Oregon rivers, stopped a spate of hydro dams on the Deschutes River, and spent seven years working for the Wisconsin Department of Natural Resources.*

Ranching on public lands is a social ailment, not a cultural crisis. The West is neither Bosnia nor northern Ireland. The ruin of its resources and the social tragedy in its wake have more in common with the modern Republic of the Congo or the colonization of Cuba. Like many Americans who live off the largesse of our federal welfare state, public lands ranchers will continue to receive handouts only to the extent the public and their elected (and nonelected) officials support a 130-year-old post–Civil War experiment gone astray. With enough money and willpower, you can grow bananas in Alaska, but, like grazing livestock in the arid lands of the West, it makes no sense.

That support is eroding quickly as more people understand the burdens to the land, American taxpayers, and future generations. Unlike past exploiters of the wilderness—the buffalo hunters, the fur trappers, and the whalers—the public lands ranching industry has benefited from owning an icon of our American West, the lone cowboy on his horse, a mythical figure cut out of our imagination and perpetuated by Owen Wister's *The Virginian* and Hollywood hawkers.

But the romantic underpinnings supporting the icon are about the place, not the rugged individual: wide open spaces, big blue skies, untamed nature —all beckoning the young and restless to adventure and danger in a land of enchantment. The cowboy is a stand-in for an American psyche that longs for the great outdoors—for wilderness. We crave nature and live vicariously on the back of the horse riding into the sunset. Louis L'Amour understood that.

I will not argue the ecological and economic failings of grazing livestock on public lands; others more adept have done so. I will argue that the long-term good of ranching on public lands crumbles under the weight of honest observation. The problem has been and continues to be our field of view and our point of observation. Our market-driven culture, coupled with the perversities in our shortsighted political system, prevent any collective ability to step back far enough, in time and space, to realize that livestock grazing on public lands is not sustainable, or desirable, in terms of its cost to the land or to society.

Cowboys round up cattle, High Rock Canyon, Black Rock Desert National Conservation Area, Nevada.

To wit: If the livestock industry were proposing, for the first time ever, to graze livestock on our public lands today, under current law it would not be allowed. Period. Which brings me back to my first assertion: we are dealing with a social sticky wicket that must accommodate the individuals affected by the gradual end of public lands grazing.

Ranchers, along with proponents of grazing on public lands, have largely failed to wean themselves from the myopic notion of agrarian stewards, having never fully realized or appreciated the accidents by which their lifestyle exists. In truth, the vast majority of grazing permits are now controlled by nouveau land barons—corporate cowboys more likely to have corrals filled with wild stock portfolios than with brutish bovines. One might even term the demise of ranching in the arid West as collateral damage on the treadmill of Manifest Destiny: another experiment in Jefferson's vision of an agrarian lifestyle gone awry. We cannot fault those who preceded, nor should we fault those who, by circumstance or predilection, find themselves in the middle of the present dilemma: between those who advocate the restoration of nature and those who abhor restraint.

We are presently mired in this conflict of values. Our challenge is to anticipate the future. Can we end livestock grazing on public lands in the foreseeable future and restore biodiversity to the landscape? Yes. But can we do it in a manner that saves face, that respects the legitimate, if not futile, toil of the yeoman rancher on our public lands, and do it with gentle firmness? One-third of Americans, according to one poll, already favor such a ban. With increased public awareness, it is only a matter of time before that number reaches 51 percent.

The following actions are mere markers on the trail, a path that will, eventually, take us close to the point of beginning. We can never start over. But we must endeavor to go back. It will take many deeds over many years to deal with the plight of the 20,000 public lands ranchers who have adopted or inherited a lifestyle out of synch with the West's 300 million acres of public lands. With so many ranchers close to retirement (and most ranch children not interested in the business), the time to begin our journey is now.

The ideas outlined below have been advocated by various individuals and organizations as potential means for helping to resolve the public lands livestock grazing conflict. These proposals are not mutually exclusive, nor would the successful implementation of any one approach guarantee an end to commercial livestock grazing. However, all these proposed actions would move us closer to achieving the removal of livestock from our public lands.

- *Adopt a policy calling for the gradual phase-out of livestock grazing on federal land.* Proponents of this action believe that Congress should declare a phase-out of commercial livestock production on public lands. Such a phase-out would be done in a manner that respects the legitimate investment of persons engaged in the livestock industry and affected local communities.

- *Allow nonranchers to hold or retire grazing permits for conservation purposes.* Current law requires persons holding grazing permits to be engaged in the livestock business. Proponents of this change in policy would allow conservation-minded buyers to acquire permits from ranchers for nonuse, conservation use, or retirement of the permit.

- *Require the government to buy out voluntarily surrendered permits.* Proponents of this tactic suggest that the federal government should pay ranchers to retire their grazing permits. Some argue for a voluntary program involving only willing sellers, whereas others believe permit retirement should be mandated after a specific grace period. As in the bailout of the insurance industry or any other social program gone astray with enormous financial consequences, federal dollars could be authorized to make transition payments to ranchers for grazing permits surrendered to the government. This program would be similar to previous federal programs addressing economic and social calamities like the whole herd dairy buyout in the Midwest, the purchase of federal tobacco allotments in the Southeast, and the proposition of a federal buyout for one-third of commercial fishers' groundfish fleet on the Pacific coast.

- *Allow ranchers or others who hold a grazing permit not to use it.* Believe it or not, there are ranchers who don't want to run livestock on their public grazing allotment. Current law grants ranchers only three years' nonuse of their permit, after which time they risk losing grazing privileges on their allotment. Proponents argue that ranchers should be able to hold permits in nonuse status indefinitely.

- *Retire grazing permits when voluntarily surrendered.* Proponents argue that the federal government should permanently retire any permit surrendered by a rancher to the government, regardless of whether any grant or payment to the permittee was made. Such permit retirements would be automatic, with no agency discretion.

- *Retire grazing allotments currently vacant and in poor condition, and cancel all permits in the hot deserts of the Southwest.* Proponents argue that Congress should take immediate action on those federal lands where grazing has been a clear and obvious ecological failure.

- *Supplement county payments for all grazing permits retired.* Equally important in the final equation is the obligation to support local schools and governments that may be financially burdened during this transition. Any solution must account for real or perceived losses in revenue. In fact, many conservationists argue that Congress should decouple and stabilize county payments irrespective of changes in grazing levels on public lands. Last year, Congress adopted a similar decoupling program for rural counties that were once dependent on national forest revenues. That legislation provides guaranteed funding of $1.1 billion over five years, regardless of whether a single tree is cut, for over eight hundred counties in forty-two states. Proponents of this solution argue that a similar program could be implemented that provides funding equal to current revenues generated from public lands grazing fees—an amount that would be considerably less than the funds provided from the sale of timber.

- *Prohibit banks from using federal grazing permits as collateral on loans to ranchers.* Current banking practice allows banks to base loans to ranchers in part on the added value of the federal grazing permit attached to a base property. Banks in turn advocate for maximum grazing returns off the public lands to ensure loan payments. This perverse relationship between banks, ranchers, and the federal land managing agencies must be reined in immediately.

The end of livestock grazing on public lands is not about the death of an industry or the end of a way of life. Most ranchers in the West (seven out of ten) don't even graze livestock on public lands, so the cowboy is not riding off into the sunset just yet! The greatest challenge to our society in the present century is the restoration of nature. If we stumble, history will remember the present generation of Americans as having failed its descendants.

It was Aldo Leopold who said, "The major premise of civilization is that the attainments of one generation shall be available to the next."[1] Like the restoration of freedom to the slaves of the South, rewilding our western public lands ultimately demands abolition of an oppressive system, not mere reform. We must work for ending livestock grazing on our public lands; reform has been tried, with little success. When nature is freed from the burden of bovines on public lands in the West, we will not be merely restoring the health of the land but nurturing our passion for freedom and wilderness. Aspiring to such lofty goals underscores the need, if not duty, to end grazing on public lands with gentle conviction, understanding, and civility.

Understanding a landscape no one alive has felt or seen makes the task at hand formidable. *Lonesome Dove* author Larry McMurtry, who spent a lifetime observing his father's deep and peculiar fascination with grass, summed it up best: "I now think it's likely that a lot of my writing about the cowboy was an attempt to understand my father's essentially tragic take on his own—and human—experience . . . [A]s we begin our long descent toward the country we won't be back from, our memory seeks to go back to where it started."[2]

In the process of exposing the seamy side of the western cattle industry, one must differentiate between the people and the industry. Ranchers have a reputation for being good neighbors—friendly and generous. . . . People may be friendly, generous, interesting, good companions, or anything else, but that does not mean that one must approve of their business. Good people can be involved in terribly destructive vocations.

—Denzel and Nancy Ferguson,
Sacred Cows at the Public Trough, 1983

UPHOLDING THE LAW
Litigation and Public Lands Livestock

Stephanie M. Parent

Litigation is an effective and important tool in the effort to protect public lands from the damaging effects of livestock grazing. The goal of legal action is to achieve compliance with existing laws and standards. Several key laws provide the basis for seeking protection of land, water, and species. Litigation is, and should be, only one of various approaches to conserving the ecological health of public lands.

Stephanie M. Parent *is a staff attorney with the Pacific Environmental Advocacy Center (PEAC), the environmental legal clinic of the Northwestern School of Law of Lewis and Clark College. She is an alumna of the school, receiving her J.D. degree in 1992. Formerly, she was a trial attorney with the U.S. Department of Justice, defending the United States in environmental litigation, primarily in actions challenging public lands management decisions. She also has taught courses in natural resources law and policy and in environmental law for the U.S. Department of Agriculture Graduate School, an adult continuing education program in Washington, D.C.*

The Diamond Bar Allotment in the Aldo Leopold Wilderness, New Mexico. Environmental groups successfully utilized legal means to stop livestock damage to wilderness values.

Wouldn't it be nice if we could all just get along? If we could reach a consensus on how to heal the public lands? Unfortunately, much change in land management has come only after the public takes its battle to court. Examples of effective litigation abound, from removal of livestock damaging the values of the Donner und Blitzen Wild River in Oregon to the protection of Desert Tortoise habitat in the California Desert Conservation Area. Livestock grazing must comply with a number of standards found in federal laws. Enforcement of these standards, through litigation if necessary, will aid in the overall strategy to restore public lands from grazing impacts.

Outside pressure through litigation is particularly important in the sphere of range management. Many federal land managers live in the local culture of remote areas. The livestock industry has disproportionate influence over the agencies and legislators. In addition, range managers have an inherent interest in maintaining, rather than diminishing, their turf. Therefore, improved management often is a consequence of court action, either because citizen enforcement of the law provides cover for the agency to meet its legal obligations without industry disapproval or because the agency really has no intention of upholding environmental standards and must be forced to do so.

As an initial step, one must ask whether land in the arid western states is suitable to livestock grazing. Both the Bureau of Land Management (BLM) and the Forest Service are required to assess whether it is appropriate to graze livestock on certain lands pursuant to the Federal Land Policy and Management Act and the National Forest Management Act.

The Federal Land Policy and Management Act of 1976 mandates that the BLM shall "take any action necessary to prevent unnecessary or undue degradation of the lands."[1] In addition, this law directs that land use plans must comply with the principle of multiple use, should "best meet the present and future needs of the American people," and should include consideration of "relative values of the resources." Land use plans should preclude "permanent impairment of the productivity of the land and the quality of the environment."[2] Thus, the BLM may permit livestock grazing only where the benefits—environmental, economic, and social—outweigh the harms. Evidence of noncompliance with these standards can provide a basis for legal action.

Importantly, the BLM's grazing regulations provide legally binding standards directed at the ecological status of public land. The "fundamentals of rangeland health" require a change in grazing management no later than the start of the next grazing year whenever the BLM finds unsatisfactory conditions of watersheds, ecological processes, water quality, and habitats for threatened, endangered, candidate, and other special status species.[3] The regulatory requirement that the BLM take action promptly is an implement for change. The BLM has been conducting range assessments for compliance with these standards and guidelines. Public involvement in this process, as well as ensuring the BLM takes required action,[4] is essential to progress in the health and restoration of the lands.

On national forests, the Forest Service must provide for multiple use as well,[5] and the National Forest Management Act provides for regulations that "require the identification of the suitability of lands for resource management."[6] The Forest Service's regulations have defined suitability as the "appropriateness of applying [grazing] to a particular area of land, as determined by an analysis of the economic and environmental consequences and the alternative uses foregone."[7] Although a 1999 case resulted in a finding that the Forest Service adequately considered the suitability of the land for grazing,[8] in a 2001 administrative appeal the chief of the Forest Service and the secretary of agriculture directed a new grazing suitability determination because the forest plan had failed to comply with the suitability regulations.[9]

Congress mandates conservation and recovery of imperiled species and their habitat in the Endangered Species Act of 1973.[10] The Endangered Species Act is a powerful tool to change and even eliminate livestock grazing where it is shown to degrade the habitat on which threatened or endangered species depend. Ensuring that species in need of protection from extinction are listed as threatened or endangered is the first step in the process and does not usually occur without litigation.[11] Once a species is listed, the federal agencies have the duty of conserving the species and ensuring that actions do not jeopardize the continued existence of the species or destroy or adversely modify the species' critical habitat.[12] Citizens must take an active role to ensure species are provided the protections required by the act.

Another federal law that provides a high standard for continued livestock grazing is the Wild and Scenic Rivers Act.[13] This law provides that federal agencies must develop plans to manage rivers designated as wild, scenic, or recreational. Where the agency fails to have a plan within three years of designation, the courts can provide relief and order completion of the plan within a certain time.[14] The Wild and Scenic Rivers Act requires that the agencies manage the rivers to "protect and enhance the values" for which the rivers were designated in the first place. Where livestock grazing is not compatible with protecting and enhancing a river's values, the government has the authority to change or completely eliminate that grazing, and the public is able to enforce this obligation in the courts.[15]

The Clean Water Act calls for the restoration of our nation's waters.[16] Livestock grazing on federal land must comply with state water quality standards, such as standards setting temperature, fecal coliform, and sedimentation limits established pursuant to the Clean Water Act.[17] Enforcement of this duty has not been tested in court, but there are many streams whose water quality has been impaired as a result of livestock grazing that could provide a factual basis for judicial review. In addition, states are in the process of developing total maximum daily loads for water quality-limited streams, including for nonpoint sources (nonpoint being any source of a pollutant that is not a discrete conveyance), such as grazing allotments. Conservationist participation in and enforcement of these provisions will ensure that livestock grazing no longer fouls our waterways.

Finally, the National Environmental Policy Act (NEPA)[18] requires that the federal agencies assess the impacts of any federal action, including leasing lands for livestock grazing, when the impacts will be significant. NEPA compliance in the context of grazing management has been the subject of much litigation as well as several appropriations riders. NEPA continues to be a valuable instrument to inform the public and reform grazing practices on public lands.

All this is not to say that litigation is the only answer. Public participation in grazing decisions is absolutely imperative if lands are to be restored and wildlife protected. Seeking creative solutions through dialogue, mediation, and other means will enhance the chances for restoration. Decisions to seek recourse from the courts should not be made lightly; even well-reasoned lawsuits are lengthy, resource-intensive, and often frustrating. Still, litigation is integral to accomplishing conservation goals by serving as a backstop—a safeguard in the most urgent and intractable cases—and by putting activists in a position of relative strength in other arenas, such as working for administrative and legislative change.

Cow pooping by stream. The Clean Water Act and other laws offer the potential, if enforced, to significantly rehabilitate many western lands presently degraded by livestock.

OUR

OUR VISION

So why end public lands livestock grazing? There are a host of reasons already enumerated throughout this book. It's bad public policy. It's a bad program. It wastes government funds. It gives certain individuals and corporations an economic advantage over other livestock producers. It gives undue political clout to a small minority in decisions affecting public resources and lands that belong to all Americans. It causes incalculable environmental damage to our natural heritage. Livestock are a major source of water pollution. They negatively affect recreational use, and so forth. Any of these are reason enough to end public lands grazing.

There is yet another reason. Our native plants and animals are under tremendous and growing pressures from development, pollution, competition with exotic species (including cows), habitat fragmentation, and all the well-documented impacts that a growing human population is having on the planet.

Our public lands, however, are a potential refuge. They are large, continuous blocks of habitat that should be managed to maintain native species and ecosystems. The 300 million acres of public lands in the West could serve as major core areas for biodiversity preservation. That does not mean all human use of these lands must end. Some activities are compatible with this goal. But we cannot allot the majority of forage, water, and space on our public lands to cattle and sheep and believe this doesn't significantly reduce the lands' overall carrying capacity for native species. The public lands pie is only so big.

Simply removing livestock won't mean that all these lands will suddenly become more productive and support all native species. Some lands have slipped below an ecological threshold. They will require active restoration or at least a long period of recovery from 150 years of livestock abuses.

We cannot limit ecosystem recovery and species protection to only public lands. Most wild animals and plants will require sensitive and sensible management on private lands if they are to flourish or even survive. Nevertheless, these 300 million acres will make a good start toward providing native species a reasonable chance of maintaining themselves into the future.

It is truly a "wise use" to remove livestock from these lands. If beef production remains a goal of our society, there are many other places where cows can be raised with less ecological damage and economic waste. Indeed, as has been pointed out, the majority of beef production already takes place on private lands, primarily in the well-watered Midwest and eastern United States. An investment almost anyplace where it rains regularly will result in far greater meat output than what one can get from trying to raise cows in the arid West—particularly on the public lands.

Although we can easily grow more beef in Georgia or Missouri (though for a variety of reasons, including diet and issues of cropland use, we may question why we should raise more livestock anywhere), we cannot readily grow grizzlies or elk in such places. We need to ask, What can our public lands do best? And with few exceptions, they are far superior as fountainheads of natural biological diversity and inspiration for the human heart than they are as large-scale feed grounds for livestock. These 300 million acres of western public lands should be managed for values that aren't preserved or protected adequately elsewhere.

This is our vision. We dream of a landscape where bison, pronghorn antelope, wolves, and grizzlies are free to roam; where streams flow clear and clean; where fences don't break up the horizon. We imagine what it would be like to hear the grunts of bison and the howl of wolves, instead of the bawling of cows or the bleating of sheep. We hope for a time when fish, frogs, snails, butterflies, and birds are secure in their futures and their homes.

We are not talking about turning back the clock to the days of Lewis and Clark or some other previous point in history. There have been too many changes, too many ecological bridges crossed, and today there are probably too many people for that to happen. But we can envision a future in which our public lands are in far better ecological condition than at present; in which there are far fewer species on the edge of extinction; in which landscape-scale ecological processes can operate with a minimum of human interference. This is an achievable goal. The elimination of livestock production from our public lands will set us on that pathway.

And so, it is time to begin.

320–321: Bison, Yellowstone National Park, Wyoming.

We encamped close to the river. The night was dark, and as we lay down we could hear, mingled with the howlings of wolves, the hoarse bellowing of the buffalo, like the ocean beating upon a distant coast.
— Francis Parkman, *The Oregon Trail*, 1849.

ACKNOWLEDGMENTS

Many dedicated, creative, and highly knowledgeable people helped bring this book to fruition. Doug Tompkins saw the need for a fresh and ambitious treatment of the topic — his energy and vision, and the vital support of his Foundation for Deep Ecology, propelled the project along its winding path from idea to reality. Foundation staff members offered their own keen perspectives, cheerful assistance, and sympathetic ears, including John Davis, earliest champion of the book and ever champion of all things wild; Sharon Donovan, agile navigator of deadlines and details; Quincey Tompkins Imhoff, font of kind and encouraging comments; and Melanie Adcock, reminder of viewpoints we sometimes forgot. Thanks to everyone at the Foundation, and also to Kris Tompkins, whose great food and good humor sustained us during those long, early meetings, and who also offered many wise suggestions of her own.

Roberto Carra was our book designer par excellence. His ability to transform stacks of images and reams of text into a visually stunning and coherent opus is breathtaking. Also making invaluable contributions to the book's accuracy and good order were Doug Bevington, pinch-hitting fact checker; Janet Vail, determined wrestler of wayward endnotes; Mary Anne Stewart, gentle but eagle-eyed copyeditor; Amy Evans McClure, typesetting virtuoso; proofreader Vicki Botnick; and indexer Ellen Davenport.

We are greatly indebted to the following people for sharing their expertise on particular aspects of western ranching and its impacts: Sue Bellagamba, Jayne Belnap, Katie Fite, Peter Galvin, Steve Herman, Jerry Holechek, John Horning, Mike Hudak, Lynn Jacobs, Jon Marvel, Brett Matzke, Ralph Maughan, Beth Painter, Tom Ribe, Alan Sands, Mark Sterns, Martin Taylor, and Larry Walker. Andy Kerr read the entire manuscript; Jerry Freilich, Scott Kronberg, and Helen Wagenknecht reviewed specific essays and made suggestions for improvement. Greg Schneider and Dale Turner also offered their help.

Finally, this book would not have been possible without the work of our authors and photographers. They unselfishly donated their time, words, and images because of their caring for this Earth. We are in awe of their commitment, as well as their knowledge and talent. Thank you very, very much.

— George Wuerthner and Mollie Matteson

Photo Credits

All photos are by George Wuerthner, except as noted below.

Jess Alford, *pages 16 (middle left), 270*

Bureau of Land Management, *page 77 (top and bottom)*

Bob Crabtree, *page 246*

Katie Fite, *pages 78–79*

Steve Herman, *pages 18–19, 154–155*

David Hollingsworth/U.S. Fish and Wildlife Service, *page 255 (9)*

Mike Hudak, *pages 16 (bottom left), 40, 134–135, 158–159, 160–161, 183, 316*

Lynn Jacobs, *pages 32, 193, 216, 276 (top), 277 (top)*

Steve Johnson, *pages 6, 112–113, 249 (middle right)*

Sandy Lonsdale, *pages 92–93, 209, 236, 249 (top right), 257 (10), 319*

Steve Maslowski/U.S. Fish and Wildlife Service, *page 257 (15)*

Brett Matzke, *pages 142–143*

Kim McMaster/U.S. Fish and Wildlife Service, *page 255 (11)*

Kim Mello/U.S. Fish and Wildlife Service, *page 255 (8)*

Dick Randall, *page 249 (top left and bottom left)*, courtesy of the Humane Society of the United States.

John Rihne/U.S. Fish and Wildlife Service, *page 255 (2)*

Bruce Rosenlund/U.S. Fish and Wildlife Service, *page 255 (4)*

Scott T. Smith, *pages 10, 76 (top and bottom), 249 (middle left), 255 (12), 304, 312*

U.S. Department of Agriculture, *page 282*

U.S. Fish and Wildlife Service, *pages 255 (1, 3, 6, 7)*

WHAT YOU CAN DO

- Reduce or eliminate beef from your diet. It's good for your health, as well as for the land.

- Visit your public lands. Find out who is grazing where. Monitor grazing allotment plans and know the landscapes firsthand.

- Join one or more of the groups working to end public lands livestock grazing.

- If you are a member of an organization that is not supportive of ending grazing on public lands, help the organization to reconsider its position.

- Sponsor speakers to come to your community and discuss livestock production issues.

- Write letters to the editor of your newspaper about livestock production and its negative effects. Remember that the most popular section of every newspaper is the op-ed page. If you are thinking of writing your congressional representative, governor, or other politician or government agency, use that letter as a basis for a letter to the editor. By also addressing a public forum, you will greatly magnify the impact of your comments.

- Do your research, increase your knowledge, and sharpen your critique of public lands livestock grazing. A thorough reading of this book is a good way to begin.

- Bring attention to the public lands livestock grazing debate among friends, family, and colleagues.

- Remember, and remind as many others as you can, that there is no other single conservation opportunity for rewilding and restoring the health and beauty of such an immense area—300 million acres—as ending livestock grazing on all public lands.

GROUPS TO CONTACT

A large number of organizations work on issues related to public lands livestock grazing. Some are explicitly committed to the elimination of livestock production on public lands. Getting involved with one of the groups below is a way to become part of the effort to protect and restore America's western lands and wildlife.

ALLIANCE FOR THE WILD ROCKIES
P.O. Box 8731
Missoula, MT 59807
www.wildrockiesalliance.org

AMERICAN LANDS ALLIANCE
726 7th Street SE
Washington, DC 20003
www.americanlands.org

AMERICAN WILDLANDS
40 East Main, Suite 2
Bozeman, MT 59715
www.wildlands.org

BIODIVERSITY ASSOCIATES
P.O. Box 1512
Laramie, WY 82073
www.biodiversityassociates.org

BIODIVERSITY LEGAL FOUNDATION
P.O. Box 278
Louisville, CO 80027
(303) 926-7606

BUFFALO FIELD CAMPAIGN
P.O. Box 957
West Yellowstone, MT 59758
www.wildrockies.org/Buffalo

CALIFORNIA TROUT
870 Market Street, Suite 1185
San Francisco, CA 94102
www.caltrout.org

CALIFORNIA WILDERNESS COALITION
2655 Portage Bay East, Suite 5
Davis, CA 95616
www.calwild.org

CENTER FOR BIOLOGICAL DIVERSITY
P.O. Box 710
Tucson, AZ 85702
www.biologicaldiversity.org

COMMITTEE FOR IDAHO'S HIGH DESERT
P.O. Box 2863
Boise, ID 83701
www.cihd.org

EARTHSAVE INTERNATIONAL
1509 Seabright Avenue, Suite B1
Santa Cruz, CA 95062
www.earthsave.org

ESCALANTE WILDERNESS PROJECT
P.O. Box 652
Escalante, UT 84726
(435) 826-4778
toripat@scinternet.net

FOREST GUARDIANS
312 Montezuma, Suite A
Santa Fe, NM 87501
www.fguardians.org

FUND FOR ANIMALS
World Building
8121 Georgia Avenue, Suite 301
Silver Spring, MD 20910
www.fund.org

GILA WATCH
P.O. Box 309
Silver City, NM 88062
(505) 388-2854

GREATER YELLOWSTONE COALITION
P.O. Box 1874
Bozeman, MT 59771
www.greateryellowstone.org

HELLS CANYON PRESERVATION COUNCIL
P.O. Box 2768
La Grande, OR 97850
www.hellscanyon.org

HUMANE SOCIETY OF THE UNITED STATES
2100 L Street, NW
Washington, DC 20037
www.hsus.org

IDAHO CONSERVATION LEAGUE
P.O. Box 844
Boise, ID 83701
www.wildidaho.org

IDAHO RIVERS UNITED
P.O. Box 633
Boise, ID 83701
www.idahorivers.org

JACKSON HOLE CONSERVATION ALLIANCE
P.O. Box 2728
Jackson, WY 83001
www.jhalliance.com

ASSESSING LIVESTOCK IMPACTS ON PUBLIC LANDS

Those opposed to eliminating livestock production from all public lands often assert that livestock operations can be environmentally beneficial, or at least conducted in such a way as to do no harm. Such an argument demonstrates a very limited notion of the types of activities raising livestock entails; adherents to this perspective generally have a narrow understanding of the ways in which livestock production in the West affects ecological systems.

When one takes a more holistic approach to the question of livestock production on western lands, it becomes extremely difficult to find any stock-growing operation that can be said to be environmentally innocuous, much less advantageous to native species and natural ecological processes. It is possible that under particular circumstances, in which conditions are carefully controlled and *in which only one specific environmental goal is sought*, grazing with livestock can be helpful for management purposes. However, if the goal is to maintain or recover native biodiversity and to conserve fully functioning ecosystems, livestock production in the West is simply not benign.

The following checklist is presented as a way to aid interested citizens in addressing the issue of whether livestock in a particular place are environmentally harmful. The more yes answers one gets to the questions provided here, the more likely it is that the operation being evaluated is significantly damaging native species and natural systems.

Keep in mind that although mitigation is possible for some impacts of livestock production, mitigation—or a lessening of harm—is not the same as an absence of impacts. For example, fencing riparian areas, a common mitigation action, transfers grazing effects to uplands, which can be detrimental to water flow regimes and species dependent on small seeps and springs. The production of so-called predator-friendly livestock ignores the fact that domestic animals are eating forage that could otherwise support higher native prey densities, and therefore support a greater number of predators. There is, as they say, no free lunch.

Proceed through these questions as they are applicable (most of them will be) to the particular area or allotment in which you are interested. It is up to you to judge the number of yes responses that indicate a significantly detrimental livestock operation. Certainly, you will be a more effective and better-informed advocate for public lands if you take the time to work through these questions.

❑ *Is there any overgrazing of plants?* Are native plants in decline in any part of the allotment? Are plant succession levels kept at early stages because of livestock grazing?

❑ *Is the natural fire regime changed or is fire precluded?* Has livestock grazing reduced fuels so that fires cannot burn, or is the presence of livestock causing agencies to put out fires so as not to reduce forage for livestock permittees?

❑ *Is there a decline in water quality due to livestock?* Is there nutrient loading of watersheds resulting from livestock manure? Is sediment loading higher due to the presence of livestock?

❑ *Are livestock hooves trampling biological crusts?* Intact soil crusts are important for preventing soil erosion, adding nitrogen and carbon to soils, and helping prevent establishment of invasive annuals, such as cheatgrass.

❑ *Is there forage competition?* Are livestock removing vegetation that is food for native herbivores (e.g., elk, small mammals, insects)? There is no "surplus" of forage. Significant appropriation of forage by livestock results in fewer native animals.

❑ *Are livestock degrading riparian areas?* Are stream banks broken down? Is streamside vegetation reduced? Degradation of these critical areas by livestock can harm many native species and also lead to greater flooding.

❑ *Have livestock compacted soils?* Compacted soils reduce water infiltration, creating droughty conditions, degrading hydrological systems, and increasing flood magnitudes.

❑ *Have livestock aided the spread of exotic plants?* Livestock distribute weed seeds through feces or on their hides. They disturb the soil by trampling and eat favored native plant species, giving the competitive advantage to weeds.

❑ *Have livestock transmitted diseases to native species?* Does the presence of livestock preclude reintroduction efforts or hamper recovery of a native species because of fears of disease transmission? Bighorn sheep are particularly susceptible to livestock diseases. Other native animals are also vulnerable.

❑ *Is the diversion of water for irrigation of livestock feed and fodder harming aquatic habitat?* Are there reductions in stream flow or changes in water quality (e.g., warmer stream temperatures)? Are there dams and reservoirs that fragment river ecosystems? Is there a reduction in the amount and width of riparian or wet meadow habitat?

❑ *Is there predator control?* That is, are predators being killed to appease livestock interests?

❑ *Is there "pest" control?* Ground squirrels, prairie dogs, and grasshoppers are commonly targeted as competitors with livestock for forage, yet such species play important ecological roles, including as prey for other native species.

❑ *Are there spring developments?* The removal of water from springs diminishes habitat quality and quantity for a variety of native species.

❑ *Are there fences?* Fences fragment habitat and block animal migrations and can provide perches for raptors that increase their predation on other species, such as sage grouse.

❑ *Are headwater streams and springs exposed to trampling hooves?* First-order streams often provide more than half the late season water flow to larger streams. Trampling can significantly reduce flows, exacerbating the effects of drought on aquatic ecosystems.

❑ *Are livestock socially displacing native species?* Elk will not graze where cattle are present. They can therefore be displaced to other, less suitable habitat. Wolves and grizzlies are "displaced" by people not willing to let them roam where there is livestock.

- *Has hiding cover been removed?* Even on rangelands rated as "good," livestock grazing can reduce vegetation heights so there is no longer effective hiding cover for some species. Sharptail grouse require eight inches of plant height to hide from predators, yet even on "excellent" condition ranges livestock may crop grasses much shorter than this.
- *Have livestock contributed to a change in water quality or quantity so as to favor nonnative aquatic species?* For instance, degraded waterways contribute to the increase of invertebrates that cause whirling disease in trout.
- *Is the presence of livestock driving wildlife management decisions?* For example: Bison are shot on public lands after they wander out of Yellowstone National Park because of perceived conflicts with livestock. Elk are not allowed to roam on national grasslands in North Dakota because of opposition from livestock permittees. Fish and game departments will set hunting quotas high in some places to appease ranchers who complain that game species "depredate" their livestock forage.

- *Are soil surfaces disturbed by livestock hooves?* Is mineral soil exposed to solar radiation and air? Disturbed soils are warmer than those covered with microbiotic crusts or plant litter, and they therefore lose more carbon in the form of carbon dioxide (a greenhouse gas), they lose more nitrogen through denitrification (reducing soil fertility), and the composition of soil microbes is altered. Soil disturbance also increases evaporation, which leads to drier soils, lower plant productivity, and altered plant composition.
- *Does livestock grazing prevent or retard recovery of damaged lands?* Does frequent livestock disturbance perpetuate soil erosion? Reduce competitive ability of native plants so that they cannot out-compete exotic species? Damage native early successional plants so they cannot participate in ecological recovery?
- *Are cows ruminating?* Livestock rumen microfauna are one of the major sources of methane gas released into the atmosphere, and a factor in global warming.

- *Are nutrient flows truncated?* When a native animal such as an elk or bison dies, it is consumed by predators and/or a host of scavengers. Most of the nutrients contained in its body are retained in the local landscape. When a cow or sheep consumes most of the above-ground biomass and then is removed from the ecosystem, natural nutrient flows are disrupted and diminished.
- *Is money going to mitigate the impacts of livestock that would otherwise be used to benefit wildlife?* By diverting funds from other projects, livestock production is indirectly harming the environment. For example, most of the money being spent on wolf restoration is devoted to monitoring and controlling wolves to avoid conflicts with livestock producers. In the Gila Box National Riparian Conservation Area of Arizona, livestock eliminated most of the cottonwood and willow. Managers then planted, fenced, and irrigated young cottonwood trees, using money out of the wildlife budget instead of livestock fees.

KEY REFERENCES AND RESOURCES

BOOKS

BEYOND BEEF: THE RISE AND FALL OF THE CATTLE CULTURE. Jeremy Rifkin. New York: Dutton, 1992.

A global perspective on the livestock industry and its diverse impacts. The book begins with a discussion of cultural attitudes about meat and then looks at issues of global meat production and its effects on people and ecological systems.

DIET FOR A NEW AMERICA. John Robbins. Walpole, N.H.: Stillpoint, 1987.

Robbins focuses mainly on animal welfare, food safety, and heath issues but also makes some thoughtful connections between diet and ecological degradation.

ECOLOGICAL IMPLICATIONS OF LIVESTOCK HERBIVORY IN THE WEST. Martin Vavra, William Laycock, and Rex Pieper, eds. Denver: Society for Range Management, 1994.

A collection of essays by a variety of authors—all of them professors and researchers in university range departments or with the U.S. Department of Agriculture. As a consequence, the book has a pro–livestock industry perspective, and the authors assume that livestock grazing will occur. Nevertheless, there is helpful information on range ecology and the effects of livestock on a variety of resources.

ECOLOGY AND ECONOMICS OF THE GREAT PLAINS. Daniel Licht. Lincoln: University of Nebraska Press, 1997.

Licht examines the multiple ways in which agriculture—both livestock production and farming—have degraded the Great Plains, once one of the most productive regions for wildlife in the world. The author's ecological insights alone make the book a valuable read, but his overview of how government policies continue to support the destruction of the Great Plains are instructive for anyone trying to understand subsidized degradation of land anywhere, including the public lands.

A GEOGRAPHY OF HOPE: AMERICA'S PRIVATE LAND. Washington, D.C.: U.S. Department of Agriculture, 1997.

An overview of the amount of land used for agriculture versus other uses, including urbanization.

GRAZING TO EXTINCTION: ENDANGERED, THREATENED AND CANDIDATE SPECIES IMPERILED BY LIVESTOCK GRAZING ON WESTERN PUBLIC LANDS. Washington, D.C.: National Wildlife Federation, 1994.

This short booklet reviews how livestock production has contributed directly or indirectly to the decline and even the extirpation of hundreds of species. It includes a list of all species on public lands thought to have been negatively affected by livestock production.

LOST LANDSCAPES AND FAILED ECONOMIES: THE SEARCH FOR A VALUE OF PLACE. Thomas Michael Power. Washington, D.C.: Island Press, 1996.

Power deals with all aspects of natural resource exploitation in the West, including agriculture. His analysis offers insight into what is really driving western economies and into how industries—including the livestock industry—distort and otherwise obfuscate their limited economic contributions to hold on to subsidies and political power.

MAD COWBOY: PLAIN TRUTH FROM THE CATTLE RANCHER WHO WON'T EAT MEAT. Howard Lyman with Glen Merzer. New York: Scribner's, 1998.

Lyman is a former Montana stock grower, now a leading vegetarian and anti–animal agriculture activist. In down-home style, Lyman explains how he changed from a rancher who "never met a chemical he didn't like" to a leading spokesperson for plant-based diets and organic agriculture. This is an insider's look at the no-win situation most ranchers face in a world of global marketing and widespread hormone and artificial chemical use, and at the wide array of problems created by our meat-oriented diets.

OVERTAPPED OASIS: REFORM OR REVOLUTION FOR WESTERN WATER. Marc Reisner and Sarah Bates. Washington, D.C.: Island Press, 1990.

A look at western agricultural water use. Although this book does not deal with public lands, it provides insights into how much western water goes into growing crops and forage consumed by livestock.

OWNING IT ALL. William Kittredge. St. Paul, Minn.: Graywolf Press, 1987.

An autobiographical account of a ranching family in eastern Oregon—owners of the MC Ranch, once the largest in the state. Kittredge, who gave up raising livestock to become an author, writes with sympathy about the rural West, putting a human face on the livestock industry; he also provides insights into the culture of ranching and the attitudes that have bankrupted the natural capital of the West.

PUBLIC RANGELANDS: SOME RIPARIAN AREAS RESTORED BUT WIDESPREAD IMPROVEMENT WILL BE SLOW. Washington, D.C.: General Accounting Office, 1988. GAO/RCED-88-105.

This report to Congress analyzes how riparian areas are damaged by livestock and how little recovery has occurred on the public lands, despite scientific recognition of their ecological value.

RANGE MANAGEMENT: PRINCIPLES AND PRACTICES. Jerry Holechek, Rex Pieper, and Carlton Herbel. Upper Saddle River, N.J.: Prentice Hall, 2001.

A good, basic overview of range management and ecology, despite the pro–livestock industry leanings of its authors, who are range professors. Lead author Jerry Holechek is one of the most published, creative, and innovative academicians in the field of range management; anything written by him is worth reading.

RANGELAND HEALTH: NEW METHODS TO CLASSIFY, INVENTORY AND MONITOR RANGELANDS. National Research Council. Washington, D.C.: National Academy Press, 1994.

A technical overview of rangeland conditions and ways to measure the health of ecosystems.

RANGELANDS MANAGEMENT: PROFILE OF THE BUREAU OF LAND MANAGEMENT GRAZING ALLOTMENTS AND PERMITS. Washington, D.C.: General Accounting Office, 1992. GAO/RCED 92-213FS.

A review of grazing permits that shows how the majority of public lands forage is allotted to a small percentage of large ranching operations. The GAO has published numerous other reports relevant to public lands livestock issues.

SACRED COWS AT THE PUBLIC TROUGH. Denzel and Nancy Ferguson. Bend, Ore.: Maverick Publications, 1983.

While managing the Malheur Biological Field Station in eastern Oregon, Denzel and Nancy Ferguson were so appalled by the abuse of public lands and the negative effects of livestock grazing on plants and wildlife that they were driven to write an exposé. Their book is filled with examples of how our public lands are managed primarily for the benefit of the livestock industry. Although the particulars of how government subsidizes ranching operations are somewhat different today, the fundamental assertions of this nearly twenty-year-old publication unfortunately still hold true. A good reference for anyone who wants some historical perspective. The authors also do a good job of showing how to approach the issue of livestock subsidies.

SAVING NATURE'S LEGACY: PROTECTING AND RESTORING BIODIVERSITY. Reed Noss and Allen Cooperrider. Washington, D.C.: Island Press, 1994.

Biologists Noss and Cooperrider explain in accessible language the basic principles of conservation biology. Although the book examines various threats to biodiversity, from logging to sprawl, it includes a good chapter on the effects of livestock grazing.

TAKING STOCK: ANIMAL FARMING AND THE ENVIRONMENT. Alan B. Durning and Holly B. Brough. Washington, D.C.: Worldwatch Institute, 1991.

The authors survey the environmental effects of animal agriculture, particularly its role in world hunger and in threats to global biodiversity.

WASTE OF THE WEST: PUBLIC LANDS RANCHING. Lynn Jacobs. Tucson: Self-published, 1991; available at www.apnm.org/waste_of_west/.

Jacobs presents an exhaustive review of the entire livestock industry, covering everything related to the issue, from statistics on public lands grazing to history, policy, ecology, and the cowboy myth. Although lengthy, at more than 600 pages, the book is user-friendly, including more than a thousand photographs, diagrams, and charts. If you have a question about livestock production, this book probably has the answer.

THE WESTERN RANGE REVISITED: REMOVING LIVESTOCK FROM PUBLIC LANDS TO CONSERVE NATIVE BIODIVERSITY. Debra Donahue. Norman: University of Oklahoma Press, 1999.

A former wildlife biologist and now a professor of law at the University of Wyoming, Donahue offers a well-researched and thorough overview of the legal, cultural, and ecological arguments for ending public lands grazing. In particular, her strong biological training makes the book a valuable resource for those attempting to understand the multiple ways in which livestock affect the West.

ARTICLES

"APPLICATION OF HERBIVORE OPTIMIZATION THEORY TO RANGELANDS OF THE WESTERN UNITED STATES." Elizabeth L. Painter and A. Joy Belsky. Ecological Applications 3, no. 1 (1993): 2–9.

An examination of the question of whether plants need to be grazed.

"THE BLACK-TAILED PRAIRIE DOG: HEADED FOR EXTINCTION?" George Wuerthner. Journal of Range Management 50, no. 5 (1997): 459–466.

An overview of the factors contributing to prairie dog decline around the West, and a summation of the ecological importance of prairie dogs to ecosystems.

"DOES HERBIVORY BENEFIT PLANTS? A REVIEW OF THE EVIDENCE." A. Joy Belsky. American Naturalist 127 (1986): 870–892.

This excellent review piece looks at the effects of grazing on plants. It basically argues that plants do not require grazing; rather, grazing is tolerated, and some costs to the plants may still be incurred.

"ECOLOGICAL COSTS OF LIVESTOCK GRAZING IN WESTERN NORTH AMERICA." Thomas Fleischner. Conservation Biology 8, no. 3 (1994): 629–644.

Still the best overview of the ecological impacts of livestock production.

"EFFECTS OF LIVESTOCK GRAZING ON STAND DYNAMICS AND SOILS OF UPLAND FORESTS OF THE INTERIOR WEST." A. Joy Belsky and Dana M. Blumenthal. Conservation Biology 11, no. 2 (1997): 315–327.

Reviews the connection between grazing and the invasion of trees into forests and meadows formerly in a more open condition.

"IMPACTS OF LIVESTOCK GRAZING ACTIVITIES ON STREAM INSECT COMMUNITIES AND THE RIVERINE ENVIRONMENT." R. Mac Strand and Richard Merritt. American Entomologist 45 (1999): 13–29.

An illuminating overview of livestock influences on streams, particularly with regard to effects on insects.

"GRAZING THE WESTERN RANGE: WHAT COSTS, WHAT BENEFITS." George Wuerthner. Western Wildlands 16, no. 2 (1990): 27–29.

A holistic look at the ecological costs of livestock grazing.

"SHOULD WE SADDLE UP WITH THE COWBOYS?" George Wuerthner. Wild Earth 8, no. 3 (1998): 68–72.

Argues that there are sharp philosophical differences between proponents of resource extraction industries and those supporting ecological processes.

"SOME ECOLOGICAL COSTS OF LIVESTOCK." George Wuerthner. Wild Earth 2, no. 1 (1992): 10–14.

Discusses how to assess the ecological costs of livestock production more fully.

"SUBDIVISIONS AND EXTRACTIVE INDUSTRIES." George Wuerthner. Wild Earth 7, no. 3 (1997): 57–62.

Looks at the "condos versus cows" issue.

"SUBDIVISIONS VERSUS AGRICULTURE." George Wuerthner. Conservation Biology 8, no. 3 (1994): 905–908.

The geographical imprint of subdivisions is compared to that associated with agriculture—particularly livestock production.

"SURVEY OF LIVESTOCK INFLUENCES ON STREAM AND RIPARIAN ECOSYSTEMS IN THE WESTERN UNITED STATES." A. Joy Belsky, Andrea Matzke, and Shauna Uselman. Journal of Soil and Water Conservation 54 (1999): 419–431.

An extensive review of the scientific literature on livestock impacts to streams and riparian areas. The authors conclude that livestock damage to riparian ecosystems began with the introduction of livestock 100–200 years ago and continues today wherever grazing is ongoing. Riparian recovery is contingent on total rest from livestock grazing.

"WHITEHORSE BUTTE ALLOTMENT: POOR PUBLIC RANGE POLICY?" George Wuerthner. Rangelands 12, no. 6 (1992): 300–304.

Reviews the multiple economic and environmental subsidies on a large public lands allotment in eastern Oregon.

SCIENTIFIC JOURNALS

JOURNAL OF RANGE MANAGEMENT and RANGELANDS, both published by the Society for Range Management, are good sources of basic information about the livestock industry and range science. Their web sites, respectively, are http://uvalde.tamu.edu/jrm/jrmhome.htm, and http://uvalde.tamu.edu/rangel/home.htm.

No other journals focus exclusively on livestock and range science issues, but CONSERVATION BIOLOGY, ECOLOGICAL APPLICATIONS, ECOLOGY, JOURNAL OF WILDLIFE MANAGEMENT, WILDLIFE BULLETIN, and other academic publications occasionally feature articles relevant to livestock production and public lands grazing.

A GUIDE TO LIVESTOCK-FREE LANDSCAPES

This annotated list is composed of areas where domestic livestock have never grazed, or where livestock grazing no longer occurs. These places serve as benchmarks for what the West could look like in the absence of domestic livestock production.

We do not claim to have done a comprehensive inventory, yet we had some difficulty coming up with even this relatively brief list. The scarcity of livestock-free sites in the West is shocking—even more so, the small number of places never grazed by livestock. Belying their designations, many national parks, wildlife refuges, wilderness areas, national recreation areas, Nature Conservancy preserves, and other places most

people assume are set aside for their natural features and ecological values are often grazed by livestock. It is indeed ironic that some of the best native grasslands remaining in the West are found within nuclear research facilities, bomb test sites, and military reservations. In most cases, land has suffered more from livestock grazing than from the pummeling of bombs or from the storage of nuclear waste!

In addition to these listed sites, many existing livestock allotments have small areas that are inaccessible or otherwise unusable by domestic stock. These include grassy slopes enclosed by cliffs or very steep terrain, areas far from water,

or the tops of badlands and buttes. The fenced margins of rural roads and highways, where no livestock grazing, mowing, or sowing of exotic grasses has occurred, and where road construction has not disrupted plant communities, can also serve as potential ecological benchmarks.

ARIZONA

AUDUBON APPLETON-WHITTELL RESEARCH RANCH. Southeast of Tucson. Ungrazed by livestock for more than 30 years. There is an extensive research record on changes since cows were removed.

BUENOS AIRES NATIONAL WILDLIFE REFUGE. Not grazed by livestock since about 1985. Currently the largest grassland not grazed by livestock in southern Arizona, about 116,000 acres.

COTTONWOOD CREEK. Near Prescott. Steep cliffs guard a stretch of creek about 1.5 miles south of Camp Wood in the Santa Maria Mountains. The stream features a cottonwood/ ash/willow association with young trees as well as old.

DUTCHWOMAN BUTTE, TONTO NATIONAL FOREST. The top of Dutchwoman Butte has never been grazed by livestock.

ORGAN PIPE CACTUS NATIONAL MONUMENT. Low-elevation Sonoran Desert site on the Mexican border, south of Ajo. Since cows were removed, regeneration of large cacti has improved.

POWELL PLATEAU, NORTH RIM OF THE GRAND CANYON. Has never been grazed by livestock to any degree, has never been logged, and has a fairly natural fire regime.

SAND TANK MOUNTAINS, SONORAN DESERT NATIONAL MONUMENT. Until recently, part of the Barry Goldwater Bombing Range; the area hasn't been grazed by livestock in 50 years.

SYCAMORE CANYON, CORONADO NATIONAL FOREST. Cows were recently removed from Sycamore Creek. The stream runs about five miles from where the road ends to the international border with Mexico. Many rare bird species are known from this area.

CALIFORNIA

FORT HUNTER LIGGETT. Monterey County. Many parts were heavily grazed from the founding of Mission San Antonio in the late eighteenth century through the late twentieth century. Most of the area was part of William Randolph Hearst's holdings. Livestock grazing permits ended in about 1992 (though some trespass still occurs). Since 1992 there has been an amazing recovery, including reestablishment of oaks.

HENRY COÈ STATE PARK. One of the largest state parks in California, this area near San Jose is no longer grazed by livestock and features wonderful oak woodlands.

JOSHUA TREE NATIONAL PARK. Within the older boundaries of this recently enlarged park, livestock grazing has not occurred for decades.

KINGS CANYON/SEQUOIA NATIONAL PARKS. Although backcountry horse use still takes its toll in some places, most of the grasslands are not touched by domestic animals. The lower-elevation oak woodlands in Sequoia National Park are particularly impressive in the spring.

LAVA BEDS NATIONAL MONUMENT. One of the nicest livestock-free bunchgrass sites in northern California.

MOJAVE NATIONAL PRESERVE. Cattle were recently removed from an extensive segment of the preserve.

MOUNT DIABLO STATE PARK. In the San Francisco Bay Area. Livestock grazing is banned in most of the park.

PENNINGTON CREEK BIOLOGICAL PRESERVE. California Polytechnic State University, San Luis Obispo. Livestock grazing ended about 25 to 30 years ago.

YOSEMITE NATIONAL PARK. The park where John Muir coined the phrase "hooved locusts" to describe domestic sheep and their effects on the land. Sheep haven't grazed since the turn of the nineteenth century, though signs of the previous abuse are still evident to the astute observer. Yosemite's Tuolumne Meadows is the largest meadow in the Sierra Nevada Range.

COLORADO

BROWNS PARK NATIONAL WILDLIFE REFUGE. Browns Park has more than 13,000 livestock-free acres, including a substantial riparian area along the Green River.

GREAT SAND DUNES NATIONAL MONUMENT. The boundaries of this park unit were recently enlarged; some

38,000 acres within the old monument boundaries are livestock-free, but livestock grazing still occurs within the newly expanded boundaries.

HOLY CROSS WILDERNESS. White River and San Isabel National Forests. Most of this wilderness is no longer grazed by livestock.

ROCKY FLATS ARSENAL. Near Denver. Has been livestock-free for decades and is one of the more easily accessible, no-livestock zones in Colorado.

ROCKY MOUNTAIN NATIONAL PARK. Was extensively grazed prior to its establishment, but some parts have been livestock-free for nearly a century.

WEMINUCHE WILDERNESS. San Juan and Rio Grande National Forests. Most of the wilderness is no longer grazed by livestock, although there are domestic sheep in some areas.

IDAHO

CRATERS OF THE MOON NATIONAL MONUMENT. In the midst of lava flows, small, isolated grasslands—known as kipukas—have never been grazed by livestock.

FRANK CHURCH-RIVER OF NO RETURN WILDERNESS. This huge wilderness, more than 2 million acres in size, is cow-free over 90 percent of its area. Although heavily forested, grasslands in the canyons of the Middle Fork Salmon and the main Salmon River, along with the extensive subalpine meadows in the Chamberlain Basin, have never been or are no longer grazed by domestic livestock.

IDAHO NUCLEAR ENGINEERING AND ENVIRONMENTAL LAB. INEEL is located northeast of Idaho Falls. Although some of the site was and is still grazed on the fringes, most of the area hasn't been grazed by livestock in decades and is one of the most intact low-elevation bunchgrass sites in the state.

LITTLE BLUE TABLE. BLM land near Grasmere. An isolated mesa top with a shrub/steppe plant community. Approximately 20,000 acres. It is the only large part of the Bruneau Resource Area that is rated as being in excellent range condition.

MINK CREEK, BANNOCK RANGE. Caribou National Forest. This is the watershed for the city of Pocatello, and has not been grazed—except for a few stray, illegal cows—for 70 years. To protect this unique place, the Forest Service established the Gibson Jack Creek and West Mink Creek Research Natural Areas about a decade ago. A few miles to the north, the City Creek watershed, with different portions managed by the city of Pocatello, the BLM, and the Caribou National Forest, has been similarly without livestock grazing, but has no permanent protection from it.

MONTANA

GLACIER NATIONAL PARK. Some large, livestock-free bunchgrass prairies may be found along the North Fork of the Flathead River on the western side of the park. Extensive rough fescue prairies on the eastern side have not been grazed by livestock in nearly a century. The east side plains are truly spectacular in late June when filled with wild rose and other blooming wildflowers.

MAKOSHIKA STATE PARK. Glendive. Great example of livestock-free Great Plains shortgrass prairie.

NATIONAL BISON RANGE. Moiese. Grazed by bison, elk, pronghorn, and bighorn sheep, but no livestock.

RED ROCK LAKES NATIONAL WILDLIFE REFUGE. Most of the refuge is no longer grazed by livestock and offers a good contrast to adjacent BLM and Forest Service lands, where livestock still trample riparian zones and eat grass down to stubble.

SQUARE BUTTE NATURAL AREA. By Geraldine. Surrounded by steep cliffs, which protect some 2,000 acres of grasslands on top. The summit is accessible by trail.

WILD HORSE ISLAND STATE PARK. Flathead Lake, by Dayton. The largest island in the lake, with handsome Palouse prairie grasslands, a herd of bighorn sheep, and no wild horses, despite the name.

NEVADA

GREAT BASIN NATIONAL PARK. Grazing was permitted in the national park at the time of its designation, but in the late 1990s cattle were removed (some domestic sheep grazing still occurs).

HAWTHORNE ARMY AMMUNITION DEPOT. In the Wassuk Range above Walker Lake. Most of the higher elevations of the Wassuk Range are part of this ammunition depot, where livestock have been excluded for 60 years. The area around Mt. Grant is a botanical paradise.

JARBIDGE WILDERNESS. Humboldt National Forest. The West Fork of the Mary's River has not been grazed by livestock in four decades.

SCOFIELD CANYON. Grant Range, Humboldt National Forest. The riparian area and creek have not been grazed by livestock for decades.

SHELDON NATIONAL WILDLIFE REFUGE. Had been heavily grazed for a century, but in the late 1990s all livestock were removed. The riparian areas are showing remarkable recovery. At 575,000 acres, it is now the largest livestock-free site in the Great Basin region.

NEW MEXICO

BANDELIER NATIONAL MONUMENT. Though mostly forested, there are some riparian areas and large, scattered midelevation meadows that have been livestock-free for decades.

GILA WILDERNESS. Gila National Forest. Most of this area was grazed at one time, but livestock were removed from some parts 20 to 30 years ago, and more recently—because of litigation by environmentalists—cattle were removed from the East Fork of the Gila River. This created a livestock-free zone larger than 700,000 acres in the center of New Mexico's largest wildlands complex.

SEVILLETA NATIONAL WILDLIFE REFUGE. One of the largest livestock-free sites in New Mexico at 229,700 acres.

WHITE SANDS MISSILE TEST SITE. Allegedly, this area has some of the highest percentage of rangeland rated "good" to "excellent" in New Mexico. Being grazed by cattle is apparently far more destructive than being hit by a nuclear bomb!

NORTH DAKOTA

THEODORE ROOSEVELT NATIONAL PARK. Aside from a few longhorns used for "historic demonstration" and a small herd of wild horses, this is a rare, livestock-free Great Plains reserve, with elk, bison, and prairie dogs.

OREGON

HART MOUNTAIN NATIONAL ANTELOPE REFUGE. Cattle were removed from this 270,000 acre refuge in 1991; it is now one of the largest livestock-free areas in the entire Great Basin. The refuge provides a great example of a bunchgrass-sagebrush community. A good place to observe pronghorn antelope and bighorn sheep.

HELLS CANYON WILDERNESS. (Part is in Idaho.) Contains some of the finest remaining interior Columbia Basin native grasslands. Although heavy cattle and sheep grazing has occurred throughout the wilderness, nearly 100,000 acres have been livestock-free for 15 to 22 years.

THE ISLAND RESEARCH NATURAL AREA. A cliff-fringed plateau in the Deschutes River Canyon, north of Bend. It has never been grazed by livestock, and contains some of the best undisturbed grasslands in the Columbia Basin. It is closed to the public but may be available to researchers.

KLAMATH MARSH NATIONAL WILDLIFE REFUGE. Formerly a ranch, the area was extensively grazed in the past. However, portions of the refuge are now livestock-free.

MOUNT PISGAH/BUFORD COUNTY PARK. Near Eugene. Over 2,000 acres in size, the park features a relict Willamette Valley oak woodland and extensive meadows dotted with wildflowers in the spring.

STEENS MOUNTAIN WILDERNESS. Near Burns. A portion of the wilderness is the first to be designated by Congress as officially livestock-free.

TOM MCCALL PRESERVE. In the Columbia Gorge near Hood River, well known for its spectacular wildflower displays. The preserve is operated by the Nature Conservancy.

SOUTH DAKOTA

BADLANDS NATIONAL PARK. Six hundred bison and a few reintroduced black-footed ferrets in this mixed-grass prairie.

CUSTER STATE PARK. Black Hills. Famous for its bison, though the herd is artificially culled every year.

WIND CAVE NATIONAL PARK. Black Hills. Mixed-grass prairie with prairie dogs, pronghorn, deer, elk, and bison.

TEXAS

BIG BEND NATIONAL PARK. Most of the park was created by buying up former ranchland in the 1930s, and was

severely overgrazed at that time. However, some upland grasslands have shown remarkable recovery.

UTAH

ARCH, FISH, MULE, OWL CREEK, AND ROAD CANYONS. East of Grand Gulch. Cows were removed from all five canyons by court order in 1993.

CANYONLANDS NATIONAL PARK. The canyons of the Maze have been ungrazed by livestock since 1975. Jasper Canyon, Pete's Mesa, and Virginia Park are said to be ungrazed relicts. Grays Pasture has been livestock-free for 30 years.

GRAND GULCH PRIMITIVE AREA. Since the removal of cows in the 1970s, there has been dramatic recovery of riparian vegetation.

NO MAN'S MESA, GRAND STAIRCASE–ESCALANTE NATIONAL MONUMENT. This area was grazed by livestock many years ago, but only by goats (no cows or sheep).

RED BUTTE CANYON RESEARCH NATURAL AREA. Wasatch Range. This site features a livestock-free riparian area.

WASHINGTON

HANFORD REACH NATIONAL MONUMENT. Formerly a federal nuclear site near Richland. Contains one of the largest livestock-free shrub-steppe grasslands in the

Columbia Basin. It also contains one of the last free-flowing stretches of the Columbia River.

MEEKS TABLE RESEARCH NATURAL AREA. Wenatchee National Forest. Never grazed by livestock.

MOUNT RAINIER NATIONAL PARK. Most of the larger subalpine meadows were heavily grazed by domestic sheep until the 1930s. The famous wildflower displays in the livestock-free meadows are now a trademark of the park.

WYOMING

DEVILS TOWER NATIONAL MONUMENT. Small in area but features typical Great Plains vegetation and a population of prairie dogs.

FOSSIL BUTTE NATIONAL MONUMENT. Small area near Kemmerer that has been livestock-free for several decades.

SEEDSKADEE NATIONAL WILDLIFE REFUGE. North of Rock Springs. Since the early 1980s, its 27,000 acres along the Green River have been livestock-free.

YELLOWSTONE NATIONAL PARK. Livestock-free since its establishment in 1872 (except for some milk cows in the early days). Part of one of the wildest, most complete, natural ecosystems in the West. The northern range between Gardiner and Cooke City is home to a wealth of wildlife, with bison, pronghorn, elk, wolves, and grizzlies regularly seen. Bechler Meadows, in the park's lush, remote, southwest corner, has never been grazed by livestock.

ENDNOTES

PART II

IN THE BEGINNING: COW

1. P. Shepard, *Coming Home to the Pleistocene* (Washington, D.C.: Island Press, 1998), p. 125.
2. D. Ferguson and N. Ferguson, *Sacred Cows at the Public Trough* (Bend, Ore.: Maverick Publications, 1983).
3. E. Lawrence, *Hoofbeats and Society* (Bloomington: University of Indiana Press, 1985).
4. Shepard, *Coming Home*, pp. 109–130.
5. Ibid., 109–110.
6. B. Lincoln, *Priests, Warriors and Cattle: A Study of the Ecology of Religion* (Berkeley and Los Angeles: University of California Press, 1981), p. 68.
7. Ibid., p. 131.
8. Shepard, *Coming Home*, p. 2.
9. L. Shlain, *The Alphabet Versus the Goddess: The Conflict Between Word and Image* (New York: Viking, 1998), pp. 10–11.
10. Brigitte Greenberg, "Cowboy Boots and Ball Gowns Make for a Hootin' and Hollerin' Party," *Detroit Free Press*, 20 January 2001.
11. C. Manes, *Other Creations: Rediscovering the Spirituality of Animals* (New York: Doubleday, 1997), pp. 93–103.

BEEF, COWBOYS, THE WEST

1. A. B. Durning and H. B. Brough, *Taking Stock: Animal Farming and the Environment*, Worldwatch Paper 103 (Washington, D.C.: Worldwatch Institute, 1991); D. Ferguson and N. Ferguson, *Sacred Cows at the Public Trough* (Bend, Ore.: Maverick Publications, 1983); T. Fleischner, "Ecological Costs of Livestock Grazing in Western North America," *Conservation Biology* 8, no. 3 (1994): 629–644; L. Jacobs, *Waste of the West: Public Lands Ranching* (Tucson: Self-published, 1991), www.apnm.org/waste_of_west/; M. Reisner, *Cadillac Desert* (New York: Penguin, 1986); J. Rifkin, *Beyond*

Beef: The Rise and Fall of the Cattle Culture (New York: Penguin, 1992); J. Robbins, *Diet for a New America* (Walpole, N.H.: Stillpoint Press, 1987); G. Wuerthner, "How the West Was Eaten," *Wilderness* (Spring 1991): 28–36; id., "Some Ecological Costs of Livestock," *Wild Earth* 2, no. 2 (1992): 10–14.
2. D. Brothwell and P. Brothwell, *Food in Antiquity: A Survey of the Diet of Early People* (New York: Praeger, 1969).
3. Rifkin, *Beyond Beef* (see note 1 above).
4. T. G. Jordan, *North American Cattle Ranching Frontiers: Origins, Diffusion, and Differentiation* (Albuquerque: University of New Mexico Press, 1993).
5. Rifkin, *Beyond Beef* (see note 1 above).
6. P. Shepard, *The Others: How Animals Made Us Human* (Washington, D.C.: Island Press, 1996).
7. E. Ross, "An Overview of Trends in Dietary Variation from Hunter-Gatherer to Modern Capitalist Societies," in *Food and Evolution: Towards a Theory of Human Food Habits*, edited by M. Harris and E. Ross (Philadelphia: Temple University Press, 1987).
8. Ibid.
9. C. J. Adams, *The Sexual Politics of Meat: A Feminist-Vegetarian Critical Theory* (New York: Continuum, 1991).
10. R. J. Hooker, *Food and Drink in America: A History* (New York: Bobbs-Merrill, 1981).
11. Adams, *Sexual Politics* (see note 9 above).
12. Ibid.
13. Ibid.
14. Ross, "Overview" (see note 7 above).
15. Rifkin, *Beyond Beef* (see note 1 above).
16. Ibid.
17. Ibid.
18. D. Cozzens, "History and Louis L'Amour's Cowboy," *Journal of American Culture* 14, no. 2 (1991): 42–52; D. Dary, *Cowboy Culture: A Saga of Five Centuries* (New

York: Knopf, 1981); O. Najera-Ramirez, "Engendering Nationalism: Identity, Discourse, and the Mexican Charro," *Anthropological Quarterly* 67, no. 1 (1994): 1–10.
19. Cozzens, "History" (see note 18 above); W. W. Savage, *The Cowboy Hero: His Image in American History and Culture* (Norman: University of Oklahoma Press, 1979);
20. R. V. Hines, *The American West: An Interpretive History* (Boston: Little, Brown, 1984); Najera-Ramirez, "Engendering Nationalism" (see note 18 above).
21. Dary, *Cowboy Culture* (see note 18 above); Jordan, *North American* (see note 4 above).
22. Dary, ibid.; Jordan, ibid.; Najera-Ramirez, "Engendering Nationalism" (see note 18 above).
23. Najera-Ramirez, ibid.
24. Jordan, *North American* (see note 4 above).
25. Dary, *Cowboy Culture* (see note 18 above); L. A. McFarlane, "British Remittance Men as Ranchers: The Case of Coutts Majoribanks and Edmund Thursby, 1884–95," *Great Plains Quarterly* 11, no. 2 (1991): 53–69.
26. Jordan, *North American* (see note 4 above).
27. U.S. General Accounting Office, *Rangeland Management: Profile of the Bureau of Land Management's Grazing Allotments and Permits*, RCED-92-213FS (Washington, D.C.: GAO, 1992).
28. Jacobs, *Waste of the West* (see note 1 above).
29. W. Cronon, G. Miles, and J. Gitlin, eds., *Under an Open Sky: Rethinking America's Western Past* (New York: W. W. Norton, 1993).
30. Jordan, *North American* (see note 4 above).
31. Ibid.
32. R. L. Knight, G. N. Wallace, and W. E. Riebsame, "Ranching the View: Subdivisions Versus Agriculture," *Conservation Biology* 9 (1995): 459–461.
33. Jordan, *North American* (see note 4 above).
34. Ferguson and Ferguson, *Sacred Cows* (see note 1 above); Cronon, Miles, and Gitlin, *Under an Open Sky* (see note

29 above); Jacobs, *Waste of the West* (see note 1 above); Wuerthner, "Some Ecological Costs" (see note 1 above).

35. L. Logan, "The Geographical Imagination of Frederic Remington: The Invention of the Cowboy," *Western Journal of Historical Geography* 18, no. 1 (1992): 75–90.

36. B. J. Stoeltje, "Rodeo: From Custom to Ritual," *Western Folklore* 48 (1989): 244–260.

37. R. Martin, *Cowboy: The Enduring Myth of the Wild West* (New York: Stewart, Tabori and Chang, 1983).

38. Dary, *Cowboy Culture* (see note 18 above); Logan, "Geographical Imagination" (see note 35 above).

39. F. J. Turner, *The Frontier in American History* (New York: Henry Holt, 1921; reprint, Mineola, N.Y.: Dover, 1996).

40. J. B. Frantz and J. E. Choate, *The American Cowboy: The Myth and the Reality* (Norman: University of Oklahoma Press, 1955).

41. Cozzens, "History" (see note 18 above); Logan, "Geographical Imagination" (see note 35 above).

42. Logan, ibid.

43. Frantz and Choate, *American Cowboy* (see note 40 above).

44. Logan, "Geographical Imagination" (see note 35 above).

45. Knight, Wallace, and Riebsame, "Ranching the View" (see note 32 above).

46. Wuerthner, "Some Ecological Costs" (see note 1 above); Wuerthner, "Subdivisions and Extractive Industries," *Wild Earth* 7, no. 3 (1997): 57–62.

LAND HELD HOSTAGE

1. D. D. Brand, "The Early History of the Range Cattle Industry in Northern Mexico," *Agricultural History* 35 (1961): 132–139.

2. G. Stewart, "History of Range Use," in *The Western Range*, 74th Cong., 2d sess., S. Doc. 199, 1936; L. A. Stoddart and A. D. Smith, *Range Management* (New York: McGraw-Hill, 1943).

3. Stewart, ibid.

4. W. C. Barnes, "Outlines Arizona's Cattle History Since Days of Kino," *Prescott (Arizona) Courier*, 17 December 1935, pp. 14–15; H. E. Bolton, *The Padre on Horseback: A Sketch of Eusebio Francisco Kino*, The American West Reprint Series (Chicago: Loyola University Press, 1963); id., *Rim of Christendom: A Biography of Eusebio Francisco Kino, Pacific Coast Pioneer* (New York: Macmillan, 1936); B. Haskett, "Early History of the Cattle Industry in Arizona," *Arizona Historical Review* 6 (1935): 3–42; N. M. Loomis, "Early Cattle Trails in Southern Arizona," *Arizoniana* 3 (1962): 18–24; Stewart, "History" (see note 2 above).

5. Stewart, "History" (see note 2 above).

6. Brand, "Early History" (see note 1 above).

7. Stewart, "History" (see note 2 above).

8. J. G. Bell, "A Log of the Texas–California Cattle Trail, 1854," *Southwestern Historical Quarterly* 35 (1932): 208–237; G. Cureton, "The Cattle Trail to California, 1840–1860," *Historical Society of Southern California Quarterly* 35 (1953): 99–109; Haskett, "Early History" (see note 4 above).

9. Stewart, "History" (see note 2 above).

10. P. A. Schlegel, "A History of the Cattle Industry in Northern Arizona, 1863–1912" (master's thesis, Northern Arizona University, 1992).

11. In this book, reference is usually made to the "eleven western states"—Arizona, California, Colorado, Idaho, Montana, Nevada, New Mexico, Oregon, Utah, Washington, and Wyoming. Most public land acreage in the lower forty-eight states occurs in these eleven westernmost states. The majority of public lands livestock grazing also occurs there. However, a small amount of public lands ranching takes place in six additional states—Kansas, Nebraska, North Dakota, Oklahoma, South Dakota, and Texas.

12. Stewart, "History" (see note 2 above).

13. K. L. Cole, N. Henderson, and D. S. Shafer, "Holocene Vegetation and Historic Grazing Impacts at Capitol Reef National Park Reconstructed Using Packrat Middens," *Great Basin Naturalist* 57 (1997): 315–326.

14. Stewart, "History" (see note 2 above); Stoddart and Smith, *Range Management* (see note 2 above).

15. J. W. Powell, *Report on the Lands of the Arid Region of the United States, with a More Detailed Account of the Lands of Utah* (Washington, D.C.: U.S. Department of the Interior, 1878).

16. V. W. Scott, "The Range Cattle Industry: Its Effect on Western Land Law," *Montana Law Review* 28 (1967): 155–183; Lyle F. Watts, "Unsuitable Land Policy," in *The Western Range*, 74th Cong., 2d sess., S. Doc. 199, 1936.

17. P. O. Foss, *Politics and Grass: The Administration of Grazing on the Public Domain* (Seattle: University of Washington Press, 1960; reprint, New York: Greenwood, 1969).

18. Foss, ibid.; J. A. Stout Jr., "Cattlemen, Conservationists, and the Taylor Grazing Act," *New Mexico Historical Review* 45 (1970): 311–332.

19. Foss, ibid.; J. Muhn and H. R. Stuart, *Opportunity and Challenge: The Story of BLM* (Washington, D.C.: U.S. Department of the Interior, Bureau of Land Management, 1988).

20. Foss, ibid.

21. Ibid.

22. Foss, ibid.; Stout, "Cattlemen" (see note 18 above).

23. Foss, ibid.; C. M. Klyza, *Who Controls Public Lands? Mining, Forestry, and Grazing Policies, 1870–1990* (Chapel Hill: University of North Carolina Press, 1996); Muhn and Stuart, *Opportunity and Challenge* (see note 19 above).

24. S. K. Fairfax and C. E. Yale, *Federal Lands: A Guide to Planning, Management, and State Revenues* (Washington, D.C.: Island Press, 1987).

25. R. L. Glicksman and G. C. Coggins, *Modern Public Land Law in a Nutshell* (St. Paul: West Publishing, 1995).

26. Muhn and Stuart, *Opportunity and Challenge* (see note 19 above).

27. Ibid.

28. G. C. Coggins, "Of Succotash Syndromes and Vacuous Platitudes: The Meaning of 'Multiple Use, Sustained Yield' for Public Land Management," *University of Colorado Law Review* 53 (1982): 229–280; Glicksman and Coggins, *Modern Public Land Law* (see note 25 above); M. E. Mansfield, "A Primer of Public Land Law," *Washington Law Review* 68 (1993): 801–857.

29. Craig W. Allin, *The Politics of Wilderness Preservation* (Westport, Conn.: Greenwood, 1982); Coggins, "Succotash" (see note 28 above); Coggins, "The Law of Public Rangeland Management IV: FLPMA, PRIA, and the Multiple Use Mandate," *Environmental Law* 14 (1983): 1–131; Glicksman and Coggins, *Modern Public Land Law* (see note 25 above); Klyza, *Who Controls?* (see note 23 above); Mansfield, "Primer" (see note 28 above); Muhn and Stuart, *Opportunity and Challenge* (see note 19 above); I. Senzel, "Genesis of a Law: Part 1," *American Forests* (January 1978): 30–32, 61–64; ibid., "Part 2" (February 1978): 32–39.

30. K. Barton, "Bureau of Land Management," in *Audubon Wildlife Report 1987*, edited by R. L. DiSilvestro (Orlando: Academic Press, 1987); Klyza, *Who Controls?* (see note 23 above); Muhn and Stuart, *Opportunity and Challenge* (see note 19 above); Dyan Zaslowsky, *These American Lands* (New York: Henry Holt, 1986).

31. Foss, *Politics* (see note 17 above); Klyza, *Who Controls?* (see note 23 above).

32. J. Woolf, "How the West Was Won, and Won, and . . . ," *High Country News*, 16 October 1995, p. 6.

33. Barton, "Bureau of Land Management (see note 30 above).

34. Foss, *Politics* (see note 17 above).

35. Klyza, *Who Controls?* (see note 23 above).

36. Coggins, "Succotash" (see note 28 above); Mansfield, "Primer" (see note 28 above).

37. Glicksman and Coggins, *Modern Public Land Law* (see note 25 above).

38. P. Fradkin, "The Eating of the West," *Audubon*, January 1979, pp. 94–121; J. N. Miller, "The Nibbling Away of the West," *Readers Digest*, December 1972, pp. 107–111; L. Williamson, "Where the Grass Is Greenest," *Outdoor Life*, February 1985, pp. 30–31.

39. Stoddart and Smith, *Range Management* (see note 2 above).

40. R. J. Smith, "Conclusions," in *Proceedings of a Seminar on Improving Fish and Wildlife Benefits in Range Management*, FWS/OBS-77/1 (Washington, D.C.: U.S. Fish and Wildlife Service, 1977).

41. Oregon-Washington Interagency Wildlife Committee, "Managing Riparian Ecosystems for Fish and Wildlife in Eastern Oregon and Eastern Washington, 1979" (available from Washington State Library, Olympia, Washington).

42. E. Chaney, W. Elmore, and W. S. Platts, *Livestock Grazing on Western Riparian Areas* (Denver: U.S. Environmental Protection Agency, 1990).

43. T. L. Fleischner, "Ecological Costs of Livestock Grazing in Western North America," *Conservation Biology* 8: 629–644; id., "Keeping the Cows Off: Conserving Riparian Areas in the American West," in *Terrestrial Ecoregions of North America: A Conservation Assessment*, edited by T. H. Ricketts et al. (Washington, D.C.: World Wildlife Fund/Island Press, 1999); R. D. Ohmart, "Historical and Present Impacts of Livestock Grazing on Fish and Wildlife Riparian Habitats," in *Rangeland Wildlife* (Denver: Society for Range Management, 1996).

44. C. L. Armour, D. A. Duff, and W. Elmore, "The Effects of Livestock Grazing on Riparian and Stream Ecosystems," *Fisheries* 16 (1991): 7–11; T. L. Fleischner et al., "Society for Conservation Biology Position Statement: Livestock Grazing on Public Lands in the United States of America," *Society for Conservation Biology Newsletter* 1, no. 4 (1994): 2–3; Wildlife Society, "The Wildlife Society Position Statement on Livestock Grazing on Federal Rangelands in the Western United States," *Wildlifer* 274 (1996): 10–13.

AN EVIL IN THE SEASON

1. M. LeSueur, *North Star Country* (New York: Book Find Club, 1946).

2. R. Lowitt and M. Beasely, eds. *One Third of a Nation: Lorena Hickok Reports on the Great Depression* (Urbana: University of Illinois Press, 1981).

3. J. Agee, "The Drought: A Post-Mortem in Pictures," *Fortune*, October 1934.

4. LeSueur, *North Star Country*.

5. E. Morris, *The Rise of Theodore Roosevelt* (New York: Coward, McCann, & Geoghegan, 1979).

6. T. Saloutos and J. D. Hicks, *Agricultural Discontent in the Middle West, 1900–1939* (Madison: University of Wisconsin Press, 1951).

7. D. Worster, *The Dust Bowl: The Southern Plains in the 1930s* (New York: Oxford University Press, 1979).

8. Anonymous, but probably Agee, "Drought."

9. K. Davis, *FDR: The New Deal Years, 1933–1937* (New York: Random House, 1986).

10. R. Lambert, "Drought Relief for Cattlemen: The Emergency Purchase Program of 1934–1935," *Panhandle-Plains Historical Review* 45 (1972): 21–35.

11. Ibid.

12. Ibid.

13. T. H. Watkins, *Righteous Pilgrim: The Life and Times of Harold L. Ickes, 1874–1952* (New York: Henry Holt, 1990).

14. E. L. Peffer, *The Closing of the Public Domain* (Palo Alto: Stanford University Press, 1951).

15. J. A. Stout Jr., "Cattlemen, Conservationists, and the Taylor Grazing Act," *New Mexico Historical Review* 45 (1970): 311–332.

16. E. B. Nixon, *Franklin D. Roosevelt and Conservation*, vol. 1 (Hyde Park: Franklin D. Roosevelt Library, 1957).

17. Ibid.

18. Ibid., p. 312.

19. Watkins, *Righteous Pilgrim*, p. 481.

20. G. Wuerthner, "How the West Was Eaten," *Wilderness*, Spring 1991, pp. 28–36.

21. B. DeVoto, "The West Against Itself," *Harper's*, January 1947, pp. 1–14.

PILLAGED PRESERVES

1. K. Davis, "General and Specific Legislative Authorities Pertaining to Domestic and Feral Livestock Grazing in the National Park Service," draft report (U.S. Department of the Interior, National Park Service, 23 July 1999).

2. 16 U.S.C. § 1 (2001).

3. 16 U.S.C. § 3 (2001).

4. R. Sellars, *Preserving Nature in the National Parks: A History* (New Haven: Yale University Press, 1997).

5. Ibid.

6. 36 C.F.R. § 2.60 (2000).

7. D. Roth, *The Wilderness Movement and the National Forests*, 2d ed. (College Station, Tex.: Intaglio, 1995).

8. C. Meine, *Aldo Leopold: His Life and Work* (Madison: University of Wisconsin Press, 1988).

9. Ibid.

10. A. Leopold, "Conservationist in Mexico," *American Forests* 43 (March 1937): 118–120; reprinted in *Wild Earth* 10 (Spring 2000): 57–60.

11. Roth, *Wilderness Movement*.

12. Ibid.

13. Ibid.

14. 16 U.S.C. § 1133(d)(4) (2001).

15. Pub. L. No. 96-560 § 108 (codified at 16 U.S.C.A. § 1133 notes [2001]).

16. House Committee on Interior and Insular Affairs, *Designating Certain National Forest System Lands in the National Wilderness Preservation System, and for Other Purposes*, 96th Cong., 1st sess., 1979, HR Rept. 617.

17. Ibid.

18. *Arizona Desert Wilderness Act of 1990*, Pub. L. No. 101-628, § 101(f)(1); *Arizona Wilderness Act of 1984*, Pub. L. No. 98-428, § 101(f)(1); Davis, "General and Specific Legislative Authorities"; *Utah Wilderness Act of 1984*, Pub. L. No. 98-428, § 301(a); *Wyoming Wilderness Act of 1984*, Pub. L. No. 98-550, § 501.

19. *Steens Mountain Cooperative Management and Protection Act*, Pub. L. No. 106-399.

20. 16 U.S.C. §§ 410aaa-5(a)-(b) (2001).

21. *Black Canyon of the Gunnison National Park and Gunnison Gorge National Conservation Area Act of 1999*, Pub. L. No. 106–76.

PART IV

SURVEYING THE WEST

Text

1. D. G. Milchunas and W. K. Lauenroth, "Quantitative Effects of Grazing on Vegetation and Soils Over a Global Range of Environments," *Ecological Monographs* 63 (1993): 327–351.

Table 1

1. D. C. Anderson, K. T. Harper, and S. R. Rushforth, "Recovery of Cryptogamic Soil Crusts from Grazing on Utah Winter Ranges," *Journal of Range Management* 35 (1982): 355–359.

2. R. J. Beymer and J. M. Klopatek, "Effects of Grazing on Cryptogamic Crusts in Pinyon-Juniper Woodlands in Grand Canyon National Park," *American Midland Naturalist* 127 (1992): 139–148.

3. C. E. Bock et al., "Responses of Birds, Rodents, and Vegetation to Livestock Exclosure in a Semidesert Grassland Site," *Journal of Range Management* 37 (1984): 239–242.

4. W. W. Brady et al., "Response of a Semidesert Grassland to 16 Years Rest from Grazing," *Journal of Range Management* 42 (1989): 284–288.

5. F. A. Branson, R. F. Miller, and I. S. McQueen, "Effects of Contour Furrowing, Grazing Intensities, and Soils on Infiltration Rates, Soil Moisture, and Vegetation Near Fort Peck, Montana," *Journal of Range Management* 15 (1962): 151–158.

6. J. D. Brotherson and W. T. Brotherson, "Grazing Impacts on the Sagebrush Communities of Central Utah," *Great Basin Naturalist* 41 (1981): 335–340.

7. J. D. Brotherson, S. R. Rushforth, and J. R. Johansen, "Effects of Long-Term Grazing on Cryptogam Crust Cover in Navajo National Monument, Arizona," *Journal of Range Management* 36 (1983): 579–581.

8. J. W. Brown and J. L. Schuster, "Effects of Grazing on a Hardland Site in the Southern High Plains," *Journal of Range Management* 22 (1969): 418–423.

9. J. C. Buckhouse and G. F. Gifford, "Sediment Production and Infiltration Rates as Affected by Grazing and Debris Burning on Chained and Seeded Pinyon-Juniper," *Journal of Range Management* 29 (1976): 83–85.

10. S. D. Conroy and T. J. Svejcar, "Willow Planting Success as Influenced by Site Factors and Cattle Grazing in Northeastern California," *Journal of Range Management* 44 (1991): 59–63.

11. W. P. Cottam and F. R. Evans, "A Comparative Study of the Vegetation of Grazed and Ungrazed Canyons of the Wasatch Range, Utah," *Ecology* 26 (1945): 171–181.

12. F. Daddy, M. J. Trlica, and C. D. Bonham, "Vegetation and Soil Water Differences Among Big Sagebrush Communities with Different Grazing Histories," *Southwestern Naturalist* 33 (1988): 413–424.

13. E. G. Dunford, "Surface Runoff and Erosion from Pine Grasslands of the Colorado Front Range," *Journal of Forestry* 52 (1954): 923–937.

14. N. D. Gamougoun et al., "Soil, Vegetation, and Hydrologic Responses to Grazing Management on Ft. Stanton, New Mexico," *Journal of Range Management* 37 (1984): 538–541.

15. J. L. Gardner, "The Effects of Thirty Years of Protection from Grazing in Desert Grassland," *Ecology* 31 (1950): 44–50.

16. W. E. Grant et al., "Structure and Productivity of Grassland Small Mammal Communities Related to Grazing-Induced Changes in Vegetative Cover," *Journal of Mammalogy* 63 (1982): 248–260.

17. L. M. Hall et al., "Effects of Cattle Grazing on Blue Oak Seedling Damage and Survival," *Journal of Range Management* 45 (1992): 503–506.

18. T. A. Hanley and J. L. Page, "Differential Effects of Livestock Use on Habitat Structure and Rodent Populations in Great Basin Communities," *California Fish and Game* 68 (1981): 160–174.

19. E. J. Heske and M. Campbell, "Effects of an 11-Year Livestock Exclosure on Rodent and Ant Numbers in the Chihuahuan Desert, Southeastern Arizona," *Southwestern Naturalist* 36 (1991): 89–93.

20. J. L. Holecheck and T. Stephenson, "Comparison of Big Sagebrush Vegetation in North Central New Mexico Under Moderately Grazed and Grazing Excluded Conditions," *Journal of Range Management* 36 (1983): 455–456.

21. D. L. Jeffries and J. M. Klopatek, "Effects of Grazing on the Vegetation of the Blackbrush Association," *Journal of Range Management* 40 (1987): 390–392.

22. K. Jepson-Innes and C. E. Bock, "Response of Grasshoppers (Orthoptera: Acrididae) to Livestock Grazing in Southeastern Arizona: Differences Between Season and Subfamilies," *Oecologia* 78 (1989): 430–431.

23. J. R. Johansen and L. L. St. Clair, "Cryptogamic Soil Crusts: Recovery from Grazing Near Camp Floyd State Park, Utah, USA," *Great Basin Naturalist* 46 (1986): 632–640.

24. S. J. Johnson, "Impacts of Domestic Livestock Grazing on Small Mammals of Forest Grazing Allotments in Southeastern Idaho," in *Proceedings of the Wildlife-Livestock Relationships Symposium, 20–22 April 1981, Coeur d'Alene, Idaho*, edited by J. M. Peek and P. D. Dalke (Moscow: University of Idaho, Forest, Wildlife, and Range Experiment Station, 1982).

25. W. M. Johnson, "The Effect of Grazing Intensity on Plant Composition, Vigor, and Growth of Pine-Bunchgrass Ranges in Central Colorado," *Ecology* 37 (1956): 790–798.

26. J. L. Kingery and R. T. Graham, "The Effect of Cattle Grazing on Ponderosa Pine Regeneration," *Forestry Chronicle* 67 (1991): 245–248.

27. G. Knoll and H. H. Hopkins, "The Effects of Grazing and Trampling upon Certain Soil Properties," *Transactions of the Kansas Academy of Science* 62 (1959): 221–231.

28. W. C. Krueger, M. Vavra, and W. P. Wheeler, "Plant Succession as Influenced by Habitat Type, Grazing Management, and Reseeding on a Northeast Oregon Clearcut," in *1980 Progress Report: Research in Rangeland Management*, Special Report No. 586 (Corvallis: Oregon State University, Agricultural Experiment Station/USDA, Agricultural Research—SEA, 1980), pp. 32–37.

29. W. A. Laycock and P. W. Conrad, "Effect of Grazing on Soil Compaction as Measured by Bulk Density on a High Elevation Cattle Range," *Journal of Range Management* 20 (1967): 136–140.

30. D. E. Medin and W. P. Clary, *Small Mammal Populations in a Grazed and Ungrazed Riparian Habitat in Nevada*, USDA Forest Service General Technical Report INT-413 (Ogden, Utah: USFS, Intermountain Research Station, 1989).

31. D. E. Medin and W. P. Clary, *Bird and Small Mammal Populations in a Grazed and Ungrazed Riparian Habitat in Idaho*, USDA Forest Service General Technical Report INT-425 (Ogden, Utah: USFS, Intermountain Research Station, 1990).

32. R. O. Meeuwig, "Effects of Seeding and Grazing on Infiltration Capacity and Soil Stability of a Subalpine Range in Utah," *Journal of Range Management* 18 (1965): 173–180.

33. D. G. Milchunas, W. K. Lauenroth, and P. L. Chapman, "Plant Competition, Abiotic, and Long- and Short-Term Effects of Large Herbivores on Demography of Opportunistic Species in a Semiarid Grassland," *Oecologia* 92 (1992): 520–531.

34. J. L. Oldemeyer and L. R. Allen-Johnson, "Cattle Grazing and Small Mammals on the Sheldon National Wildlife Refuge, Nevada," in *Management of Amphibians, Reptiles, and Small Mammals in North America*, USDA Forest Service General Technical Report RM-166, edited by R. C. Szaro, K. E. Severson, and D. R. Patton (Fort Collins, Colo.: USDA Forest Service, Rocky Mountain Forest and Range Experiment Station, 1988).

35. A. B. Orodho, M. J. Trlica, and C. D. Bonham, "Long-Term Heavy-Grazing Effects on Soil and Vegetation in the Four Corners Region," *Southwestern Naturalist* 35 (1990): 9–14.

36. M. K. Owens and B. E. Norton, "Interactions of Grazing and Plant Protection on Basin Big Sagebrush (*Artemisia tridentata* ssp. *tridentata*) Seedling Survival," *Journal of Range Management* 45 (1992): 257–262.

37. P. E. Packer, "Effects of Trampling Disturbance on Watershed Condition, Runoff, and Erosion," *Journal of Forestry* 51 (1953): 28–31.

38. L. C. Pearson, "Primary Production in Grazed and Ungrazed Desert Communities of Eastern Idaho," *Ecology* 46 (1965): 278–285.

39. R. D. Pieper, "Comparison of Vegetation on Grazed and Ungrazed Pinyon-Juniper Grassland Sites in Southcentral New Mexico," *Journal of Range Management* 21 (1968): 51–53.

40. J. J. Pluhar, R. W. Knight, and R. K. Heitschmidt, "Infiltration Rates and Sediment Production as Influenced by Grazing Systems in the Texas Rolling Plains," *Journal of Range Management* 40 (1987): 240–243.

41. L. L. Rasmussen and J. D. Brotherson, "Response of Winterfat Communities to Release from Grazing Pressure," *Great Basin Naturalist* 46 (1986): 148–156.

42. P. O. Reardon and L. B. Merrill, "Vegetative Responses Under Various Grazing Management Systems," *Journal of Range Management* 29 (1976): 195–198.

43. E. D. Rhoades et al., "Water Intake on a Sandy Range as Affected by 20 Years of Differential Stocking Rates," *Journal of Range Management* 17 (1964): 185–190.

44. L. Rich and H. Reynolds, "Grazing in Relation to Runoff and Erosion on Chaparral Watersheds in Central Arizona," *Journal of Range Management* 16 (1963): 322–326.

45. J. H. Robertson, "Changes on a Sagebrush-Grass Range in Nevada Ungrazed for 30 Years," *Journal of Range Management* 24 (1971): 397–400.

46. S. S. Rosenstock, "Shrub-Grassland Small Mammal and Vegetation Responses to Rest from Grazing," *Journal of Range Management* 49 (1996): 199–203.

47. B. A. Roundy and G. L. Jordan, "Vegetation Changes in Relation to Livestock Exclusion and Rootplowing in Southeastern Arizona," *Southwestern Naturalist* 33 (1988): 425–436.

48. D. O. Salihi and B. E. Norton, "Survival of Perennial Grass Seedlings Under Intensive Grazing in Semi-Arid Rangelands," *Journal of Applied Ecology* 24 (1987): 145–151.

49. T. L. Schmidt and J. Stubbendieck, "Factors Influencing Eastern Redcedar Seedling Survival on Rangeland," *Journal of Range Management* 46 (1993): 448–451.

50. J. L. Schuster, "Root Development of Native Plants Under Three Different Grazing Intensities," *Ecology* 45 (1964): 63–70.

51. D. A. Smith and E. M. Schmutz, "Vegetative Changes on Protected Versus Grazed Desert Grassland Ranges in Arizona," *Journal of Range Management* 28 (1975): 453–458.

52. J. R. Thompson, "Effect of Grazing on Infiltration in a Western Watershed," *Journal of Soil and Water Conservation* 23 (1968): 63–65.

53. M. K. Wood, *Impacts of Grazing Systems on Watershed Values* (Las Cruces: New Mexico State University, Department of Animal and Range Sciences, 1982).

54. M. K. Wood and W. H. Blackburn, "Sediment Production as Influenced by Livestock Grazing in the Texas Rolling Plains," *Journal of Range Management* 34 (1981): 228–231.

LIFEBLOOD OF THE WEST

1. J. B. Kauffman et al., "Wildlife of Riparian Habitats," in *Wildlife-Habitat Relationships in Oregon and Washington* (Corvallis: Oregon State University Press, 2001).

2. T. L. Fleischner, "Ecological Costs of Livestock Grazing in Western North America," *Conservation Biology* 8 (1994): 629–644; J. B. Kauffman and W. C. Krueger, "Livestock Impacts on Riparian Ecosystems and Streamside Management Implications: A Review," *Journal of Range Management* 37 (1984): 430–437.

3. L. R. Roath and W. C. Krueger, "Cattle Grazing Influences on a Mountain Riparian Zone," *Journal of Range Management* 35 (1982): 100–104.

4. K. A. Dwire, B. A. McIntosh, and J. B. Kauffman, "Ecological Influences of the Introduction of Livestock on Pacific Northwest Ecosystems," in *Northwest Lands, Northwest Peoples: Readings in Environmental History*, edited by D. D. Goble and P. W. Hirt (Seattle: University of Washington Press, 1999); J. B. Kauffman and D. A. Pyke, "Range Ecology: Global Livestock Influences," in *Encyclopedia of Biological Diversity*, vol. 5, edited by S. Levin (San Diego: Academic Press, 2001), pp. 33–52.

5. H. W. Li et al., "Cumulative Effects of Riparian Disturbances Along High Desert Trout Streams of the John Day Basin, Oregon," *Transactions of the American Fisheries Society* 123 (1994): 627–640.

6. J. B. Kauffman et al., "Ecological Approaches to Riparian Restoration in the United States," *Fisheries* 22 (1997): 12–24.

WHAT THE RIVER ONCE WAS

1. E. Chaney, W. Elmore, and W. S. Platts, *Livestock Grazing on Western Riparian Areas* (Eagle, Idaho: Northwest Resource Information Center, 1990).

2. U.S. Department of the Interior, "Rangeland Reform '94," draft environmental impact statement (Washington, D.C.: Bureau of Land Management, 1994).

3. A. J. Belsky, A. Matzke, and S. Uselman, "Survey of Livestock Influences on Stream and Riparian Ecosystems in the Western United States," *Journal of Soil and Water Conservation* 54 (1999): 419–431.

4. L. R. Roath and W. C. Krueger, "Cattle Grazing and Influence on a Forested Range," *Journal of Range Management* 35 (1982): 332–338.

5. J. W. Thomas, C. Maser, and J. E. Rodiek, *Wildlife Habitats in Managed Rangelands—The Great Basin of Southeastern Oregon: Riparian Zones*, USDA Forest Service General Technical Report PNW-80 (1979).

6. C. H. Flather, L. A. Joyce, and C. A. Bloomgarden, *Species Endangerment Patterns in the United States*, USDA Forest Service General Technical Report RM-241 (Fort Collins, Colo.: USDA Forest Service, Rocky Mountain Forest and Range Experiment Station 1994).

7. W. Elmore and B. Kauffman, "Riparian and Watershed Systems: Degradation and Restoration," in *Ecological Implications of Livestock Herbivory in the West*, edited by

M. Vavra, W. A. Laycock, and R. D. Pieper (Denver: Society for Range Management, 1994).

8. R. D. Ohmart, "Historical and Present Impacts of Livestock Grazing on Fish and Wildlife Resources in Western Riparian Habitats," in *Rangeland Wildlife*, edited by P. R. Krausman (Denver: Society for Range Management, 1996), pp. 245–279.

CATTLE AND STREAMS

1. T. J. Myers and S. Swanson, "Temporal and Geomorphic Variations of Stream Stability and Morphology: Mahogany Creek, Nevada," *Water Resources Bulletin* 32, no. 2 (1996): 253–265; S. W. Trimble and A. C. Mendel, "The Cow as a Geomorphic Agent—A Critical Review," *Geomorphology* 13 (1995): 233–253.

2. J. M. Friedman, W. R. Osterkamp, and W. M. Lewis Jr., "The Role of Vegetation and Bed Level Fluctuations in the Process of Channel Narrowing," *Geomorphology* 14 (1996): 341–351; C. R. Hupp and W. R. Osterkamp, "Riparian Vegetation and Fluvial Geomorphic Processes," *Geomorphology* 14 (1996): 227–295; C. R. Hupp and A. Simon, "Bank Accretion and the Development of Vegetated Depositional Surfaces Along Modified Alluvial Channels," *Geomorphology* 4 (1991): 111–124; R. McKenney, R. B. Jacobson, and R. C. Wertheimer, "Woody Vegetation and Channel Morphogenesis in Low-Gradient, Gravel-Bed Streams in the Ozark Plateaus, Missouri and Arkansas," *Geomorphology* 13 (1995): 175–198; Myers and Swanson, "Temporal and Geomorphic Variations" (see note 1 above); Trimble and Mendel "Cow as Geomorphic Agent" (see note 1 above).

3. C. R. Hupp and A. Simon, "Bank Accretion."

4. E. R. Montgomery et al., "Pool Spacing in Forest Channels," *Water Resources Research* 31, no. 4 (1995): 1097–1105; J. E. O'Connor, "Paleohydrology of Pool-and-Riffle Pattern, Development: Boulder Creek, Utah," *Geological Society of America Bulletin* 97 (1986): 410–420; R. D. Smith, R. C. Sidle, and P. E. Porter, "Effects on Bedload Transport of Experimental Removal of Woody Debris from a Forest Gravel-Bed Stream," *Earth Surface Processes and Landforms* 18 (1993): 455–468; E. E. Wohl, K. R. Vincent, and D. J. Merritts, "Pool and Riffle Characteristics in Relation to Channel Gradient," *Geomorphology* 6 (1993): 99–110.

5. D. M. Lawler, "The Measurement of River Bank Erosion and Lateral Channel Change: A Review," *Earth Surfaces Processes and Landforms* 18 (1993): 777–821; J. E. Pizzuto, "Bank Erodibility of Shallow Sandbed Streams," *Earth Surface Processes and Landforms* 9 (1984): 113–124; S. A. Schumm, *The Shape of Alluvial Channels in Relation to Sediment Type*, U.S. Geological Survey Professional Paper 352-B (Washington, D.C.: 1960), pp. 17–30; D. G. Smith, "Effects of Vegetation on Lateral Migration of Anastomosed Channels of a Glacier Meltwater Stream," *Geological Society of America Bulletin* 87 (1976): 857–860; C. R. Thorne, "Processes and Mechanisms of River Bank Erosion in Gravel-Bed Rivers," in *Gravel-Bed Rivers: Fluvial Processes, Engineering, and Management*, edited by R. D. Hey, J. C. Bathurst, and C. R. Thorne (New York: Wiley, 1982); C. R. Thorne and N. K. Tovey, "Stability of Composite River Banks," *Earth Surface Processes and Landforms* 6 (1981): 469–484.

6. G. M. Kondolf, "Lag in Stream Channel Adjustment to Livestock Exclusure, White Mountains, California," *Restoration Ecology* (December 1993): 226–230.

7. L. L. Apple et al., "The Use of Beavers for Riparian/Aquatic Habitat Restoration of Cold Desert, Gully-Cut Stream Systems in Southwestern Wyoming," paper presented to the American Fisheries Society/Wildlife Society Joint Chapter Meeting, Logan, Utah, February 1984; R. J. Naiman, C. A. Johnston, and J. C. Kelley, "Alteration of North American Streams by Beaver," *BioScience* 38, no. 11 (1988): 753–762.

8. S. C. Fouty, "Stream Channel Response to Changes in Cattle and Elk Grazing Pressure and Beaver Activity" (Ph.D. dissertation, University of Oregon, in progress); J. L. Retzer et al., *Suitability of Physical Factors for Beaver Management in the Rocky Mountains of Colorado*, Colorado Department of Game and Fish Technical Bulletin No. 2 (March 1956), pp. 24–30; J. D. Stock and I. J. Schlosser, "Short-Term Effects of a Catastrophic Beaver Dam Collapse on a Stream Fish Community," *Environmental Biology of Fishes* 31 (1991): 123–129.

STINK WATER

1. U.S. Environmental Protection Agency, *The Quality of Our Nation's Water: 1996—Executive Summary of the National Water Quality Inventory: Report to Congress*, EPA841-S-97-001 (Washington, D.C.: USEPA, Office of Water, 1998).

2. Environmental Defense Fund, "Animal Waste—A National Overview," *Environmental Defense Fund Scorecard*, 15 January 2000.

3. U.S. General Accounting Office, *Animal Waste Management and Water Quality Issues*, GAO/RCED-95-200BR (Washington, D.C.: USGAO, 1995).

4. U.S. Department of Agriculture/U.S. Environmental Protection Agency, *Unified National Strategy for Animal Feeding Operations* (Washington, D.C.: USEPA, 9 March 1999), www.epa.gov\owm\finafost.htm.

5. JoAnn M. Burkholder, "The Lurking Perils of *Pfiesteria*," *Scientific American*, August 1999, pp. 42–49.

6. USDA/USEPA, *Unified National Strategy* (see note 4 above).

7. Ibid.

8. J. H. Martin Jr., "The Clean Water Act and Animal Agriculture," *Journal of Environmental Quality* 26 (1997): 1198–1203.

9. USDA/USEPA, *Unified National Strategy* (see note 4 above).

10. USGAO, *Animal Waste Management* (see note 3 above).

11. USDA/USEPA, *Unified National Strategy* (see note 4 above).

12. USGAO, *Animal Waste Management* (see note 3 above).

13. A. J. Belsky, A. Matzke, and S. Uselman, "Survey of Livestock Influences on Stream and Riparian Ecosystems in the Western United States," *Journal of Soil and Water Conservation* 54 (1999): 419–431.

14. K. E. Saxon et al., *Effect of Animal Grazing on Water Quality of Non-Point Runoff in the Pacific Northwest*, project summary, EPA-600/S2-83-071 (Ada, Okla.: Robert S. Kerr Environmental Research Laboratory, 1983).

15. R. K. Hubbard et al., "Surface Runoff and Shallow Ground Water Quality as Affected by Center Pivot Applied Dairy Cattle Wastes," *Transactions of the American Society of Agricultural Engineers* 30, no. 2 (1987): 430–437.

16. J. S. Schepers, B. L. Hackes, and D. D. Francis, "Chemical Water Quality of Runoff from Grazing Land in Nebraska: II. Contributing Factors," *Journal of Environmental Quality* 11, no. 3 (1982): 355–359.

17. G. Nader et al., "Water Quality Effect of Rangeland Beef Cattle Excrement," *Rangelands* 20, no. 5 (1998): 19–25.

18. Willow Creek Ecology, *Watersheds, Livestock and Water Quality: A Case Study from the Cache National Forest, Utah and Idaho*, Publication 99-01 (Mendon, Utah: Willow Creek Ecology, 1999).

19. A. N. Pell, "Manure and Microbes: Public and Animal Health Problem?" *Journal of Dairy Science* 80 (1997): 2673–2681.

20. Ibid.

21. Ibid.

22. S. Wells et al., "*E. coli* O157 and *Salmonella*—Status on U.S. Dairy Operations" (1996), www.aphis.usda.gov/vs/ceah/cahm/Dairy_Cattle/ecosalm 98.htm; Animal and Plant Health Inspection Service, "*E. coli* O157:H7 Overview" (February 1994), www.aphis.usda.gov/vs/ceah/cahm/Dairy_Cattle/ndhep/ dhpeco1txt.htm.

23. Animal and Plant Health Inspection Service, ibid.

24. P. Brasher, "Fourth of Beef Cattle Harbor *E. coli*," *Salt Lake Tribune*, 4 July 2000.

25. Brasher, ibid.; M. Koohmaraie, W. W. Laegreid, and R. L. Hruska, "*E. coli*: Unwelcome from Farm to Fork," *Agricultural Research* 48 (October 2000).

26. Ibid.

27. Food Safety and Inspection Service, "Meat and Poultry Product Recalls: News Releases and Information for Consumers" (2001), www.fsis.usda.gov/OA/news/xrecalls.htm (September 2001).

28. U.S. Environmental Protection Agency, *Quality Criteria for Water, July 1976: Fecal Coliform Bacteria*, U.S. Environmental Protection Agency 42–50 (Washington, D.C.: USEPA, 1976).

29. G. L. Howard, S. R. Johnson, and S. L. Ponce, "Cattle Grazing Impact on Surface Water Quality in a Colorado Front Range Stream," *Journal of Soil and Water Conservation* (March–April 1983): 124–128.

30. U.S. General Accounting Office, *Rangeland Management: More Emphasis Needed on Declining and Overstocked Grazing Allotments*, GAO/RCED-88-80 (Washington, D.C.: USGAO, 1988).

31. USDA/USEPA, *Unified National Strategy* (see note 4 above).

32. Saxon et al., *Effect of Animal Grazing* (see note 14 above).

33. Willow Creek Ecology, *Watersheds* (see note 18 above).

34. U.S. Forest Service, *Wasatch-Cache National Forest Land and Resource Management Plan* (Ogden, Utah: USFS, Intermountain Region, 1985).

35. Utah Department of Environmental Quality, *Utah Water Quality Assessment Report to Congress* (Salt Lake City: Utah Division of Water Quality, 1996).

GUZZLING THE WEST'S WATER

1. M. Reisner and S. Bates, *Overtapped Oasis: Reform or Revolution for Western Water* (Washington, D.C.: Island Press, 1990).

2. Ibid.

3. *Montana GAP Analysis* (CD-ROM) (Missoula: University of Montana, Montana Cooperative Wildlife Research Unit, Wildlife Spatial Analysis Lab, 1998).

4. W. B. Solley, R. R. Pierce, and H. A. Perlman, *Estimated Use of Water in the United States in 1990*, USGS Circular 1081 (1993).

5. J. Cobourn et al., *Nevada's Water Future: Making Tough Choices* (Reno: University of Nevada, 1992).

6. W. L. Minckley and J. E. Deacon, eds., *Battle Against Extinction: Native Fish Management in the West* (Tucson: University of Arizona Press, 1991).

7. T. Palmer, *The Snake River: Window to the West* (Washington, D.C.: Island Press, 1991).

8. Montana Department of Fish, Wildlife and Parks, "Dewatered Streams List" (Helena, 1991).

9. Minckley and Deacon, *Battle Against Extinction.*

10. Reisner and Bates, *Overtapped Oasis.*

11. Minckley and Deacon, *Battle Against Extinction.*

12. U.S. General Accounting Office, *Public Rangelands: Some Riparian Areas Restored but Widespread Improvement Will Be Slow*, GAO/RCED-88-105 (Washington, D.C.: USGAO, 1988).

13. J. W. Duffield, T. C. Brown, and S. Allen, *Economic Value of Instream Flow in Montana's Big Hole and Bitterroot Rivers*, Research Paper RM-317 (Fort Collins, Colo.: Rocky Mountain Forest and Range Experiment Station, 1994).

14. J. W. Duffield, J. B. Loomis, and R. Brooks, *The Net Economic Value of Fishing in Montana* (Helena: Montana Department of Fish, Wildlife and Parks, 1987).

15. Minckley and Deacon, *Battle Against Extinction.*

16. U.S. General Accounting Office, *Bureau of Reclamation: Information on Allocation and Repayment Costs of Constructing Water Project*, GAO/RCED-96-109 (Washington, D.C.: USGAO, 1996).

17. Ibid.

18. A. Melnykovych, "In the West, Subsidy Begets Subsidy Begets Subsidy," *High Country News*, 11 April 1988.

19. See Figure 9 in U.S. Environmental Protection Agency, *National Water Quality Inventory: 1998 Report to Congress*, EPA 841-R-00-001, www.epa.gov/ow/resources/brochure/.

20. Ibid.

21. U.S. Senate Committee, *Animal Waste Pollution in America: An Emerging National Problem*, report by Minority Staff of the United States Committee on Agriculture, Nutrition and Forests (Washington, D.C., 1997).

22. M. Strand and R. W. Merritt, "Impacts of Livestock Grazing Activities on Stream Insect Communities and the Riverine Environment," *American Entomologist* 45, no. 1 (1999): 13–21.

23. R. E. Larson et al., "Water-Quality Benefits of Having Cattle Manure Deposited Away from Streams," *Bioresource Technology* 48 (1994): 113–118.

24. A. J. Belsky, A. Matzke, and S. Uselman, "Survey of Livestock Influences on Stream and Riparian Ecosystems in the Western United States," *Journal of Soil and Water Conservation* 54 (1999): 419–431.

25. Belsky, Matzke, and Uselman, ibid.; Strand and Merritt, "Impacts of Livestock Grazing Activities"; USGAO, *Public Rangelands.*

26. T. J. Frest and E. J. Johannes, *Interior Columbia Basin Mollusk Species of Special Concern*, final report (Walla Walla, Wash.: Interior Columbia Basin Ecosystem Management Project, 1995).

27. D. S. Wilcove et al., "Quantifying Threats to Imperiled Species in the United States," *Bioscience* (1998): 607–613.

28. Minckley and Deacon, *Battle Against Extinction.*

THE SOIL'S LIVING SURFACE

1. J. Belnap, "Potential Role of Cryptobiotic Soil Crust in Semiarid Rangelands," in *Proceedings—Ecology and Management of Annual Rangelands*, edited by S. B. Monsen and S. G. Kitchen, USDA Forest Service General Technical Report INT–GTR-313 (Ogden, Utah: USDA Forest Service Intermountain Research Station, 1994).

2. J. H. Kaltenecker, M. C. Wicklow-Howard, and R. Rosentreter, "Biological Soil Crusts in Three Sagebrush Communities Recovering from a Century of Livestock Trampling," in *Proceedings Shrublands Ecotones*, RMRS-P-11 (USDA Rocky Mountain Research Station, 1999).

3. J. Belnap, "Soil Surface Disturbances in Cold Deserts: Effects on Nitrogenase Activity in Cyanobacterial-Lichen Soil Crusts," *Biology and Fertility of Soils* 23 (1996): 362–367; R. J. Beymer and J. M. Kiopatek, "Effects of Grazing on Cryptogamic Crusts in Pinyon-Juniper Woodlands in Grand Canyon National Park," *American Midland Naturalist* 127 (1992): 139–148; J. H. Kaltenecker, M. C. Wicklow-Howard, and R. Rosentreter, "Biological Soil Crusts: Natural Barriers to *Bromus tectorum* Establishment in the Northern Great Basin, USA," in *Proceedings of the VI International Rangeland Congress*, vol. 1, edited by D. Eldridge and D. Freudenberger (Aitkenvale, Queensland, Australia, 1999); J. R. Marble and K. T. Harper, "Effect of Timing of Grazing on Soil-Surface Cryptogamic Communities in a Great Basin Low-Shrub Desert: A Preliminary Report," *Great Basin Naturalist* 49 (1989): 104–107.

4. R. Rosentreter, "Compositional Patterns Within a Rabbit-brush *(Chrysothamnus)* Community of the Idaho Snake River Plain," in *Proceedings—Symposium on the Biology of Artemisia and Chrysothamnus*, USDA Forest Service General Technical Report INT-200 (Ogden, Utah: USDA Forest Service Intermountain Research Station, 1986).

5. Ibid.

6. J. D. Williams, J. P. Dobrowolski, and N. E. West, "Microphytic Crust Influence on Interrill Erosion and Infiltration Capacity," *Transactions of the American Society of Agricultural Engineers* 38 (1995): 139–146.

7. Ibid.

8. J. Belnap and J. S. Gardner, "Soil Microstructure in Soils of the Colorado Plateau: The Role of the Cyanobacterium *Microcieus vaginatus*," *Great Basin Naturalist* 53 (1993): 40–47.

9. Beymer and Kiopatek, "Effects of Grazing"; R. D. Evans and J. R. Ehleringer, "A Break in the Nitrogen Cycle in Arid Lands? Evidence from Nitrogen-15 of Soils," *Oecologia* 94 (1993): 314–317.

10. Beymer and Kiopatek, ibid.

11. H. E. Dregne, *Desertification of Arid Lands* (New York: Harwood, 1983).

12. K. T. Harper and R. L. Pendleton, "Cyanobacteria and Cyanolichens: Can They Enhance Availability of Essential Minerals for Higher Plants?" *Great Basin Naturalist* 53 (1993): 59–72.

13. Kaltenecker, Wicklow-Howard, and Rosentreter, "Biological Soil Crusts"; K. D. Larsen, "Effects of Microbiotic Crusts on the Germination and Establishment of Three Range Grasses" (master's thesis, Boise State University, Boise, Idaho 1995).

14. J. Belnap, "Surface Disturbances: Their Role in Accelerating Desertification," *Environmental Monitoring and Assessment* 37 (1995): 39–57.

15. C. S. Crawford, "The Community Ecology of Macroarthropod Detritivores," in *Ecology of Desert Communities*, edited by G. Polis (Tucson: University of Arizona Press, 1991); J. T. Doyen and W. F. Tschinkel, "Population Size, Microgeographic Distribution and Habitat Separation in Some Tenebrionid Beetles," *Annals of the Entomological Society of America* 67 (1974): 617–626.

16. Rosentreter, "Compositional Patterns."

17. J. R. Johansen et al., "Recovery Patterns of Cryptogamic Soil Crusts in Desert Rangelands Following Fire Disturbance," *Bryologist* 87 (1984): 238–243.

18. D. J. Eldridge, "Trampling of Microphytic Crusts on Calcareous Soils and Its Impact on Erosion Under Rain-Impacted Flow," *Catena* 33 (1998): 221–239.

19. Kaltenecker, Wicklow-Howard, and Rosentreter, "Biological Soil Crusts."

20. K. L. Memmot, V. J. Anderson, and S. B. Monsen, "Seasonal Grazing Impact on Cryptogamic Crusts in a Cold Desert Ecosystem," *Journal of Range Management* 51 (1998): 547–550.

21. Kaltenecker, Wicklow-Howard, and Rosentreter, "Biological Soil Crusts."

22. R. N. Mack and J. N. Thompson, "Evolution in Steppe with Few, Large, Hooved Mammals," *American Midland Naturalist* 119 (1982): 757–773; G. L. Stebbins, "Coevolution of Grasses and Herbivores," *Annual of the Missouri Botanical Garden* 68 (1981): 75–86.

COMRADES IN HARM

1. W. D. Billings, "*Bromus tectorum*, a Biotic Cause of Ecosystem Impoverishment in the Great Basin," in *The Earth in Transition: Patterns and Processes of Biotic Impoverishment*, edited by G. M. Woodwell (New York: Cambridge University Press, 1990); R. N. Mack, "Temperate Grasslands Vulnerable to Plant Invasions: Characteristics and Consequences," in *Biological Invasions: A Global Perspective*, edited by J. A. Drake et al. (Chinchester, U.K.: Wiley and Sons, 1989), pp. 155–179; S. Whisenant, "Changing Fire Frequencies on Idaho's Snake River Plains: Ecological and Management Implications," in *Proceedings from the Symposium on Cheatgrass Invasion, Shrub Dieoff and Other Aspects of Shrub Biology and Management*, USDA Forest Service General Technical Report INT-276 (1990), pp. 4–10.

2. Bureau of Land Management, *Partners Against Weeds: An Action Plan for the Bureau of Land Management*, BLM/MT/ST-96/003+1020 (Billings, Mont.: Bureau of Land Management, 1996).

3. W. G. Dore and L. C. Raymond, "Viable Seeds in Pasture Soil and Manure," *Scientia Agricola* 23 (1942): 69–76.

4. G. A. Harris, "Grazing Lands of Washington State," *Rangelands* 13 (1991): 222–227; R. N. Mack, "Temperate Grasslands" (see note 1 above); R. N. Mack and J. N. Thompson, "Evolution in Steppe with Few Large, Hooved Mammals," *American Naturalist* 119 (1982): 757–773; D. G. Milchunas, O. E. Sala, and W. K. Lauenroth, "A Generalized Model of the Effects of Grazing by Large Herbivores on Grassland Community Structure," *American Naturalist* 132, no. 1 (1988): 87–106.

5. Mack and Thompson, "Evolution" (see note 4 above).

6. M. J. Crawley, *Herbivory: The Dynamics of Animal-Plant Interactions* (Berkeley and Los Angeles: University of California Press, 1983); Lacey, "The Influence of Livestock Grazing on Weed Establishment and Spread,"

Proceedings Montana Academy of Science 47 (1987): 131–146; G. Stewart and A. C. Hull Jr., "Cheatgrass *(Bromus tectorum L.)*: An Ecological Intruder in Southern Idaho," *Ecology* 30 (1949): 58–74; B. R. Watkin and R. J. Clements, "The Effects of Grazing Animals on Pastures," in *Plant Relations in Pastures*, edited by J. R. Wilson (East Melbourne, Australia: CSIRO, 1978), pp. 273–289.

7. D. J. Bedunah, "The Complex Ecology of Weeds, Grazing, and Wildlife," *Western Wildlands* (Summer 1992): 6–11; Lacey, "Influence of Livestock Grazing" (see note 6 above); S. M. Louda, K. H. Keeler, and R. D. Holt, "Herbivore Interactions on Plant Performance and Competitive Interactions," in *Perspectives on Plant Competition*, edited by J. B. Grace and D. Tilman (San Diego: Academic Press, 1990); R. L. Sheley, B. E. Olson, and L. L. Larson, "Effect of Weed Seed Rate and Grass Defoliation Level on Diffuse Knapweed," *Journal of Range Management* 50, no. 1 (1997): 39–43.

8. L. Ellison, "Influence of Grazing on Plant Succession of Rangelands," *Botanical Review* 26 (1960): 1–78; G. A. Harris, "Some Competitive Relationships Between *Agropyron spicatum* and *Bromus tectorum*," *Ecological Monographs* 37, no. 2 (1967): 90–111; W. H. Rickard, "Experimental Cattle Grazing in a Relatively Undisturbed Shrubsteppe Community," *Northwest Science* 59, no. 1 (1985): 66–72.

9. Ellison, "Influence of Grazing" (see note 8 above); R. J. Hobbs, "The Nature and Effects of Disturbance Relative to Invasions," in *Biological Invasions: A Global Perspective*, edited by J. A. Drake et al. (Chinchester, U.K.: Wiley, 1989), pp. 389–405; R. J. Hobbs and L. F. Huenneke, "Disturbance, Diversity, and Invasion: Implications for Conservation," *Conservation Biology* 6, no. 3 (1992): 324–337; Rickard, "Experimental Cattle Grazing" (see note 8 above).

10. R. A. Dahlgren, M. J. Singer, and X. Huang, "Oak Tree and Grazing Impacts on Soil Properties and Nutrients in a California Oak Woodland," *Biogeochemistry* 39 (1997): 45–64; J. F. Dormaar and W. D. Willms, "Effect of Forty-Four Years of Grazing on Fescue Grassland Soils," *Journal of Range Management* 51 (1998): 122–126; J. W. Menke, "Management Controls on Productivity," in *Grassland Structure and Function: California Annual Grassland*, edited by L. F. Huenneke and H. A. Mooney (Dordrecht, Netherlands: Kluwer, 1989), pp. 173–199; Watkin and Clements, "Effects of Grazing Animals" (see note 6 above).

11. Ellison, "Influence of Grazing" (see note 8 above); G. C. Lusby, "Hydrologic and Biotic Effects of Grazing vs. Non-Grazing Near Grand Junction, Colorado," *Journal of Range Management* 23 (1971): 256–260.

12. R. A. Evans and J. A. Young, "Microsite Requirements for Establishment of Annual Rangeland Weeds," *Weed Science* 23, no. 5 (1972): 354–357.

13. J. Belnap and O. L. Lange, *Biological Soil Crusts: Structure, Function and Management* (Berlin: Springer-Verlag, 2001).

14. J. Belnap, "Surface Disturbances: Their Role in Accelerating Desertification," *Environmental Monitoring and Assessment* 37 (1995): 39–57; Belnap and Lange, "Biological Soil Crusts" (see note 13 above); R. E. Eckert Jr. et al., "Effects of Soil-Surface Morphology on Emergence and Survival of Seedlings in Big Sagebrush Communities," *Journal of Range Management* 39, no. 5 (1986): 414–420; R. D. Evans and J. R. Ehleringer, "A Break in the Nitrogen Cycle in Aridlands? Evidence From N15 Isotope of Soils," *Oecologia* 94 (1993): 314–317; R. N. Mack, "Temperate Grasslands" (see note 1 above); T. J. Stohlgren et al., "Patterns of Plant Invasions: A Case Example in Native Species Hotspots and Rare Habitats," *Biological Invasions* 3 (2001): 37–50

15. E. B. Allen, "Mycorrhizal Limits to Rangeland Restoration: Soil Phosphorous and Fungal Species Composition," in *Proceedings of the Fifth International Rangeland Congress*, vol. 2, *Rangelands in a Sustainable Biosphere*, pp. 57–61 (Salt Lake City, 1995); E. B. Allen and M. F. Allen, "Facilitation of Succession by the Nonmycotrophic Colonizer *Salsola kali* on a Harsh Site: Effects on

Mycorrhizal Fungi," *American Journal of Botany* 75 (1988): 257–266; G. J. Bethlenfalvay and S. Dakessian, "Grazing Effects on Mycorrhizal Colonization and Floristic Composition of the Vegetation on a Semiarid Range in Northern Nevada," *Journal of Range Management* 37, no. 4 (1984): 312–316; G. J. Bethlenfalvay, R. A. Evans, and A .L. Lesperance, "Mycorrhizal Colonization of Crested Wheatgrass as Influenced by Grazing," *Agronomy Journal* 77 (1985): 233–236; T. B. Doerr, E. F. Redente, and F. B. Reeves, "Effects of Soil Disturbance on Plant Succession and Levels of Mycorrhizal Fungi in a Sagebrush-Grassland Community," *Journal of Range Management* 37 (1984): 135–139.

16. Evans and Ehleringer, "Break in the Nitrogen Cycle" (see note 14 above); Hobbs, "Nature and Effects of Disturbance" (see note 9 above); L. F. Huenneke et al., "Effects of Soil Resources on Plant Invasion and Community Structure in Californian Serpentine Grassland," *Ecology* 71 (1990): 478–491.

17. S. Archer and D. E. Smeins, "Ecosystem Level Processes," in *Grazing Management: An Ecological Perspective*, edited by R. K. Heitschmidt and J. W. Stuth (Portland, Ore.: Timber Press, 1991); W. L. Loope and G. F. Gifford, "Influence of a Soil Micro-Floral Crust on Select Property of Soils Under Pinyon Juniper in Southeastern Utah," *Journal of Soil and Water Conservation* 27 (1972): 164–167; R. L. Piemeisel, "Causes Affecting Change and Rate of Change in a Vegetation of Annuals in Idaho," *Ecology* 32, no. 1 (1951): 53–72.

18. Belnap, "Surface Disturbances" (see note 14 above); R. F. Daubenmire, "Plant Succession on Abandoned Fields, and Fire Influences, in a Steppe Area in Southeastern Washington," *Northwest Science* 49 (1975): 36–48; R. F. Daubenmire, *Steppe Vegetation of Washington*, Washington Agricultural Experimental Station Technical Bulletin No. 62 (1970); J. R. Goodwin et al., "Persistence of Idaho Fescue on Degraded Sagebrush Steppe," *Journal of Range Management* 52 (1999): 187–198; J. R. Lacey, P. Husby, and G. Handl, "Observations on Spotted and Diffuse Knapweed Invasion into Ungrazed Bunchgrass Communities in Western Montana," *Rangelands* 12 (1990): 30–32; G. D. Pickford, "The Influence of Continued Heavy Grazing and of Promiscuous Burning on Spring–Fall Ranges in Utah," *Ecology* 13 (1932): 159–171; R. L. Piemeisel, "Causes Affecting Change and Rate of Change in a Vegetation of Annuals in Idaho," *Ecology* 32, no. 1 (1951): 53–72; J. A. Young and F. L. Allen, "Cheatgrass and Range Science: 1930–1950," *Journal of Range Management* 50, no. 5 (1997): 530–535.

SILENT SPRINGS

1. A. R. Blaustein et al., "The Biology of Amphibians and Reptiles in Old-Growth Forests in the Pacific Northwest," USDA Forest Service General Technical Report PNW-GTR-337 (Portland, Ore.: USDA Forest Service, Pacific Northwest Research Station, 1995).

2. J. C. Munger et al., "Habitat Conservation Assessment for the Columbia Spotted Frog *(Rana luteiventris)* in Idaho," draft report submitted to the Idaho State Conservation Effort, 1997.

3. J. C. Engle and J. C. Munger, *Population Structure of Spotted Frogs in the Owyhee Mountains*, Idaho BLM Technical Bulletin No. 98–20 (1998).

4. I. Hanski, *Metapopulation Ecology* (New York: Oxford University Press, 1999).

5. Munger et al., "Habitat Conservation Assessment."

6. W. P. Clary, "Vegetation and Soil Responses to Grazing Simulation on Riparian Meadows," *Journal of Range Management* 48 (1995): 18–25; W. S. Platts, *Managing Fisheries and Wildlife on Rangelands Grazed by Livestock: A Guidance and Reference Document for Biologists* (Reno: Nevada Department of Wildlife, 1990); P. T. Tueller, "Vegetation Science Applications for Rangeland Analysis and Management," in *Handbook of Vegetation Science* (Dordrecht, Netherlands: Kluwer, 1988), p. 14.

7. V. S. Lamoureux and D. M. Madison, "Overwintering Habitats of Radio-Implanted Green Frogs, *Rana clamitans*," *Journal of Herpetology* 33 (1999): 430–435.

8. D. G. Penney, "Frogs and Turtles: Different Ectotherm Overwintering Strategies," *Comparative Biochemistry and Physiology* 86A (1987): 609–615; A. W. Pinder and M. E. Feder, "Effect of Boundary Layers on Cutaneous Gas Exchange," *Journal of Experimental Biology* 143 (1990): 67–80; F. L. Rose and R. B. Drotman, "Anaerobiosis in a Frog, *Rana pipiens*," *Journal of Experimental Zoology* 166 (1967): 427–432.

9. R. G. Wetzel, *Limnology*, 2d ed. (Fort Worth: Saunders College Publishing, 1983).

10. Munger et al., "Habitat Conservation Assessment."

11. E. Chaney, W. Elmore, and W. Platts, "Livestock Grazing on Western Riparian Areas," report produced for the U.S. Environmental Protection Agency by the Northwest Resource Information Center (Eagle, Idaho, 1993).

12. Lamoureux and Madison, "Overwintering."

NATIVE SNAILS

1. H. A. Pilsbry, *Land Mollusca of North America (North of Mexico)*, vol. 1, pt. 1, Academy of Natural Sciences of Philadelphia Monograph 3, no. 1 (1939): 1–574; D. M. Turgeon et al., *Common and Scientific Names of Aquatic Invertebrates from the United States and Canada: Mollusks*, 2d ed., AFS Special Publication 26 (American Fisheries Society, 1998).

2. T. J. Frest and B. Roth, "Mollusk Conservation in the Western United States," American Malacological Union, 61st Annual Meeting, 1995, *Program and Abstracts*, p. 26.

3. E. O. Wilson, *The Diversity of Life* (New York: W. W. Norton, 1992).

4. J. B. Burch, *North American Freshwater Snails* (Hamburg, Mich.: Malacological Publications, 1989).

5. Turgeon et al., *Common and Scientific Names of Aquatic Invertebrates* (see note 1 above).

6. R. Hershler, "A Systematic Review of the Hydrobiid Snails of the Great Basin, Western United States: Part I, Genus *Pyrgulopsis*," *Veliger* 41, no. 1 (1998): 1–132; ibid., "Part II, Genera *Colligyrus, Eremopyrgus, Fluminicola, Pristinicola*, and *Tryonia*," *Veliger* 42, no. 4 (1999): 306–337.

7. E. O. Wilson, "The Little Things That Run the World (The Importance and Conservation of Invertebrates)," *Conservation Biology* 1 (1987): 344–346.

8. T. J. Frest and R. S. Rhodes II, "*Oreohelix strigosa cooperi* (Binney) in the Midwest Pleistocene," *Nautilus* 95 (1981): 47–55.

9. R. Hershler, *A Review of the North American Freshwater Snail Genus* Pyrgulopsis *(Hydrobiidae)*, Smithsonian Contributions to Zoology No. 554 (1994).

10. T. J. Frest and E. J. Johannes, "Interior Columbia Basin Mollusk Species of Special Concern," final report to the Interior Columbia Basin Ecosystem Management Project (Seattle: Deixis Consultants, 1995); id., "Land Snails of the Lower Salmon River Drainage, Idaho," American Malacological Union, 63rd Annual Meeting, and Western Society of Malacologists, 30th Annual Meeting, *Program and Abstracts* (1997): 28; id., "Land Snails of the Lucile Caves ACEC," final report to the USDI Bureau of Land Management (Seattle: Deixis Consultants, 1995); id., *Land Snails of the Lucile Caves ACEC*, Technical Bulletin 97–16 (Boise: USDI Bureau of Land Management, Idaho State Office, 1997); id., *Land Snail Survey of the Lower Salmon River Drainage*, Technical Bulletin 97–18 (Boise: USDI Bureau of Land Management, Idaho State Office, 1997); id., "Land Snail Survey of the Lower Salmon River Drainage, Idaho," final report to USDI Bureau of Land Management (Seattle: Deixis Consultants, 1995).

11. T. J. Frest and E. J. Johannes, "Endemics in an Ancient Western North American Lake (Upper Klamath Lake, Oregon): Lake or Stream Origin?" *World Congress of Malacology Abstracts* (1998): 105.

12. T. L. Fleischner, "Ecological Costs of Livestock Grazing in Western North America," *Conservation Biology* 8 (1994): 629–644.

13. C. L. Armour, D. A. Duff, and W. Elmore, "The Effects of Livestock Grazing on Riparian and Stream Ecosystems," *Fisheries* 16 (1991): 7–11; A. J. Belsky, A. Matzke, and

S. Uselman, "Survey of Livestock Influences on Stream and Riparian Ecosystems in the Western United States," *Journal of Soil and Water Conservation* 54, no. 1 (1999): 419–431; Fleischner, "Ecological Costs of Livestock Grazing" (see note 12 above); E. L. Painter and A. J. Belsky, "Application of Herbivore Optimization Theory to Rangelands of the Western United States," *Ecological Applications* 3, no. 1 (1993): 2–9.

14. T. J. Frest and E. J. Johannes, "Additional Information on Certain Mollusk Species of Special Concern Occurring Within the Range of the Northern Spotted Owl," report to the Oregon State Office of the Bureau of Land Management (Seattle: Deixis Consultants, 1996); id., *Field Guide to Survey and Manage Freshwater Mollusk Species*, BLM/OR/WA/PL-99/045+1792 (Portland, Ore.: USDI, BLM, Oregon State Office; USFWS, Northwest Regional Ecosystems Office; and USDA, Forest Service, Region 6, 1999), www.or.blm.gov/surveyandmanage/; id., "Freshwater Mollusks of the Upper Klamath Drainage, Oregon," Third Klamath Basin Watershed Restoration and Research Conference, *Conference Abstracts* (Klamath Falls, Ore.: USFWS, Klamath Basin Ecosystem Restoration Office, 1999), p. 15; id., "Freshwater Mollusks of the Upper Klamath Drainage, Oregon, 1998," yearly report to the Oregon Natural Heritage Program and Klamath Project, USDI Bureau of Reclamation (Seattle: Deixis Consultants, 1998); id., "Freshwater Mollusks of the Upper Sacramento System, California, with Particular Reference to the Cantara Spill," final report to the California Department of Fish & Game (Seattle: Deixis Consultants, 1995); id., "The Hydrobiid Subfamily Amnicolinae in the Northwestern United States," *World Congress of Malacology, Abstracts*, 1998: 106; id., "Interior Columbia Basin Mollusk Species of Special Concern" (see note 10 above); id., "Land Snails of the Lower Salmon River Drainage, Idaho" (1997) (see note 10 above); id., "Land Snails of the Lower Salmon River Drainage, Idaho," *Western Society of Malacologists Annual Reports* 30 (1998): 22; id., "Land Snails of the Lucile Caves ACEC" (1995) (see note 10 above); id., *Land Snails of the Lucile Caves ACEC* (1997) (see note 10 above); id., *Land Snail Survey of the Black Hills National Forest, South Dakota and Wyoming*, final report to the USDA Forest Service and USDI Fish & Wildlife Service (Seattle: Deixis Consultants, 1993); id., *Land Snail Survey of the Lower Salmon River Drainage* (1997) (see note 10 above); id., "Land Snail Survey of the Lower Salmon River Drainage, Idaho" (1995) (see note 10 above); id., "Mollusc Species of Special Concern Within the Range of the Northern Spotted Owl," final report to the Forest Ecosystem Management Working Group, USDA Forest Service (Seattle: Deixis Consultants, 1993); id., "Mollusk Survey of Southwestern Oregon, with Emphasis on the Rogue and Umpqua River Drainage" (Seattle: Deixis Consultants, 1999); id., "Upper Sacramento System Freshwater Mollusk Monitoring, California, 1996," final report to the Cantara Trustee Council (Seattle: Deixis Consultants, 1997).

15. A. Solem, *The Shell Makers: Introducing Mollusks* (New York: Wiley-Interscience, 1974); id., "A World Model of Land Snail Diversity and Abundance," in *World-wide Land Snails*, edited by A. Solem and A. C. Van Bruggen (Leiden, Germany: E. J. Brill, 1984).

16. A. Solem, "Notes on Salmon River Oreohelicid Land Snails, with Description of *Oreohelix waltoni*," *Veliger* 18, no. 1 (1975): 16–30; A. Solem and A. H. Clarke, "Report on Status Survey for Salmon River Valley Land Snails," unpublished letter to the Office of Endangered Species, Washington, D.C., 23 August 1974.

17. Solem, "Notes on Salmon River Oreohelicid Land Snails" (see note 16 above).

18. Frest and Johannes, "Land Snails of the Lucile Caves ACEC" (1995) (see note 10 above); id., *Land Snails of the Lucile Caves ACEC* (1997) (see note 10 above); id., *Land Snail Survey of the Lower Salmon River Drainage* (1997) (see note 10 above); id., "Land Snail Survey of the Lower Salmon River Drainage, Idaho" (1995) (see note 10 above).

19. R. Hershler, "Conservation of Springsnails in the Great Basin: Simple Solutions Complicated by Political Realities," in American Malacological Union, *Program and Abstracts, 61st Annual Meeting* (1995), p. 30; id., *Review of the North American Freshwater Snail Genus Pyrgulopsis* (see note 9 above); id., "A Systematic Review of the Hydrobiid Snails of the Great Basin, Western United States: Part I, Genus *Pyrgulopsis*" (see note 6 above); ibid., "Part II, Genera *Colligyrus, Eremopyrgus, Fluminicola, Pristinicola,* and *Tryonia*" (see note 6 above).

20. Frest and Johannes, "Freshwater Mollusks of the Upper Sacramento System, California" (see note 14 above); id., "Interior Columbia Basin Mollusk Species of Special Concern" (see note 10 above); id., "Land Snails of the Lucile Caves ACEC" (1995) (see note 10 above); id., "Land Snail Survey of the Lower Salmon River Drainage, Idaho" (1995) (see note 10 above).

21. Hershler, "Conservation of Springsnails in the Great Basin" (see note 19 above); id., "A Systematic Review of the Hydrobiid Snails of the Great Basin: Part I, Genus *Pyrgulopsis*" (see note 6 above).

22. Public Employees for Environmental Responsibility, *Public Trust Betrayed: Employee Critique of Bureau of Land Management Rangeland Management* (Washington, D.C.: Public Employees for Environmental Responsibility, 1994).

23. Ibid.

BIRDS AND BOVINES

1. T. L. Fleischner, "Ecological Costs of Livestock Grazing in Western North America," *Conservation Biology* 8 (1994): 629–644.

2. C. E. Bock et al., "Effects of Livestock Grazing on Neotropical Migratory Landbirds in Western North America," in *Status and Management of Neotropical Migratory Birds*, edited by D. M. Finch and P. W. Stangel, USDA Forest Service General Technical Report RM-229 (Fort Collins, Colo.: USDA Forest Service, Rocky Mountain Forest and Range Experiment Station, 1993).

3. D. G. Milchunas, W. K. Lauenroth, and I. C. Burke, "Livestock Grazing: Animal and Plant Biodiversity of Shortgrass Steppe and the Relationship to Ecosystem Function," *Oikos* 83 (1998): 65–74.

4. B. G. Peterjohn and J. R. Sauer, "Population Status of North American Grassland Birds from the North American Breeding Bird Survey, 1966–1996," *Studies in Avian Biology* 19 (1999): 27–44.

5. F. L. Knopf, "Avian Assemblages on Altered Grasslands," *Studies in Avian Biology* 15 (1994): 247–257.

6. R. N. Mack and J. N. Thompson, "Evolution in Steppe with Few Large, Hooved Mammals," *American Naturalist* 119 (1982): 757–773.

7. Milchunas, Lauenroth, and Burke, "Livestock Grazing"; D. G. Milchunas, O. E. Sala, and W. K. Lauenroth, "A Generalized Model of the Effects of Grazing by Large Herbivores on Grassland Community Structure," *American Naturalist* 132 (1988): 87–106.

8. C. P. Ortega, *Cowbirds and Other Brood Parasites* (Tucson: University of Arizona Press, 1998).

9. R. E. Marvil and A. Cruz, "Impact of Brown-headed Cowbird Parasitism on the Reproductive Success of the Solitary Vireo," *Auk* 106 (1989): 476–480.

10. J. C. Uyehara and P. M. Narins, "Nest Defense by Willow Flycatchers to Brood-Parasitic Intruders," *Condor* 97 (1995): 361–368.

11. S. L. Lima and T. J. Valone, "Predators and Avian Community Organization: An Experiment in a Semi-Desert Grassland," *Oecologia* 86 (1991): 105–112.

12. C. E. Bock and J. H. Bock, "Response of Winter Birds to Drought and Short-Duration Grazing in Southeastern Arizona," *Conservation Biology* 13 (1999): 1117–1123; H. R. Pulliam and J. B. Dunning, "The Influence of Food Supply on Local Density and Diversity of Sparrows," *Ecology* 68 (1987): 1009–1014.

13. R. D. Ohmart, "The Effects of Human-Induced Changes on the Avifauna of Western Riparian Habitats," *Studies in Avian Biology* 15 (1994): 273–285.

14. E. M. Ammon and P. B. Stacey, "Avian Nest Success in Relation to Past Grazing Regimes in a Montane Riparian System," *Condor* 99 (1997): 7–13; D. S. Dobkin, A. C. Rich, and W. H. Pyle, "Habitat and Avifaunal Recovery from Livestock Grazing in a Riparian Meadow System of the Northwestern Great Basin," *Conservation Biology* 12 (1998): 209–221; Ohmart, "Effects of Human-Induced Changes."

15. V. A. Saab et al., "Livestock Grazing Effects in Western North America," in *Ecology and Management of Neotropical Migratory Birds*, edited by T. E. Martin and D. M. Finch (New York: Oxford University Press, 1995).

16. F. L. Knopf, "Mountain Plover *(Charadrius montanus)*," in *The Birds of North America*, No. 211, edited by A. Poole and F. Gill (Philadelphia: Academy of Natural Sciences, and Washington, D.C.: American Ornithologists' Union, 1996).

17. Milchunas, Lauenroth, and Burke, "Livestock Grazing."

18. Knopf, "Mountain Plover."

19. Peterjohn and Sauer, "Population Status."

20. P. D. Vickery, "Grasshopper Sparrow *(Ammodramus savannarum)*," in *The Birds of North America*, No. 239, edited by A. Poole and F. Gill (Philadelphia: Academy of Natural Sciences, and Washington, D. C.: American Ornithologists' Union, 1996).

21. Bock and Bock, "Response of Winter Birds to Drought."

22. E. A. Webb and C. E. Bock, "Botteri's Sparrow *(Aimophila botterii)*," in *The Birds of North America*, No. 216, edited by A. Poole and F. Gill (Philadelphia: Academy of Natural Sciences, and Washington, D.C.: American Ornithologists' Union, 1996).

23. J. B. Dunning Jr. et al., "Cassin's Sparrow *(Aimophila cassinii)*," in *The Birds of North America*, No. 471, edited by A. Poole and F. Gill (Philadelphia: The Birds of North America, Inc., 1999).

24. D. F. Bradford et al., "Bird Species Assemblages as Indicators of Biological Integrity in Great Basin Rangelands," *Environmental Monitoring and Assessment* 49 (1998): 1–22; Saab, "Livestock Grazing Effects."

25. D. L. Johnson et al., *Effects of Management Practices on Grassland Birds: Northern Harrier* (Jamestown, N.D.: Northern Prairie Wildlife Research Center, 1998); id., *Effects of Management Practices on Grassland Birds: Short-eared Owl* (Jamestown, N.D.: Northern Prairie Wildlife Research Center, 1998).

26. J. A. Dechant et al., *Effects of Management Practices on Grassland Birds: Ferruginous Hawk* (Jamestown, N.D.: Northern Prairie Wildlife Research Center, 1999).

27. Saab, "Livestock Grazing Effects."

28. T. Nelson et al., "Wildlife Numbers on Late and Mid Seral Chihuahuan Desert Rangelands," *Journal of Range Management* 50 (1997): 593–599; G. Smith, J. L. Holechek, and M. Cardenas, "Wildlife Numbers on Excellent and Good Chihuahuan Desert Rangelands: An Observation," *Journal of Range Management* 49 (1996): 489–493.

29. C. E. Bock, J. H. Bock, and H. M. Smith, "Proposal for a System of Federal Livestock Exclosures on Public Rangelands in the Western United States," *Conservation Biology* 7 (1993): 731–733; D. L. Donahue, *The Western Range Revisited: Removing Livestock from Public Lands to Conserve Native Biodiversity* (Norman: University of Oklahoma Press, 1999).

RANCHING IN BEAR COUNTRY

1. F. R. Gowans, *Mountain Man and Grizzly* (Orem, Utah: Mountain Grizzly Publications, 1986; T. I. Storer and L. P. Tevis Jr., *California Grizzly* (Lincoln: University of Nebraska Press, 1955).

2. D. E. Brown, *The Grizzly in the Southwest: Documentary of an Extinction* (Norman: University of Oklahoma Press, 1985).

3. D. J. Mattson et al., "Grizzly Bears," in *Our Living Resources* (Washington, D.C.: U.S. Department of the Interior, National Biological Service, 1995).

4. M. J. Madel, *Rocky Mountain Front Grizzly Bear Management Program: Four Year Progress Report*,

1991–1994 (Helena: Montana Department of Fish, Wildlife and Parks, 1996).

5. K. Aune and W. Kasworm, *East Front Grizzly Studies: Final Report* (Helena: Montana Department of Fish, Wildlife and Parks, 1989).

6. A. R. Dood and H. I. Pac, *The Grizzly Bear in Northwestern Montana* (Helena: Montana Department of Fish, Wildlife and Parks, 1993).

7. L. P. Horstman and J. R. Gunson, "Black Bear Predation on Livestock in Alberta," *Wildlife Society Bulletin* 10 (1982): 34–39.

8. U.S. Department of Agriculture, National Agricultural Statistics Service, *Cattle Predator Loss* (Washington, D.C.: USDA, 2000).

9. L. F. James, D. B. Nielsen, and K. E. Panter, "Impact of Poisonous Plants on the Livestock Industry," *Journal of Range Management* 45 (1992): 3–8.

10. Horstman and Gunson, "Black Bear Predation" (see note 7 above).

11. C. J. Jorgensen, "Bear-Sheep Interactions, Targhee National Forest," *International Conference on Bear Research and Management* 5 (1983): 191–200.

12. Wyoming Game and Fish Department, *Management Guidelines to Address Nuisance Grizzly Bears on Sheep and Cattle Allotments Outside the Yellowstone Recovery Zone in Wyoming* (Cheyenne: Wyoming Game and Fish Department, 1999).

13. Dood and Pac, *Grizzly Bear* (see note 6 above).

14. B. L. Horejsi, "Uncontrolled Land-Use Threatens an International Grizzly Bear Population," *Conservation Biology* 3, no. 3 (1988): 220–223.

15. Alberta Environment, Natural Resources Services, *Grizzly Translocations from Alberta Southern Mountains/Foothills Area* (Pincher Creek: Alberta Environment, Natural Resources Services, 1999).

16. Horstman and Gunson, "Black Bear Predation" (see note 7 above).

17. K. Olchowy and S. Heschl, "The Hidden Secret of the Deep Creek Bear," *Alberta Game Warden* 4, no. 3 (1992): 24–25.

18. Wyoming Game and Fish Department, *Management Guidelines* (see note 12 above).

19. C. R. Anderson et al., *Grizzly Bear–Cattle Interactions on Two Cattle Allotments in Northwest Wyoming* (Lander: Wyoming Game and Fish Department, 1997).

20. R. B. Wielgus and F. L. Bunnell, "Sexual Segregation and Female Grizzly Bear Avoidance of Males," *Journal of Wildlife Management* 58, no. 3 (1994): 405–413.

21. Wyoming Game and Fish Department, *Management Guidelines* (see note 12 above).

22. Brown, *Grizzly* (see note 2 above).

23. A. L. Harting Jr., "Relationships Between Activity Patterns and Foraging Strategies of Yellowstone Grizzly Bears" (master's thesis, Montana State University, Bozeman, 1985).

24. Wyoming Game and Fish Department, *Management Guidelines* (see note 12 above).

25. Anderson et al., *Grizzly Bear–Cattle Interactions* (see note 19 above); Interagency Grizzly Bear Study Team, *A Report to the Yellowstone Ecosystem Subcommittee on Grizzly Bear Mortalities and Conflicts in the Greater Yellowstone Ecosystem* (Bozeman, Mont.: IGBST, 2000).

26. Anderson et al., *Grizzly Bear–Cattle Interactions* (see note 19 above).

27. T. S. Stivers and L. R. Irby, "Impacts of Cattle Grazing on Mesic Grizzly Bear Habitat Along the East Front of the Rocky Mountains, Montana," *Intermountain Journal of Sciences* 3, no. 1 (1997): 17–37.

28. Madel, *Rocky Mountain Front* (see note 4 above).

29. Ibid.

30. K. Aune and T. Stivers, *Ecological Studies of the Grizzly Bear in the Pine Butte Preserve* (Helena: Montana Department of Fish, Wildlife and Parks, 1985).

31. Ibid.

32. E. S. Bacon and G. M. Burghardt, "Ingestive Behaviors of the American Black Bear," *International Conference on Bear Research and Management* 3 (1976): 13–25; id., "Learning and Color Discrimination in the American

Black Bear," *International Conference on Bear Research and Management* 3 (1976): 27–36.

33. Aune and Stivers, *Ecological Studies* (see note 30 above)

34. L. L. Irwin and F. M. Hammond, "Managing Black Bear Habitats for Food Items in Wyoming," *Wildlife Society Bulletin* 13 (1985): 477–483.

35. L. Lee and C. Jonkel, "Grizzlies and Wetlands," *Western Wildlands* 7, no. 4 (1981): 26–30.

36. D. Hamer, *Wildfire's Influence on Yellow* Hedysarum *Digging Habitat Used by Grizzly Bears in Banff National Park, Alberta* (Fort St. John, B.C.: Northern Lights College, 1996); D. J. Mattson, "Selection of Microsites by Grizzly Bears to Excavate Biscuitroots," *Journal of Mammalogy* 78, no. 1 (1997): 228–238.

37. R. B. Campbell Jr., "Ecology of *Heracleum lanatum* MICHX (Cow Parsnip) Communities in Northwestern Montana" (master's thesis, University of Montana, Missoula, 1991).

38. G. B. Hilderbrand et al., "The Importance of Meat, Particularly Salmon, to Body Size, Population Productivity, and Conservation of North American Brown Bears," *Canadian Journal of Zoology* 77 (1999): 132–138.

39. United States Forest Service, *Environmental Assessments for 36 Livestock Grazing Allotments on the Shoshone National Forest* (Cody, Wyo.: U.S. Forest Service, Shoshone National Forest, 1996).

40. R. E. Kirby et al., "Grazing on National Wildlife Refuges: Do the Needs Outweigh the Problems?" *Transactions of the North American Wildlife & Natural Resources Conference* 57 (1992): 611–626.

41. T. A. Dull, "Palynological Evidence for the 19th Century Grazing-Induced Vegetation Change in the Southern Sierra Nevada, California, U.S.A.," *Journal of Biogeography* 26 (1999): 899–912.

A WEST WITHOUT WOLVES

1. R. Maughan, "Idaho Wolf Mortalities" (updated 20 August 1999), www.forwolves.org/ralph/deadwolf-id.htm; id., "Yellowstone Wolf Mortalities" (updated 15 December 2001), www.forwolves.org/ralph/deadwolf.htm.

2. Ibid.

PRAIRIE DOG GONE

1. R. E. Marsh, "Ground Squirrels, Prairie Dogs, and Marmots as Pests on Rangeland," in *Proceedings of the Conference for Organization and Practice of Vertebrate Pest Control, 30 August–3 September 1982* (Fernherst, U.K.: ICI Plant Protection Division, 1984).

2. Ibid.

3. Biodiversity Legal Foundation, J. C. Sharps, and Predator Project, "Black-tailed Prairie Dog *(Cynomys ludovicianus)*," unpublished petition (Denver: USFWS Region 6, 1998); U.S. Fish and Wildlife Service, "Reply to C-2 Petition Submitted by the Biodiversity Legal Foundation for the Black-tailed Prairie Dog," unpublished document (Washington, D.C.: USFWS, 1995).

4. Biodiversity Legal Foundation, Sharps, and Predator Project, "Black-tailed Prairie Dog" (see note 3 above); C. J. Knowles, "Status of the Black-tailed Prairie Dog," unpublished report (Washington, D.C.: U.S. Fish and Wildlife Service, 1998); National Wildlife Federation, *Petition for Rule Listing the Black-tailed Prairie Dog* (Cynomys ludovicianus) *as Threatened Throughout Its Range* (Denver: USFWS Region 6, 1998).

5. R. List, "Ecology of Kit Fox *(Vulpes macrotis)* and Coyote *(Canis latrans)* and the Conservation of the Prairie Dog Ecosystem in Northern Mexico" (Ph.D. dissertation, University of Oxford, Oxford, U.K., 1997).

6. J. Hoogland, personal communication, 2001.

7. G. Ceballos, E. Mellink, and L. R. Hanebury, "Distribution and Conservation Status of Prairie Dogs *Cynomys mexicanus* and *Cynomys ludovicianus* in Mexico," *Biological Conservation* 63 (1993): 105–112.

8. B. Miller, R. P. Reading, and S. Forrest, *Prairie Night: Recovery of the Black-footed Ferret and Other Endangered Species* (Washington, D.C.: Smithsonian Institution Press, 1996).

9. B. Miller et al., "A Proposal to Conserve Black-footed

Ferrets and the Prairie Dog Ecosystem," *Environmental Management* 14 (1990): 763–769.

10. E. T. Seton, *Lives of Game Animals* (Garden City, N.Y.: Doubleday, 1929).

11. J. F. Cully, "Plague, Prairie Dogs and Black-footed Ferrets," in *Management of Prairie Dog Complexes for the Reintroduction of the Black-footed Ferret*, edited by J. L. Oldemeyer et al., Biological Report 13 (Washington, D.C.: USFWS, 1993).

12. Miller, Reading, and Forrest, *Prairie Night* (see note 8 above).

13. T. W. Clark, *Conservation Biology of the Black-footed Ferret* Mustela nigripes, Wildlife Preservation Trust Special Scientific Report No. 3 (Philadelphia, 1989).

14. E. Dowd Stukel et al., "A Conservation Design: Conservation Assessment and Conservation Strategy for Swift Fox in the United States," First North American Swift Fox Symposium, 18–19 February 1998, Saskatoon, Saskatchewan.

15. N. B. Kotliar et al., "A Critical Review of Assumptions About the Prairie Dog as a Keystone Species," *Environmental Management* 24 (1999): 177–192.

16. B. Miller, G. Ceballos, and R. P. Reading, "The Prairie Dog and Biotic Diversity," *Conservation Biology* 8 (1994): 677–681; G. Wuerthner, "Viewpoint: The Black-tailed Prairie Dog—Headed for Extinction?" *Journal of Range Management* 50 (1997): 459–466.

17. D. M. Roemer and S. C. Forrest, "Prairie Dog Poisoning in the Northern Great Plains: An Analysis of Programs and Policies," *Environmental Management* 20 (1996): 349–359.

18. Biodiversity Legal Foundation, Sharps, and Predator Project, "Black-tailed Prairie Dog" (see note 3 above); Roemer and Forrest, "Prairie Dog Poisoning" (see note 17 above).

19. Biodiversity Legal Foundation, Sharps, and Predator Project, "Black-tailed Prairie Dog" (see note 3 above).

20. Ibid.; C. J. Knowles, "A Summary of Black-tailed Prairie Dog Abundance and Distribution on the Central and Northern Great Plains," unpublished report (Washington, D.C.: Defenders of Wildlife, 1995); U.S. Fish and Wildlife Service, *90-Day Finding for a Petition to List the Black-tailed Prairie Dog as Threatened* (Washington, D.C.: USFWS, 1999).

21. L. B. Field and E. Hansen, *The Landowner's View of Wildlife Damage Control Techniques and Agency Programs*, USDA Forest Service General Technical Report RM-171 (1989), pp. 20–23; K. Krueger, *Prairie Dogs Overpopulation: Value Judgment or Ecological Reality?* USDA Forest Service General Technical Report RM-154 (1988), pp. 39–45; C. D. Lee and F. R. Henderson, *Kansas Attitudes on Prairie Dog Control*, USDA Forest Service General Technical Report RM-171 (1989), p. 165; G. Probasco, *Involving the Public in Prairie Dog Management on the Nebraska National Forest*, USDA Forest Service General Technical Report RM-154 (1988), pp. 113–114; R. P. Reading, B. J. Miller, and S. R. Kellert, "Values and Attitudes Toward Prairie Dogs," *Anthrozoös* 12 (1999): 43–52. See also, as an example, the following statement by anonymous rancher, "[I]t's the prairie dog. I want to kill every last one," quoted in R. P. Reading, "Toward an Endangered Species Reintroduction Paradigm: A Case Study of the Black-footed Ferret" (Ph.D. dissertation, Yale University, 1993).

22. A. V. Dicey, *Law and Public Opinion* (London: Macmillan, 1926); H. D. Lasswell and A. Kaplan, *Power and Society: A Framework for Political Inquiry* (New Haven: Yale University Press, 1950).

23. M. G. Wallace et al., "Moving Toward Ecosystem Management: Examining a Change in Philosophy for Resource Management," *Journal of Political Ecology* 3 (1996): 1–36.

24. L. White Jr., "The Historical Roots of Our Ecological Crisis," *Science* 155 (1967): 1203–1207.

25. J. Locke, *The Second Treatise on Government* (London: Awnsham Churchill, 1690); A. Smith, *The Wealth of Nations* (New York: Collier, 1776/1902).

26. D. Casey and T. Clark, *Tales of the Wolf* (Moose, Wyo.:

Homestead Publishing, 1996); B. Lopez, *Of Wolves and Men* (New York: Scribner's, 1978).

27. R. B. Keiter, "Greater Yellowstone's Bison: Unraveling of an Early American Wildlife Conservation Achievement," *Journal of Wildlife Management* 61 (1997): 1–11; M. Racicot, "Shooting Bison Tragic Reality," *West Yellowstone News*, 20 March 1997.

28. Reading, Miller, and Kellert, "Values and Attitudes" (see note 21 above). See also, as an example, the following comment by rancher Richard Peterson, Haigle, Nebraska, "I know of no good that they do. . . . They tear up your ground, and the new weeds come in. . . . If these prairie dogs are not allowed to be controlled, somebody is going to have to pay us for the grass that they eat and the damage they do," quoted in J. Anderson, "Battle Looms on Prairie Dogs: Ranchers Oppose a Wildlife Federation Request That One Species Be Declared Threatened," *Omaha World-Herald*, 30 September 1998, p. 1.

29. E.g., "Cowboys like animals. . . . But there is no such thing as a cute prairie dog. They cut our income," statement by rancher Miles Davies, Deer Trail, Colorado, quoted in D. Luzadder, "Prairie Dog Bill Goes to House Floor: 4-Hour Panel Debate Pitted Developers, Wildlife Advocates Against Rural Workers," *Rocky Mountain News*, 25 February 1999, p. 14A.

30. E.g., "[Prairie dog] numbers are there. . . . The endangered species we should be worried about is the rancher because I think he is more important. These people with the government don't understand that scratching a living out of the earth is more difficult than milking taxpayers," statement by DuWayne Miller, prairie dog shooter from Devils Lake, North Dakota, quoted in R. Wilson, "Status of Prairie Dogs Will Get Further Review," *Bismarck Tribune*, 24 March 1999, p. 1A.

31. R. M. Hansen and I. K. Gold, "Black-tailed Prairie Dogs, Desert Cottontails and Cattle Trophic Relations on Shortgrass Range," *Journal of Range Management* 30 (1977): 210–214; M. E. O'Meilia, F. L. Knopf, and J. C. Lewis, "Some Consequences of Competition Between Prairie Dogs and Beef Cattle," *Journal of Range Management* 35 (1982): 580–585.

32. A. R. Collins, J. P. Workman, and D. W. Uresk, "An Economic Analysis of Black-tailed Prairie Dog *(Cynomys ludovicianus)* Control," *Journal of Range Management* 37 (1984): 358–361.

33. L. E. Klatt and D. Hein, "Vegetative Differences Among Active and Abandoned Towns of Black-tailed Prairie Dogs *(Cynomys ludovicianus),*" *Journal of Range Management* 31 (1978): 315–317; D. W. Uresk, "Effects of Controlling Black-tailed Prairie Dogs on Plant Production," *Journal of Range Management* 38 (1985): 466–468.

34. A. D. Whicker and J. K. Detling, "Ecological Consequences of Prairie Dog Disturbances," *BioScience* 38 (1988): 778–785.

35. Colorado Department of Public Health and Environment, "Animal Species Implicated in Transmission of Plague to Humans, Colorado, 1957–1998 (N = 43)," unpublished memorandum (Denver, 1998).

36. Kotliar et al., "Critical Review" (see note 15 above); B. Miller et al., "Focal Species in Design of Reserve Networks," *Wild Earth* 8, no. 4 (1998/99): 81–92; L. S. Mills, M. E. Soulé, and D. F. Doak, "The History and Current Status of the Keystone Species Concept," *BioScience* 43 (1993): 219–224; R. T. Paine, "Food Webs: Linkage, Interaction Strength and Community Infrastructure," *Journal of Animal Ecology* 49 (1980): 667–685; M. E. Power et al., "Challenges in the Quest for Keystones," *BioScience* 466 (1996): 9–20.

37. D. L. Coppock et al., "Plant-Herbivore Interactions in a North American Mixed-Grass Prairie. I. Effects of Black-tailed Prairie Dogs on Intraseasonal Aboveground Plant Biomass and Nutrient Dynamics and Plant Species Diversity," *Oecologia* 56 (1983): 1–9; J. K. Detling and A. D. Whicker, "Control of Ecosystem Processes by Prairie Dogs and Other Grassland Herbivores," in *Proceedings of the Eighth Great Plains Wildlife Damage Control Workshop, 28–30 April 1987, Rapid City, S.D.,* USDA Forest Service General Technical Report RM-154 (1988), pp.

23–29; N. B. Kotliar, B. W. Baker, and A. D. Whicker, "Are Prairie Dogs a Keystone Species?" abstract, Seventh International Theriological Congress, Acapulco, Mexico, 1997; Kotliar et al., "Critical Review" (see note 15 above); B. Miller et al., "The Role of Prairie Dogs as Keystone Species: Response to Stapp," *Conservation Biology* 14 (2000): 318–321; R. P. Reading et al., "Attributes of Black-tailed Prairie Dog Colonies in Northcentral Montana, with Management Recommendation for Conservation of Biodiversity," in *The Prairie Dog Ecosystem: Managing for Biological Diversity,* Montana BLM Wildlife Technical Bulletin No. 2, edited by Tim W. Clark, Dan Hinckley, and Terell Rich (Billings, Mont.: Bureau of Land Management, 1989), pp. 13–27; Society for Conservation Biology, "Resolution of the Society for Conservation Biology: Conservation of Prairie Dog Ecosystems," *Society for Conservation Biology Newsletter* 1 (May 1994): 7; A. D. Whicker and J. K. Detling, "Control of Grassland Ecosystem Processes by Prairie Dogs," in *Management of Prairie Dog Complexes for the Reintroduction of the Black-footed Ferret,* edited by J. L. Oldemeyer et al., Biological Report 13 (Washington, D.C.: USFWS, 1993); id., "Ecological Consequences" (see note 34 above); id., "Modification of Vegetation Structure and Ecosystem Processes by North American Grassland Mammals," in *Plant Form and Vegetation Structure: Adaptation, Plasticity and Relation to Herbivory,* edited by M. J. A. Werger et al. (The Hague: SPB Academic Publishing, 1988); Wuerthner, "Viewpoint" (see note 16 above).

38. T. W. Clark, D. Hinckley, and T. Rich, eds., *The Prairie Dog Ecosystem: Managing for Biological Diversity,* Montana Bureau of Land Management Wildlife Technical Bulletin No. 2 (1989); Miller, Reading, and Forrest, *Prairie Night* (see note 8 above).

39. W. Agnew, D. W. Uresk, and R. M. Hansen, "Flora and Fauna Associated with Prairie Dog Colonies and Adjacent Ungrazed Mixed-Grass Prairie in Western South Dakota," *Journal of Range Management* 39 (1986): 135–139; V. A. Barko, "Effect of the Black-tailed Prairie Dog on Avifaunal Composition in Southern Shortgrass Prairie" (master's thesis, Oklahoma State University, Stillwater, 1996); T. M. Campbell and T. W. Clark, "Colony Characteristics and Vertebrate Associates of White-tailed and Black-tailed Prairie Dogs in Wyoming," *American Midland Naturalist* 105 (1981): 269–276; G. Ceballos and J. Pacheco, "The Relationship of Prairie Dogs and Regional Biodiversity in Mexico," abstract, Seventh International Theriological Congress, Acapulco, Mexico, 1997; T. W. Clark et al., "Prairie Dog Colony Attributes and Associated Vertebrate Species," *Great Basin Naturalist* 42 (1982): 572–582; R. R. Koford, "Prairie Dogs, Whitefaces, and Blue Grama," *Wildlife Monographs* 3 (1958): 1–78; Kotliar et al., "Critical Review" (see note 15 above); P. Manzano, "Avian Communities Associated with Prairie Dog Towns in Northwestern Mexico" (master's thesis, University of Oxford, Oxford, U.K., 1996); E. Mellink and H. Madrigal, "Ecology of Mexican Prairie Dogs, *Cynomys mexicanus,* in El Manantial, Northeastern Mexico," *Journal of Mammalogy* 74 (1993): 631–635; O'Meilia, Knopf, and Lewis, "Some Consequences" (see note 31 above); Reading et al., "Attributes" (see note 37 above); J. C. Sharps and D. W. Uresk, "Ecological Review of Black-tailed Prairie Dogs and Associated Species in Western South Dakota," *Great Basin Naturalist* 50 (1990): 339–345; J. D. Tyler, "Distribution and Vertebrate Associates of the Black-tailed Prairie Dog in Oklahoma" (Ph.D. dissertation, University of Oklahoma, Norman, 1968).

40. Kotliar et al., "Critical Review" (see note 15 above); Manzano, "Avian Communities" (see note 39 above); Reading et al., "Attributes" (see note 37 above).

41. M. E. Berry, C. E. Bock, and S. L. Haire, "Abundance of Diurnal Raptors on Open Space Grasslands in an Urbanized Landscape," *Condor* 100 (1998): 601–608; Kotliar et al., "Critical Review" (see note 15 above); Miller, Reading, and Forrest, *Prairie Night* (see note 8 above); D. L. Plumpton and D. E. Andersen, "Habitat Use and Time Budgeting by Wintering Ferruginous

Hawks," *Condor* 99 (1997): 888–893; Reading et al., "Attributes" (see note 37 above).

42. Clark, *Conservation Biology* (see note 13 above); Miller, Reading, and Forrest, *Prairie Night* (see note 8 above).

43. Berry, Bock, and Haire, "Abundance of Diurnal Raptors" (see note 41 above); J. M. Goodrich and S. W. Buskirk, "Spacing and Ecology of North American Badgers *(Taxidae taxus)* in a Prairie-Dog *(Cynomys leucurus)* Complex," *Journal of Mammalogy* 79 (1998): 171–179; D. L. Plumpton and D. E. Andersen, "Anthropogenic Effects on Winter Behavior of Ferruginous Hawks," *Journal of Wildlife Management* 62 (1998): 340–346; id., "Habitat Use" (see note 41 above); Sharps and Uresk, "Ecological Review" (see note 39 above); D. W. Uresk and J. C. Sharps, "Denning Habitat and Diet of the Swift Fox in Western South Dakota," *Great Basin Naturalist* 46 (1986): 249–253.

44. Kotliar et al., "Critical Review" (see note 15 above); Reading et al., "Attributes" (see note 37 above).

45. M. J. Desmond, J. A. Savidge, and T. F. Seibert, "Spatial Patterns of Burrowing Owls *(Speotyto cunicularia)* Nests Within Black-tailed Prairie Dog *(Cynomys ludovicianus)* Towns," *Canadian Journal of Zoology* 73 (1995): 1375–1379; D. L. Plumpton and R. S. Lutz, "Nesting Habitat Use by Burrowing Owls in Colorado," *Journal of Raptor Research* 27 (1993): 175–179; Reading et al., "Attributes" (see note 37 above); Sharps and Uresk, "Ecological Review" (see note 39 above).

46. W. Agnew, D. W. Uresk, and R. M. Hansen, *Arthropod Consumption by Small Mammals on Prairie Dog Colonies and Adjacent Ungrazed Mixed Grass Prairie in Western South Dakota,* USDA Forest Service General Technical Report RM-154 (1988); L. Dano, "Cottontail Rabbit *(Sylvilagus audubonii baileyi)* Populations in Relation to Prairie Dog *(Cynomys ludovicianus ludovicianus)* Towns (master's thesis, Colorado State University, Fort Collins, 1952); O'Meilia, Knopf, and Lewis, "Some Consequences" (see note 31 above); P. Stapp, "A Reevaluation of the Role of Prairie Dogs in the Great Plains Grasslands," *Conservation Biology* 12 (1998): 1253–1259.

47. Ceballos and Pacheco, "Relationship of Prairie Dogs" (see note 39 above); G. Ceballos, J. Pacheco, and R. List, "Influence of Prairie Dogs *(Cynomys ludovicianus)* on Habitat Heterogeneity and Mammalian Diversity in Mexico," *Journal of Arid Environments* 41 (1999): 161–172; Manzano, "Avian Communities" (see note 39 above).

48. Coppock et al., "Plant-Herbivore Interactions. I" (see note 37 above); Detling and Whicker, "Control of Ecosystem Processes" (see note 37 above); Stapp, "Reevaluation" (see note 46 above); J. F. Weltzin, S. L. Dowhower, and R. K. Heitschmidt, "Prairie Dog Effects on Plant Community Structure in Southern Mixed-Grass Prairie," *Southwestern Naturalist* 42 (1997): 251–258; Whicker and Detling, "Control of Grassland Ecosystem Processes" (see note 37 above); id., "Ecological Consequences" (see note 34 above); id., "Modification of Vegetation Structure" (see note 37 above).

49. C. D. Bonham and A. Lerwick, "Vegetation Changes Induced by Prairie Dogs on Shortgrass Range," *Journal of Range Management* 29 (1976): 221–225; Coppock et al., "Plant-Herbivore Interactions. I" (see note 37 above); J. K. Detling, "Mammalian Herbivores: Ecosystem-Level Effects in Two Grassland National Parks," *Wildlife Society Bulletin* 26 (1998): 438–448; Detling and Whicker, "Control of Ecosystem Processes" (see note 37 above); Weltzin, Dowhower, and Heitschmidt, "Prairie Dog Effects" (see note 48 above); Whicker and Detling, "Control of Grassland Ecosystem Processes" (see note 37 above); id., "Ecological Consequences" (see note 34 above); id., "Modification of Vegetation Structure" (see note 37 above).

50. Bonham and Lerwick, "Vegetation Changes " (see note 49 above); Coppock et al., "Plant-Herbivore Interactions. I" (see note 37 above); Detling, "Mammalian Herbivores" (see note 49 above); Weltzin, Dowhower, and Heitschmidt, "Prairie Dog Effects" (see note 48 above).

51. Bonham and Lerwick, "Vegetation Changes" (see note 49

above); Coppock et al., "Plant-Herbivore Interactions. I" (see note 37 above); List, "Ecology of Kit Fox and Coyote" (see note 5 above); J. F. Weltzin, S. Archer, and R. K. Heitshmidt, "Small Mammal Regulation of Vegetation Structure in a Temperate Savanna," *Ecology* 78 (1997): 751–763.

52. J. A. Miller, "Research Updates," *BioScience* 41 (1991): 750–753.

53. Whicker and Detling, "Control of Grassland Ecosystem Processes" (see note 37 above).

54. R. E. Ingham and J. K. Detling, "Plant-Herbivore Interactions in a North American Mixed-Grass Prairie. III. Soil Nematode Populations and Root Biomass on *Cynomys ludovicianus* Colonies and Adjacent Uncolonized Areas," *Oecologia* 63 (1984): 307–313; L. C. Munn, "Effects of Prairie Dogs on Physical and Chemical Properties of Soils," in *Management of Prairie Dog Complexes for the Reintroduction of the Black-footed Ferret*, edited by J. L. Oldemeyer et al., Biological Report 13 (Washington, D.C.: USFWS, 1993); A. Outwater, *Water: A Natural History* (New York: Basic Books, 1996); Whicker and Detling, "Ecological Consequences" (see note 34 above); id., "Modification of Vegetation Structure" (see note 37 above).

55. Detling, "Mammalian Herbivores" (see note 49 above); Ingham and Detling, "Plant-Herbivore Interactions. III" (see note 54 above).

56. Whicker and Detling, "Control of Grassland Ecosystem Processes" (see note 37 above).

57. Coppock et al., "Plant-Herbivore Interactions. I" (see note 37 above); Detling, "Mammalian Herbivores" (see note 49 above); Detling and Whicker, "Control of Ecosystem Processes" (see note 35 above); K. Krueger, "Feeding Relationships Among Bison, Pronghorn, and Prairie Dogs: An Experimental Analysis," *Ecology* 67 (1986): 760–770.

58. A. V. R. Vanderhyde, "Interspecific Nutritional Facilitation: Do Bison Benefit from Feeding on Prairie Dog Towns?" (master's thesis, Colorado State University, Fort Collins, 1985); Whicker and Detling, "Control of Grassland Ecosystem Processes" (see note 37 above).

59. Kotliar et al., "Critical Review" (see note 15 above).

60. H. D. Lasswell and M. S. McDougal, *Jurisprudence for a Free Society: Studies in Law, Science and Policy* (Dordrecht, Netherlands: Kluwer, 1992).

61. S. L. Yaffee, *Prohibitive Policy: Implementing the Federal Endangered Species Act* (Cambridge: M.I.T. Press, 1982).

62. J. Treviño-Villarreal, "Mexican Prairie Dog," in *Endangered Animals: A Reference Guide to Conflicting Issues*, edited by R. P. Reading and B. J. Miller (Westport, Conn.: Greenwood Press, 2000).

63. R. P. Reading et al., "Recent Directions in Black-footed Ferret *(Mustela nigripes)* Recovery," *Endangered Species Update* 13, nos. 10–11 (1996): 1–6; Reading et al., "Black-footed Ferret *(Mustela nigripes)* Conservation Update," *Small Carnivore Conservation* 17 (1997): 1–6.

64. A. Mathews, *Where the Buffalo Roam* (New York: Grove Weidenfeld, 1992); D. Popper and F. Popper, "The Buffalo Commons: A Metaphor as Method," *Geographical Review* 89, no. 4 (1999): 491–510; id., "The Great Plains: From Dust to Dust," *Planning* 53, no. 12 (1987): 12–18.

65. Miller et al., "Focal Species" (see note 36 above); R. F. Noss and A. Y. Cooperrider, *Saving Nature's Legacy: Protecting and Restoring Biodiversity* (Washington, D.C.: Island Press, 1994); M. Soulé and R. Noss, "Rewilding and Biodiversity: Complementary Goals for Conservation," *Wild Earth* 8 (1998): 19–28; J. Terborgh, "The Big Things That Run the World—A Sequel to E. O. Wilson," *Conservation Biology* 2 (1988): 402–403.

66. Miller, Ceballos, and Reading, "Prairie Dog" (see note 16 above); National Wildlife Federation, *Petition* (see note 4 above).

67. U.S. Fish and Wildlife Service, *Federal and State Endangered Species Expenditures for Fiscal Year 1991* (Washington, D.C.: USFWS, 1992).

68. Miller, Reading, and Forrest, *Prairie Night* (see note 8 above); U.S. Fish and Wildlife Service, *Black-footed Ferret Recovery Plan* (Washington, D.C.: USFWS, 1988).

69. Reading et al., "Recent Directions" (see note 63 above); Reading et al., "Black-footed Ferret" (see note 63 above); A. Vargas et al., "Black-footed Ferret Reproduction and Reintroduction in 1998," *Small Carnivore Conservation* 20 (1999): 32.

70. Miller, Ceballos, and Reading, "Prairie Dog" (see note 16 above); Miller, Reading, and Forrest, *Prairie Night* (see note 8 above); Reading, Miller, and Kellert, "Values and Attitudes" (see note 21 above).

71. G. D. Brewer and P. de Leon, *The Foundations of Policy Analysis* (Homewood, Ill.: Dorsey, 1983); Yaffee, *Prohibitive Policy* (see note 61 above).

72. H. D. Lasswell, *Psychopathology and Politics* (New York: Viking, 1930); Lasswell and Kaplan, *Power and Society* (see note 22 above).

73. D. L. Flores, "A Long Love Affair with an Uncommon Country: Environmental History and the Great Plains," in *Prairie Conservation: Preserving North America's Most Endangered Ecosystem*, edited by F. B. Samson and F. L. Knopf (Washington, D.C.: Island Press, 1996).

SAGE GROUSE

1. A. C. Bent, "Life Histories of North American Gallinaceous Birds (Orders Galliformes and Columbiformes)," *U.S. National Museum Bulletin* 162 (1932): 300–310; R. L. Patterson, *The Sage Grouse in Wyoming* (Denver: Sage, 1952).

2. C. E. Braun, "Historic and Present Distribution/Status of Sage Grouse in North America," in *Summary of Proceedings, Western Sage Grouse Status Conference, 14 January 1999, Boise, Idaho* (Portland, Ore.: American Lands Alliance, 1999).

3. American Lands, *The Sagebrush Sea* (booklet) (Portland, Ore.: American Lands Alliance, 2001), p. 5.

4. R. E. Autenrieth, *Sage Grouse Management in Idaho*, Federal Aid Wildlife Restoration Project W-125-R, W-160-R (1981); C. E. Braun, T. Britt, and R. O. Wallestad, "Guidelines for Maintenance of Sage Grouse Habitats," *Wildlife Society Bulletin* 5 (1977): 99–106; M. A. Gregg, "Use and Selection of Nesting Habitat by Sage Grouse in Oregon" (master's thesis, Oregon State University, Corvallis, 1992); J. A. Roberson, "Sage Grouse–Sagebrush Relationships: A Review," in *Proceedings—Symposium on the Biology of Artemisia and* Chrysothamnus, *9–13 July 1984, Provo, Utah*, compiled by E. McArthur and B. L. Welch, USDA Foreest Service General Technical Report INT-200 (Ogden, Utah: USDA Forest Service, Intermountain Research Station, 1984); W. L. Wakkinen, "Nest Site Characteristics and Spring–Summer Movements of Migratory Sage Grouse in Southeastern Idaho" (master's thesis, University of Idaho, Moscow, 1990).

5. A. K. DeLong, J. A. Crawford, and D. C. DeLong, "Relationships Between Vegetational Structure and Predation of Artificial Sage Grouse Nests," *Journal of Wildlife Management* 59 (1995): 88–92; Gregg, "Use and Selection of Nesting Habitat" (see note 4 above); D. R. Webb, *Effects of Cattle Grazing on Sage Grouse: Indirect Biophysical Effects*, final report to the Wyoming Department of Game and Fish (Cheyenne, 1993).

6. J. W. Connelly, "Population Ecology and Habitat Needs," in *Summary of Proceedings, Western Sage Grouse Status Conference, 14 January 1999, Boise, Idaho* (Portland, Ore.: American Lands Alliance, 1999).

7. B. V. Hulet et al., "Seasonal Movements and Habitat Selection of Sage Grouse in Southern Idaho," in *Proceedings—Symposium on the Biology of* Artemisia *and* Chrysothamnus, *9–13 July 1984, Provo, Utah*, compiled by E. McArthur and B. L. Welch, USDA Forest General Technical Report INT-200 (Ogden, Utah: USDA Forest Service, Intermountain Research Station, 1986).

8. Braun, "Historic and Present Distribution/Status of Sage Grouse in North America" (see note 2 above).

9. R. F. Noss, E. T. La Roe, and J. M. Scott, *Endangered Ecosystems of the United States: A Preliminary Assessment of Loss and Degradation*, Biological Report 28 (Washington D.C.: USDI, National Biological Service, 1995); R. F. Noss and R. L. Peters, *Endangered Ecosystems: A Status Report on America's Vanishing Habitat and Wildlife* (Washington, D.C.: Defenders of Wildlife, 1995).

10. Noss, La Roe, and Scott, *Endangered Ecosystems* (see note 9 above).

11. R. E. Autenrieth, W. Molini, and C. Braun, eds., *Sage Grouse Management Practices*, Technical Bulletin No. 1 (Twin Falls, Idaho: Western States Sage Grouse Committee, 1982); A. J. Belsky, A. Matzke, S. Uselman, "Survey of Livestock Influences on Stream and Riparian Ecosystems in the Western United States," *Journal of Soil and Water Conservation* 54, no. 1 (1999): 419–431; M. W. Call and C. Maser, *Wildlife Habitats in Managed Rangelands—The Great Basin of Southeastern Oregon: Sage Grouse*, USDA Forest Service General Technical Report PNW-187 (Portland, Ore.: USDA Forest Service, Pacific Northwest Forest and Range Experimental Station, 1985); D. A. Klebenow, "Habitat Management for Sage Grouse in Nevada," *World Pheasant Association* 10 (1985): 34–46; Patterson, *Sage Grouse in Wyoming* (see note 1 above); id., "Livestock Grazing Interactions with Sage Grouse," in *Proceedings of the Wildlife-Livestock Relationships Symposium, 20–22 April 1981, Coeur d'Alene, Idaho*, edited by J. M. Peek and P. D. Dalke (Moscow: University of Idaho, Forest, Wildlife, and Range Experiment Station, 1982); D. I. Rasmussen and L. A. Griner, "Life History and Management Studies of the Sage Grouse in Utah, with Special Reference to Nesting and Feeding Habits," *Transactions of the North American Wildlife Conference* 3 (1938): 852–864.

12. N. E. West, "Strategies for Maintenance and Repair of Biotic Community Diversity on Rangelands," in *Biodiversity in Managed Landscapes*, edited by R. C. Szaro and D. W. Johnston (New York: Oxford University Press, 1996).

13. A. J. Belsky, and D. M. Blumenthal, "Effects of Livestock Grazing on Stand Dynamics and Soils in Upland Forests of the Interior West," *Conservation Biology* 11 (1997): 315–327.

14. R. F. Noss and A. Cooperrider, *Saving Nature's Legacy: Protecting and Restoring Biodiversity* (Washington, D.C.: Defenders of Wildlife and Island Press, 1994); D. Sheridan, *Desertification of the United States* (Washington, D.C.: Council on Environmental Quality, 1981).

15. Klebenow, "Habitat Management" (see note 11 above).

16. "Gunnison Sage Grouse Conservation Plan, 1998, Gunnison Basin, Colorado," draft report (GBCP, June 1997).

17. Ibid.

18. U.S. Fish and Wildlife Service, electronic mail string, last author Jill Parker, 15 December 1998 (two-page printout of electronic mail communications, with a two-page printed attachment entitled "Sage Grouse—(Candidate Conservation?)" obtained in response to FOIA request to USDI—Fish and Wildlife Service, 1998).

19. R. H. Braun, "Emerging Limits on Federal Land Management Discretion: Livestock, Riparian Ecosystems, and Clean Water Law," *Environmental Law* 17 (1986): 43–79; B. S. Low and J. A. Berlin, "Natural Selection and the Management of Rangelands," in *Developing Strategies for Rangeland Management* (Boulder: Westview, 1984).

20. Autenrieth, Molini, and Braun, *Sage Grouse Management Practices* (see note 11 above); R. J. Oakleaf, "Relationship of Sage Grouse to Upland Meadows in Nevada" (master's thesis, University of Nevada, Reno, 1971); D. E. Savage, "The Relationship of Sage Grouse to Upland Meadows in Nevada," *Proceedings of the Western States Sage Grouse Workshop* 6 (1969): 134–141.

21. T. L. Fleischner, "Ecological Costs of Livestock Grazing in Western North America," *Conservation Biology* 8 (1994): 629–644.

22. L. L. St. Clair et al., "Cryptogamic Soil Crusts: Enhancement of Seedling Establishment in Disturbed and Undisturbed Areas," *Reclamation and Vegetation Research* 3 (1984): 129–136.

23. Fleischner, "Ecological Costs" (see note 21 above).

24. J. Belnap, "Potential Role of Cryptobiotic Soil Crusts in Semiarid Rangelands," in *Proceedings Ecology and Management of Annual Rangelands*, compiled by S. B. Monsen and S. G. Kitchen, USDA Forest Service General

Technical Report INT-313 (Ogden, Utah: USDA Forest Service, Intermountain Research Station, 1994); id., "Recovery Rates of Cryptobiotic Crusts: Inoculant Use and Assessment Methods," *Great Basin Naturalist* 53 (1993): 89–95; J. H. Kaltenecker, "The Recovery of Microbiotic Crusts Following Post-Fire Rehabilitation on Rangelands of the Western Snake River Plain" (master's thesis, Boise State University, Boise, Idaho, 1997); L. L. St. Clair and J. R. Johansen, "Introduction to the Symposium on Soil Crust Communities," *Great Basin Naturalist* 53 (1993): 1–4.

25. Ibid.

26. R. F. Daubenmire, "An Ecological Study of the Vegetation of Southeastern Washington and Adjacent Idaho," *Ecological Monographs* 12, no. 1 (1942): 55–79.

27. Ibid.

28. Webb, *Effects of Cattle Grazing on Sage Grouse* (see note 5 above); id., *Sage Grouse Nest Site Characteristics and Microclimates in Wyoming*, report to the Wyoming Department of Game and Fish (Cheyenne, 1993).

29. Ibid.

30. Autenrieth, *Sage Grouse Management in Idaho* (see note 4 above); Call and Maser, *Wildlife Habitats in Managed Rangelands* (see note 11 above); J. A. Crawford and A. K. Delong, *Relationships Between Vegetative Structure and Predation Rates of Artificial Sage Grouse Nests*, final report submitted to the BLM (Corvallis: Oregon State University, Department of Fisheries and Wildlife, 30 September 1993); M. A. Gregg et al., "Vegetational Cover and Predation of Sage Grouse Nests in Oregon," *Journal of Wildlife Management* 58 (1994): 162–166; D. Hein et al., *Evaluation of the Effects of Changes in Hunting Regulations on Sage Grouse Populations: Evaluation of Census of Females*, Job Completion Report W-37-R-33, WP-3, J-9c (Colorado Division of Wildlife, 1980); D. A. Klebenow, "Sage Grouse Nesting and Brood Habitat in Idaho," *Journal of Wildlife Management* 33, no. 3 (1969): 649–662; C. M. Sveum, *Habitat Selection by Sage Grouse Hens During the Breeding Season in South-Central Eastern Washington* (Corvallis: Batelle Northwest and Oregon State University, 1995); Wakkinen, "Nest Site Characteristics" (see note 4 above).

31. Northwest Ecosystem Alliance, "Petition for a Rule to List the Washington Population of Western Sage Grouse, *Centrocercus urophasianus phaios*, as 'Threatened' or 'Endangered' Under the Endangered Species Act," petition from the Northwest Ecosystem Alliance (Bellingham, Wash.) to the Office of Endangered Species of the United States Fish and Wildlife Service (14 May 1999).

32. Ibid.

33. R. F. Miller, T. Svejcar, and N. West, "Implications of Livestock Grazing in the Intermountain Sagebrush Region: Plant Composition," in *Ecological Implications of Livestock Herbivory in the West*, edited by M. Vavra, W. A. Laycock, and R. D. Pieper (Denver: Society for Range Management, 1993).

34. Fleischner, "Ecological Costs" (see note 21 above).

35. R. F. Miller and L. L. Eddleman, *Spatial and Temporal Changes of Sage Grouse Habitat in the Sagebrush Biome*, Technical Bulletin No. 151 (Corvallis: Oregon State University Agricultural Experiment Station, 2000).

36. J. Christensen, "Fire and Cheatgrass Conspire to Create a Weedy Wasteland," *High Country News*, 22 May 2000, pp. 8–9.

37. C. E. Braun, "Sage Grouse Declines in Western North America: What Are the Problems?" *Proceedings of the Western Association of State Fish and Wildlife Agencies* 78 (1998): 139–156; Call and Maser, *Wildlife Habitats in Managed Rangelands* (see note 11 above); "Gunnison Sage Grouse Conservation Plan, 1998, Dove Creek, Colorado," final report (DCCP, 23 November 1998).

38. Braun, ibid.

39. "Gunnison Sage Grouse Conservation Plan, 1998, Dove Creek, Colorado" (see note 37 above).

40. C. E. Braun, "Conservation Planning for Sage Grouse in Colorado," in *Western Sage and Columbian Sharp-tailed Grouse Workshop*, abstract (Billings, Montana, 1998).

41. J. E. Swenson, C. A. Simmons, and C. D. Eustace,

"Decrease of Sage Grouse, *Centrocercus urophasianus*, After Ploughing of Sagebrush Steppe," *Biological Conservation* 41 (1987): 125–132.

42. N. S. Martin, "Life History and Habitat Requirements of Sage Grouse in Relation to Sagebrush Treatment," *Proceedings of the Western Association of State Game and Fish Commissions* 56 (1976): 289–294; D. B. Pyrah, "Sage Grouse Habitat Research in Central Montana," *Proceedings Western Association State Game and Fish Commissions* 51 (1971): 293–300; R. O. Wallestad, "Summer Movements and Habitat Use by Sage Grouse Broods in Central Montana," *Journal of Wildlife Management* 35 (1971): 129–136.

43. D. W. Hays, M. J. Tihri, and D. K. Stinson, *Washington State Status Report for the Sage Grouse* (Olympia: Washington Department of Fish and Wildlife, Wildlife Management Program, 1998); M. A. Schroeder, *Minimum Viable Populations for Sage Grouse in Washington*, job progress report (Olympia: Washington Department of Fish and Wildlife, 1998).

44. J. F. Pechanec et al., *Controlling Sagebrush on Rangelands*, Farmer's Bulletin No. 2072 (Washington, D.C.: U.S. Department of Agriculture, 1954).

45. Miller and Eddleman, *Spatial and Temporal Changes of Sage Grouse Habitat* (see note 35 above).

46. M. S. Drut, *Status of Sage Grouse with Emphasis on Populations in Oregon and Washington* (Portland, Ore.: Portland Audubon Society, 1994).

47. Braun, Britt, and Wallestad, "Guidelines for Maintenance of Sage Grouse Habitats" (see note 4 above); D. A. Klebenow, "Sage Grouse Versus Sagebrush Control in Idaho," *Journal of Range Management* 23 (1970): 396–400; N. S. Martin, "Sagebrush Control Related to Habitat and Sage Grouse Occurrence," *Journal of Wildlife Management* 34, no. 2 (1970): 313–320; D. B. Pyrah, "Effects of Chemical and Mechanical Sagebrush Control on Sage Grouse," in *Ecological Effects of Chemical and Mechanical Sagebrush Control*, edited by E. F. Schlatterer and D. B. Pyrah (Helena: Montana Department of Fish and Game, 1970); Pyrah, "Sage Grouse Habitat Research in Central Montana" (see note 42 above); G. E. Rogers, *Sage Grouse Investigations in Colorado*, Technical Publication No. 16 (Denver: Colorado Game, Fish and Parks Department, Game Research Division, 1964); R. O. Wallestad, *Life History and Habitat Requirements of Sage Grouse in Central Montana* (Helena: Montana Department of Fish and Game, Game Management Division, in cooperation with the U.S. Department of the Interior, Bureau of Land Management, 1975); id., "Male Sage Grouse Responses to Sagebrush Treatment," *Journal of Wildlife Management* 39 (1975): 482–484; id., "Summer Movements and Habitat Use" (see note 42 above).

48. T. D. I. Beck, "Winter Ecology of Sage Grouse in North Park, Colorado," *Journal of Colorado-Wyoming Academy of Science* 7, no. 6 (1975): 43.

49. C. E. Braun, electronic mail transmission re: value of crested wheatgrass, to Dr. Randy Webb, NetWork Associates Ecological Consulting, 5 May 2000.

50. Ibid.

51. E. W. Tisdale and M. Hironaka, *The Sagebrush-Grass Region: A Review of the Ecological Literature*, Bulletin No. 33 (Moscow: University of Idaho, Forest, Wildlife, and Range Experiment Station, 1981).

52. M. L. Commons, R. K. Baydack, and C. E. Braun, "Sage Grouse Response to Pinyon-Juniper Management," in *Proceedings Ecology and Management of Pinyon-Juniper Communities Within the Interior West, 15–18 September 1997, Provo, Utah*, compiled by S. B. Monsen and R. Stevens, RMRS-P-9 (Ogden, Utah: USDA Forest Service, Rocky Mountain Research Station, 1999); J. M. Hanf, P. A. Schmidt, and E. B. Groshens, *Sage Grouse in the High Desert of Central Oregon—Results of a Study, 1968–1993* (Prineville, Ore.: USDI BLM, Prineville District Office, 1994).

53. Braun, e-mail (see note 49 above).

WHERE BISON ONCE ROAMED

1. B. Willers, "Animals Wild and Domestic: A Comment on Ratios," *Wild Earth* (Spring 1995): 6.

2. R. J. Mackie, "Impacts of Livestock Grazing on Wild Ungulates," in *Transactions of the 43rd North American Wildlife and Natural Resources Conference, 18–22 March 1978, Phoenix, Ariz.* (Washington, D.C.: Wildlife Management Institute, 1978).

3. J. E. Townsend and R. J. Smith, *Improving Fish and Wildlife Benefits in Range Management: Proceedings of a Seminar Held in Conjunction with the 41st North American Wildlife and Natural Resources Conference, 20 March 1976*, FWS/OBS-77/01 (Washington, D.C.: USDA Fish and Wildlife Service, 1977).

4. F. Wagner, "Livestock Grazing and the Livestock Industry," in *Wildlife and America: Contributions to an Understanding of American Wildlife and Its Conservation*, edited by H. P. Brokaw (Washington, D.C.: Council of Environmental Quality, 1978).

5. E. Chaney et al., *Livestock Grazing on Western Riparian Areas* (Eagle, Idaho: U.S. Environmental Protection Agency/Northwest Resource Information Center, 1990).

6. U.S. General Accounting Office, *Public Rangelands: Some Riparian Areas Restored but Widespread Improvement Will Be Slow*, GAO/RCED-88-105 (Washington, D.C.: USGAO, 1988).

7. G. Wuerthner, "Are Cows Just Domestic Bison? Behavioral and Habitat Use Differences Between Cattle and Bison," in *International Symposium on Bison Ecology and Management in North America*, edited by L. Irby, L. Knight, and J. Knight (Bozeman: Montana State University, 1998).

8. G. E. Plumb and J. L. Dodd, "Foraging Ecology of Bison and Cattle on a Mixed Prairie: Implications for Natural Area Management," *Ecological Applications* 3, no. 4 (1996): 631–643.

9. D. C. Hartnett et al., "Comparative Ecology of Native and Introduced Ungulates," in *Ecology and Conservation of Great Plains Vertebrates*, edited by F. L. Knopf and F. B. Sampson (New York: Springer-Verlag, 1996).

10. D. A. Frank, "The Ecology of the Earth's Grazing Ecosystems," *BioScience* 48 (1998): 513–521.

11. A. C. Isenberg, "The Returns of the Bison: Nostalgia, Profit, and Preservation," *Environmental History* 2, no. 2 (1997): 179–196.

12. Wagner, "Livestock Grazing."

13. Mackie, "Impacts of Livestock Grazing."

14. Wagner, "Livestock Grazing."

15. Mackie, "Impacts of Livestock Grazing."

16. Ibid.

17. Wagner, "Livestock Grazing."

18. P. Shepard, *The Tender Carnivore and the Sacred Game* (New York: Scribner's, 1973).

19. P. S. Martin and C. R. Szuter, "Megafauna of the Columbia Basin, 1800–1840: Lewis and Clark in a Game Sink," in *Northwest Lands, Northwest Peoples*, edited by D. Goble and P. Hirt (Seattle: University of Washington Press, 1999).

20. Wagner, "Livestock Grazing."

A WAR AGAINST PREDATORS

1. Predator Conservation Alliance, *Wildlife "Services"? A Presentation and Analysis of the USDA Wildlife Services Program's Expenditures and Kill Figures for Fiscal Year 1999* (Bozeman, Mont.: PCA, 2001); Wildlife Services, *Annual Report* (Washington, D.C.: USDA/APHIS Wildlife Services, 1999), www.aphis.usda.gov/ws.

2. Ibid.

3. Wildlife Services, *Annual Report* (Washington, D.C.: USDA/APHIS Wildlife Services, 1998).

4. General Accounting Office, *Animal Damage Control Program: Efforts to Protect Livestock from Predators*, GAO/RCED-96-3 (Washington, D.C.: USGAO, 1995).

5. Wildlife Services, *Annual Report* (see note 3).

6. Predator Conservation Alliance, *Wildlife "Services"?*

7. R. L. Crabtree, personal communication, 2001; R. L. Crabtree and J. W. Sheldon, "Coyotes and Canid Coexistence in Yellowstone," in *Carnivores in Ecosystems: The Yellowstone Experience*, edited by T. W. Clark et al. (New Haven: Yale University Press, 1999), pp. 127–163.

8. W. Keefover-Ring, "The U.S. Department of Agriculture's

Star Wars Program: Airborne Hunting Kills More Than Wildlife," *Predator Press* (newsletter of the Predator Defense Institute, Eugene, Ore.) 7, no. 1 (2000): 12–13; National Agricultural Statistics Service, *Cattle Predator Loss* (Washington, D.C.: USDA/NASS, 1996), www.usda.gov/nass.

9. Ibid.

10. Keefover-Ring, "U.S. Department of Agriculture's Star Wars Program"; National Agricultural Statistics Service, *Sheep and Goats Predator Loss* (Washington, D.C.: USDA/NASS, 2000), www.usda.gov/nass.

11. K. Easthouse, "Agency Spends Lots of Time Killing Coyotes for Newscaster," *Santa Fe New Mexican*, 14 March 1997.

12. Environmental News Network, "USDA Researchers on Trail of Coyote Control" (15 May 2000), www.cnn.com; L. A. Miller, "Induced Infertility: A Wildlife Management Tool," www.aphis.usda.gov/ws/nwrc/InducedInfertilityProjectPage.htm (16 February 2001).

13. Environmental News Network, "Coyote Control Breeds Small Predators" (17 November 2000), www.cnn.com.

PART V

TAKING STOCK OF PUBLIC LANDS GRAZING

1. U.S. Department of Commerce, Bureau of Economic Analysis, *Regional Economic Information System, 1996*, CD-ROM.

2. These estimates are obtained by comparing the forage supplied by Bureau of Land Management and Forest Service grazing leases with the feed required for the beef cow and ewe inventories in each of the eleven western states for 1997. See U.S. Department of Agriculture, Forest Service, Range Management, *Grazing Statistical Summary, FY 1997* (Washington, D.C.: Superintendent of Documents, 1998) and U.S. Department of the Interior, *Public Land Statistics*, vol. 183, *Statistical Appendix to the Annual Report of the Director, Bureau of Land Management, to the Secretary of the Interior* (Washington, D.C.: Superintendent of Documents, 1998). In both cases, authorized AUMs (animal unit months—one AUM being the amount of forage required by a cow-calf pair, or five ewes with lambs, for a month) for cattle and sheep were used. These AUMs were divided by 12 months for cattle and by 12/5 for sheep to establish the annual feed provided for an animal. The resulting figures were compared with the inventory of beef cows and ewes to establish the percentage of total feed provided by federal grazing leases.

3. U.S. Department of Agriculture, National Agricultural Statistics Service, *1997 Census of Agriculture*, vol. 1, *Geographic Area Series*, www.nass.usda.gov/census.

4. Ibid.

5. Employment, income, and agricultural sales data are from U.S. Department of Commerce, Bureau of Economic Analysis, *Regional Economic Information System, 1996* (CD-ROM). It was assumed that the percentage of total agricultural income and employment associated with cattle and sheep operations was directly proportional to the share of total agricultural marketings that were cattle and sheep. Five years of data ending in 1996 (deflated using the consumer price index) were averaged to smooth out fluctuations in agricultural income. 1996 was the latest year available on the CD-ROM database. For employment, the average of the five years 1992–1996 was also used. The days of economic growth necessary to equal the income and employment associated with federal grazing was based on the average growth in each measure between 1990 and 1996.

6. L. Frewing-Runyon, "Importance and Dependency of the Livestock Industry on Federal Lands in the Columbia River Basin," unpublished report prepared at the Oregon State Office of the Bureau of Land Management for the Interior Columbia Basin Ecosystem Management Project, Walla Walla, Wash. (10 April 1995).

7. The results reported here are this author's modification of the BLM analysis (see note 6 above). The BLM used total earnings instead of the total income used here. Since as

much as 40 percent of income received in these counties comes from sources other than earnings, that seemed an important adjustment. The BLM analysis also deviated in two other ways from that used in the analysis of the eleven western states provided here. First, it assumed that all livestock production was cattle production. As discussed earlier, this is not the case. In Washington, for instance, less than half of livestock production is cattle and sheep production. Second, the BLM tables indicate that the total of cattle and calves was used to represent the livestock population to be supported. Since an AUM of forage is intended to support a cow with a nursing calf, this may overstate the cattle population's feed needs. These deviations from the appropriate analysis have not been corrected. Their impacts tend to offset each other.

8. The federal forage and cattle inventory data come from BLM and Forest Service sources in New Mexico. They are taken from an unpublished report written by Karyn Moskowitz (K. Moskowitz, personal communication, 21 January 1999). This analysis also did not seek to distinguish cattle from other livestock production.

9. C. K. Gee, L. A. Joyce, and A. G. Madsen, *Factors Affecting the Demand for Grazed Forage in the United States*, USDA Forest Service General Technical Report RM-210 (Fort Collins, Colo.: USDA Forest Service, Rocky Mountain Forest and Range Experiment Station, 1992).

10. U.S. Department of Agriculture, "Farm Operator Household Income and Wealth Compare Favorably with All U.S. Households," *Rural Conditions and Trends* 8, no. 2 (1997): 79–84.

11. T. M. Power, *Environmental Protection and Economic Well-Being: The Economic Pursuit of Quality* (Armonk, N.Y.: M. E. Sharpe, 1996); id., *Lost Landscapes and Failed Economies: The Search for a Value of Place* (Washington, D.C.: Island Press, 1996).

MORTGAGING PUBLIC ASSETS

1. 43 U.S.C. § 315b (2001).

2. 36 C.F.R. § 222.3(b) (2001).

3. USDI BLM/USDA Forest Service, *Rangeland Reform '94 Final Environmental Impact Statement* (Washington, D.C., 1995), p. 125.

4. D. Donahue, *The Western Range Revisited: Removing Livestock from Public Lands to Conserve Native Biodiversity* (Norman: University of Oklahoma Press, 1999), p. 38.

5. *Public Lands Council v. Babbitt*, 529 U.S. 728, 741 (2000).

6. J. M. Fowler and J. R. Gray, *Market Values of Federal Grazing Permits in New Mexico* (Las Cruces: New Mexico State University Cooperative Extension Service, Range Improvement Task Force, 1980).

7. L. A. Torell and J. P. Doll, "Public Land Policy and the Value of Grazing Permits," *Western Journal of Agricultural Economics* 16, no. 1 (1991): 174–184.

8. J. R. Winter and J. K. Whittaker, "The Relationship Between Private Ranchland Prices and Public Land Grazing Permits," *Land Economics* 57, no. 3 (1981): 414–421.

9. Brief of Amicus Curiae, State Bank of Southern Utah, in Support of Petition for a Writ of Certiorari, Public Lands Council v. Babbitt (98-1191), p. 4.

10. 43 U.S.C. § 315b (2001).

11. 36 C.F.R. § 222.3(c)(1 (2001).

12. USDA Forest Service, Grazing and Livestock Use Permit System, 2230.3 (Policy).

13. USDA Forest Service, Grazing Permit Administration Handbook, 2209.13 § 18.32 (Exhibit 01).

14. USDA Forest Service, Grazing Permit Administration Handbook, 2209.13 § 18.21 (Execution of escrow waiver).

15. Memorandum: Curtis Anderson, Farm Credit Administration, to Dale Robertson, Chief, Forest Service (1 May 1990); Memorandum: Clayton Yeutter, Department of Agriculture, to Curtis Anderson, Secretary to the Board, FCA (15 June 1990).

16. Memorandum of Understanding Between the Farm Credit Banks and the Forest Service, United States Department of Agriculture (21 December 1990).

17. 58 F.R. 22074-01, 22081 (1993).

18. Brief, Farm Credit Bank of Texas, Appeal of 30 August 1996, Decision re. Diamond Bar Grazing Permit (16 October 1996).

19. USDI BLM/USDA Forest Service, *Rangeland Reform '94*, p. 128.

20. J. Bovard, *The Farm Credit Quagmire*, Policy Analysis 122 (Washington, D.C.: Cato Institute, 1989).

IT'S WHAT'S FOR DINNER

1. Z. Djuric et al., "Oxidative DNA Damage Levels in Blood from Women at High Risk for Breast Cancer Are Associated with Dietary Intakes of Meats, Vegetables, and Fruits," *Journal of the American Dietetic Association* 98, no. 5 (1998): 524–528.

2. E. De Stefani et al., "Meat Intake, Heterocyclic Amines, and Risk of Breast Cancer: A Case-Control Study in Uruguay," *Cancer Epidemiology, Biomarkers, and Prevention* 6 (1997): 573–581.

3. S. A. Bingham et al., "Does Increased Endogenous Formation of N-Nitroso Compounds in the Human Colon Explain the Association Between Red Meat and Colon Cancer?" *Carcinogenesis* 17 (1996): 515–523.

4. B. M. Brenner, T. W. Meyer, and T. H. Hostetter, "Dietary Protein Intake and the Progressive Nature of Kidney Disease: The Role of Hemodynamically Medicated Glomerular Injury in the Pathogenesis of Progressive Glomerular Sclerosis in Aging, Renal Ablation, and Intrinsic Renal Disease," *New England Journal of Medicine* 307 (1982): 652–659; H. M. Linkswiler et al., "Protein-Induced Hypercalciuria," *Federation Proceedings* 40 (1981): 2429–2433.

5. S. H. A. Holt, J. C. Brand Miller, and P. Petocz, "An Insulin Index of Foods: The Insulin Demand Generated by 1000-Kj Portions of Common Foods," *American Journal of Clinical Nutrition* 66 (1997): 1264–1276; R. J. Wurtman and J. J. Wurtman, "Brain Serotonin, Carbohydrate Craving, Obesity, and Depression," *Obesity Research* 3 (1995): 477S–480S.

6. M. J. Hill, "Cereals, Cereal Fibre and Colorectal Cancer Risk: A Review of the Epidemiological Literature," *European Journal of Cancer* 7, no. 2 (1998): S5–10.

7. S. R. Glore et al., "Soluble Fiber and Serum Lipids: A Literature Review," *Journal of the American Dietetic Association* 94 (1994): 425–436.

8. F. G. De Waart, U. Moser, and F. J. Kok, "Vitamin E Supplementation in Elderly Lowers the Oxidation Rate of Linoleic Acid in LDL," *Atherosclerosis* 133 (1997): 255–263; D. Harats et al., "Citrus Fruit Supplementation Reduces Lipoprotein Oxidation in Young Men Ingesting a Diet High in Saturated Fat: Presumptive Evidence for an Interaction Between Vitamins C and E in Vivo," *American Journal of Clinical Nutrition* 67, no. 2 (1998): 240–245; O. P. Heinonen et al., "Prostate Cancer and Supplementation with Alpha–Tocopherol and Beta-Carotene: Incidence and Mortality in a Controlled Trial," *Journal of the National Cancer Institute* 90, no. 6 (1998): 440–446.

9. W. J. Craig, "Phytochemicals: Guardians of Our Health," *Journal of the American Dietetic Association* 97 (suppl. 2) (1997): S199–204; K. A. Steinmetz and J. D. Potter, "Vegetables, Fruit, and Cancer. II. Mechanisms," *Cancer Causes and Control* 2 (1991): 427–442.

10. T. J. Key et al., "Mortality in Vegetarians and Nonvegetarians: Detailed Findings from a Collaborative Analysis of Five Prospective Studies," *American Journal of Clinical Nutrition* 70 (1999): 516S–524S.

11. G. E. Fraser, "Associations Between Diet and Cancer, Ischemic Heart Diseases, and All-Cause Mortality in Non-Hispanic White California Seventh-Day Adventists," *American Journal of Clinical Nutrition* 70 (1999): 532S–538S.

12. A. Tavani et al., "Red Meat Intake and Cancer Risk: A Study in Italy," *International Journal of Cancer* 86 (2000): 425–428.

13. A. N. Donaldson, "The Relation of Protein Foods to Hypertension," *California West Medicine* 24 (1926): 328–331.

14. I. L. Rouse et al., "Blood-Pressure-Lowering Effect of a

Vegetarian Diet: Controlled Trial in Normotensive Subjects," *Lancet* 1, no. 8314–8315 (1983): 5–10.

15. M. J. Wiseman et al., "Dietary Composition and Renal Function in Healthy Subjects," *Nephron* 46 (1987): 37–42.

16. W. G. Robertson, M. Peacock, and D. H. Marshall, "Prevalence of Urinary Stone Disease in Vegetarians," *European Urology* 8 (1982): 334–339.

17. M. J. Messina and V. K. Messina, *Dietitian's Guide to Vegetarian Diets* (Gaithersburg, Md.: Aspen, 1996).

18. V. R. Young and P. L. Pellet, "Plant Proteins in Relation to Human Protein and Amino Acid Nutrition," *American Journal of Clinical Nutrition* 59 (1994): 1203S–1212S.

19. V. K. Messina and K. I. Burke, "Position of the American Dietetic Association: Vegetarian Diets," *Journal of the American Dietetic Association* 97 (1997): 1317–1321.

EATING IS AN AGRICULTURAL ACT

1. A. B. Durning and H. B. Brough, *Taking Stock: Animal Farming and the Environment*, Worldwatch Paper 103 (Washington, D.C.: Worldwatch Institute, 1991).

2. J. N. Abramovitz et al., *Vital Signs 2000: The Environmental Trends That Are Shaping Our Future* (Washington, D.C.: Worldwatch Institute, 2000); Durning and Brough, *Taking Stock* (see note 1 above).

3. G. Gardner and B. Halweil, *Underfed and Overfed: The Global Epidemic of Malnutrition*, Worldwatch Paper 150 (Washington, D.C.: Worldwatch Institute, 2000).

4. D. H. Wright, "Human Impacts on Energy Flows Through Natural Ecosystems, and Implications for Species Endangerment," *Ambio* 19, no. 4 (1990): 189–194.

5. H. L. Mortimer Jr., "Laying Out Land Loss," *Environment* 34, no. 4 (1992): 23–24; R. F. Noss and A.Y. Cooperrider, *Saving Nature's Legacy: Protecting and Restoring Biodiversity* (Washington, D.C.: Island Press, 1994).

6. R. B. Primack, *Essentials of Conservation Biology* (Sunderland, Mass.: Sinauer Associates, 1993).

7. World Commission on Environment and Development, *Our Common Future*, the Brundtland Commission Report (Oxford, UK: Oxford University Press, 1987).

8. Noss and Cooperrider, *Saving Nature's Legacy* (see note 5 above).

9. J. Rifkin, *Beyond Beef: The Rise and Fall of the Cattle Culture* (New York: Plume, 1992).

10. M. Oesterheld, O. E. Sala, and S. J. McNaughton, "Effects of Animal Husbandry on Herbivore Carrying-Capacity at a Regional Scale," *Nature* 356 (1992): 234–236.

11. D. S. Wilcove et al., "Quantifying Threats to Imperiled Species in the U.S.," *BioScience* 48, no. 8 (1998): 607–615.

12. Durning and Brough, *Taking Stock* (see note 1 above).

13. Primack, *Essentials* (see note 6 above).

14. E. C. Wolf, *On the Brink of Extinction: Conserving the Diversity of Life*, Worldwatch Paper 78 (Washington, D.C.: Worldwatch Institute, 1987).

15. J. Enslow and C. Padoch, *People of the Tropical Rainforest* (Berkeley and Los Angeles: University of California Press, 1988).

16. Durning and Brough, *Taking Stock* (see note 1 above).

17. L. R. Brown, *The Agricultural Link: How Environmental Deterioration Could Disrupt Economic Progress*, Worldwatch Paper 136 (Washington, D.C.: Worldwatch Institute, 1997).

18. Durning and Brough, *Taking Stock* (see note 1 above).

19. M. Webb and J. Jacobsen, *U.S. Carrying Capacity: An Introduction* (Washington, D.C.: Carrying Capacity, Inc., 1982).

20. P. R. Ehrlich et al., *Ecoscience: Population, Resources, Environment* (San Francisco: W. H. Freeman, 1977).

21. Calculated from statistics in Brian Halweil, "United States Leads World in Meat Stampede: A Worldwatch Institute Press Briefing" (2 July 1998), www.worldwatch.org/alerts/pr980704.html.

22. Brown, *Agricultural Link* (see note 17 above).

23. L. R. Brown, *Tough Choices: Facing the Challenge of Food Scarcity*, Worldwatch Environmental Alert Series (New York: W. W. Norton, 1996).

24. Durning and Brough, *Taking Stock* (see note 1 above).

25. Ibid.

26. Rifkin, *Beyond Beef* (see note 9 above).

27. A. Gore, Introduction to *Silent Spring*, by Rachel Carson, new ed. (Boston: Houghton Mifflin, 1994).

28. N. Myers, *Gaia: An Atlas of Planet Management*, rev. ed. (New York: Anchor, 1993).

29. Durning and Brough, *Taking Stock* (see note 1 above).

30. F. M. Lappé, *Diet for a Small Planet*, 20th anniv. ed. (New York: Ballantine, 1991).

31. Brown, *Agricultural Link* (see note 17 above).

32. Lappé, *Diet for a Small Planet* (see note 30 above).

33. D. Pimentel, "Livestock Production: Energy Inputs and Environment," in *Proceedings of the 47th Annual Meeting of the Canadian Society of Animal Science, 24–26 July 1997, Montreal, Quebec*, edited by S. L. Scott and Xin Zhao.

34. Durning and Brough, *Taking Stock* (see note 1 above).

35. J. R. Gillespie, *Modern Livestock and Poultry Production* (San Francisco: Delmar Publishing,1997).

36. G. Borgstrom, *The Food and People Dilemma* (Belmont, Calif.: Duxbury Press, 1973).

37. C. Flavin, "Italia Verde," *World Watch*, September/October 1989; E. Hagerman, "Can Europe Save Its Beloved Baltic?" *World Watch*, November/December 1990.

38. D. Pimentel, "Waste in Agriculture and Food Sectors: Environmental and Social Costs," paper for Gross National Waste Product, Arlington, Va. (1989); Rifkin, *Beyond Beef* (see note 9 above).

39. M. Gibbs and K. Hogan, "Methane," *EPA Journal* 16 (March/April 1990): 23–25.

40. L. R. Brown et al., *State of the World 1990* (Washington, D.C.: Worldwatch Institute, 1990).

41. G. F. Hallberg, "Nitrate in Groundwater in the United States," in *Nitrogen Management and Groundwater Protection*, edited by R. F. Follett (Amsterdam: Elsevier, 1989).

42. S. Postel, *Last Oasis: Facing Water Scarcity*, Worldwatch Environmental Alert Series (New York: W. W. Norton, 1992).

43. Rifkin, *Beyond Beef* (see note 9 above).

44. Lappé, *Diet for a Small Planet* (see note 30 above).

45. D. Pimentel and C. W. Hall, *Food and Natural Resources* (San Diego: Academic Press, 1989).

46. Lappé, *Diet for a Small Planet* (see note 30 above).

47. A. Kimbrell, "On the Road," in *The Green Lifestyle Handbook*, edited by J. Rifkin (New York: Henry Holt, 1990).

48. Durning and Brough, *Taking Stock* (see note 1 above).

49. Halweil, "United States Leads World Meat Stampede" (see note 21 above).

PART VI

THE DONUT DIET

1. Savory Center for Holistic Management web site, www.holisticmanagement.org.

2. C. J. Hadley, "The Wild Life of Allan Savory," *Rangelands* 22, no. 1 (2000): 6–10.

3. A. Savory, *Holistic Management* (Washington, D.C.: Island Press, 1999); id., *Holistic Resource Management* (Washington, D.C.: Island Press, 1988); id., "Reversing Desertification," www.holisticmanagement.org/art_desert.cfm (3 December 2001).

4. Hadley, "Wild Life of Allan Savory" (see note 2 above); Savory, "Reversing Desertification" (see note 3 above).

5. For examples, see the Savory Center for Holistic Management web site, www.holisticmanagement.org.

6. D. Brown, "Out of Africa," *Wilderness Magazine* (Winter 1994): 24–33; J. L. Holecheck et al., "Short-Duration Grazing: The Facts in 1999," *Rangelands* 22, no. 1 (2000): 18–22; L. Jacobs, *Waste of the West: Public Lands Ranching* (Tucson: Self-published, 1991), www.apnm.org/waste_of_west/; S. Johnson, "Allan Savory: Guru of False Hopes and an Overstocked Range," *High Country News*, 27 April 1987.

7. Savory, *Holistic Management* (see note 3 above); id., "Reversing Desertification" (see note 3 above).

8. B. Sindelar, "Myths in Grazing Management," in *Grazing Management: It's More Than Grass! a symposium presented by the New Pasture and Grazing Technologies Project, 19–20 January 1989, Saskatoon, Saskatchewan*, pp. 1–23; Savory, "Reversing Desertification" (see note 3 above).

9. D. Dagget, *Beyond the Rangeland Conflict* (Layton, Utah: Gibbs Smith Books, 1997); Savory, "Reversing Desertification" (see note 3 above).

10. Dagget, ibid.; Savory, *Holistic Resource Management* (see note 3 above); id., "Reversing Desertification" (see note 3 above).

11. Hadley, "Wild Life of Allan Savory" (see note 2 above).

12. J. Skovlin, "Southern Africa's Experience with Intensive Short Duration Grazing," *Rangelands* 9, no. 4 (1987): 162–164.

13. Holecheck et al., "Short-Duration Grazing" (see note 6 above).

14. Jacobs, *Waste of the West* (see note 6 above); Johnson, "Allan Savory," (see note 6 above).

15. J. Belsky, "Does Herbivory Benefit Plants? A Review of the Evidence," *American Midland Naturalist* 127 (1986): 870–892.

16. N. Ambos, G. Robertson, and J. Douglas, "Dutchwoman Butte: A Relict Grassland in Central Arizona," *Rangelands* 22, no. 2 (2000): 3–8.

17. J. E. Anderson and K. E. Holte, "Vegetation Development Over 25 Years Without Grazing on Sagebrush-Dominated Rangeland in Southeastern Idaho," *Journal of Range Management* 34 (1981): 25–29; C. E. Bock and J. H. Bock, "Cover of Perennial Gasses in Southeastern Arizona in Relation to Livestock Grazing," *Conservation Biology* 7 (1993): 371–377; W. W. Brady et al., "Response of a Semidesert Grassland to 16 Years of Rest from Grazing," *Journal of Range Management* 42, no. 4 (1989): 284–288; R. S. Driscoll, "A Relict Area in Central Oregon Juniper Zone," *Ecology* 45 (1964): 345–353; J. H. Robertson, "Changes on a Sagebrush-Grass Range in Nevada Ungrazed for 30 Years," *Journal of Range Management* 24 (1971): 397–400.

18. R. N. Mack and J. N. Thompson, "Evolution in Steppe with Few Large, Hooved Mammals," *American Midland Naturalist* 119 (1982): 757–773.

19. G. Wuerthner, "Are Cows Just Domestic Bison? Behavioral and Habitat Use Differences Between Cattle and Bison," in *International Symposium on Bison Ecology and Management in North America*, edited by L. Irby and J. Knight (Bozeman: Montana State University, 1998), pp. 374–383.

20. R. K. Heitschmidt, S. L. Dowhower, and J. W. Walker, "14- vs. 42-Paddock Rotational Grazing: Aboveground Biomass Dynamics, Forage Production, and Harvest Efficiency," *Journal of Range Management* 40, no. 3 (1987): 216–223.

21. Hadley, "Wild Life of Allan Savory" (see note 2 above); Savory, "Reversing Desertification" (see note 3 above).

22. Dagget, *Beyond the Rangeland Conflict* (see note 9 above); Hadley, "Wild Life of Allan Savory" (see note 2 above).

23. M. Meagher and D. B. Houston, *Yellowstone and the Biology of Time* (Norman: University of Oklahoma Press, 1998).

24. Belsky, "Does Herbivory Benefit Plants?" (see note 15 above).

25. Belsky, ibid.; E. L. Painter and A. J. Belsky, "Application of Herbivore Optimization Theory to Rangelands of the Western United States," *Ecological Applications* 3 (1993): 2–9.

26. L. E. Hughes, "A Drought and 2 Grazing Systems," *Rangelands* 13, no. 5 (1991): 229–231.

27. Holecheck et al., "Short-Duration Grazing" (see note 6 above).

28. J. F. Dormaar, S. Smoliak, and W. D. Willms, "Vegetation and Soil Responses to Short-Duration Grazing on Fescue Grasslands," *Journal of Range Management* 42, no. 3 (1989): 252–256; G. R. McCalla, W. H. Blackburn, and L. B. Merrill, "Effects of Livestock Grazing on Infiltration Rates: Edwards Plateau of Texas," *Journal of Range*

Management 37 (1984): 265–269; J. J. Pluhar, R. W. Knight, and R. K. Heitschmidt, "Infiltration Rates and Sediment Production as Influenced by Grazing Systems in the Texas Rolling Plains," Journal of Range Management 40, no. 3 (1987): 240–246; S. D. Warren, W. H. Blackburn, and C. A. Taylor, "Effects of Season and Stage of Rotation Cycle on Hydrologic Condition of Rangelands Under Intensive Rotation Grazing," Journal of Range Management 39, no. 6 (1986): 486–490.

29. Dormaar, Smoliak, and Willms, "Vegetation and Soil Responses" (see note 28 above).

30. Holecheck et al., "Short-Duration Grazing" (see note 6 above).

31. Warren, Blackburn, and Taylor, "Effects of Season and Stage of Rotation Cycle" (see note 28 above).

32. F. C. Bryant et al., "Does Short-Duration Grazing Work in Arid and Semiarid Regions?" Journal of Soil and Water Conservation 44 (1989): 290–296.

33. J. H. Kaltenecker, M. Wicklow-Howard, and R. Rosentreter, "Biological Soil Crusts in Three Sagebrush Communities Recovering from a Century of Livestock Trampling," in Proceedings: Shrubland Ecotones, 12–14 August 1998, Ephraim, Utah, compiled by E. D. McArthur, W. K. Ostler, and C. L. Wambolt, RMRS-P-11 (Ogden, Utah: USDA Forest Service, Rocky Mountain Research Station, 1999), pp. 222–226.

34. Dagget, Beyond the Rangeland Conflict (see note 9 above); Savory, "Reversing Desertification" (see note 3 above).

35. R. Noss and A. Cooperrider, Saving Nature's Legacy (Washington D.C.: Island Press, 1994).

JUST A DOMESTIC BISON?

1. H. W. Reynolds, R. D. Glahot, and A. W. Hawley, "Bison," in Wild Mammals of North America: Biology, Management, and Economics, edited by J. A. Chapman and G. A. Feldhamer (Baltimore: John Hopkins University Press, 1982).

2. J. K. Detling and E. L. Painter, "Defoliation Responses of Western Wheatgrass Populations with Diverse Histories of Prairie Dog Grazing," Oecologia 57 (1983): 65–71; E. L. Painter, J. K. Detling, and D. A. Steingraeber, "Grazing History, Defoliation and Frequency-Dependent Competition: Effects on Two North American Grasses," American Journal of Botany 76 (1989): 1368–1379.

3. V. Geist, Buffalo Nation: History and Legend of the North American Bison (Stillwater, Minn.: Voyageur Press, 1996); Reynolds, Glahot, and Hawley, "Bison" (see note 1 above); P. J. Urness, "Why Did Bison Fail West of the Rockies?" Utah Science 50 (1989): 175–179.

4. M. M. Meagher, The Bison of Yellowstone National Park, National Park Service Science Monograph 1 (1973).

5. H. Goetz, Letter to the editor, Rangelands 16, no. 2 (1994); W. K. Lauenroth, "Effects of Grazing on Ecosystems of the Great Plains," in Ecological Implications of Livestock Herbivory in the West, edited by M. Vavra, W. A. Laycock, and R. D. Pieper (Denver: Society for Range Management, 1994); A. Savory, "The Savory Grazing Method or Holistic Resource Management," Rangelands 5 (1983): 155–159; H. S. Thomas, "Buffalo: Early Range Users," Rangelands 13, no. 6 (1991.): 285–287.

6. Thomas, ibid.

7. Ibid.

8. Savory, "Savory Grazing Method" (see note 5 above); U.S. Department of Agriculture, Livestock Grazing Successes on Public Range, pamphlet (USDA Forest Service, 1989).

9. A. Tohill and J. Dollerschell, "Livestock: The Key to Resource Improvement on Public Lands," Rangelands 12, no. 6 (1990): 329–336.

10. W. A. Laycock, "Implications of Grazing Versus No Grazing on Today's Rangelands," in Ecological Implications of Livestock Herbivory in the West, edited by M. Vavra, W. A. Laycock, and R. D. Pieper (Denver: Society for Range Management, 1994).

11. S. Archer and F. E. Smeins, "Ecosystem-Level Processes," in Grazing Management: An Ecological Perspective,

edited by R. K. Heitschmidt and J. W. Stuth (Portland, Ore.: Timber Press, 1991).

12. R. K. Heitschmidt, "The Role of Livestock and Other Herbivores in Improving Rangeland Vegetation," Rangelands 12, no. 2 (1990): 112–115.

13. A. J. Belsky, "Does Herbivory Benefit Plants? A Review of the Evidence," American Midland Naturalist 127 (1986): 870–892; A. J. Belsky et al., "Overcompensation by Plants: Herbivore Optimization or Red Herring?" Evolutionary Ecology 7 (1993): 109–121.

14. Geist, Buffalo Nation (see note 3 above); J. N. McDonald, North American Bison: Their Classification and Evolution (Berkeley and Los Angeles: University of California Press, 1981).

15. R. D. Guthrie, "Bison and Man in North America," Canadian Journal of Anthropology 1 (1980): 55–73.

16. Geist, Buffalo Nation (see note 3 above); McDonald, North American Bison (see note 14 above).

17. McDonald, ibid.

18. Geist, Buffalo Nation (see note 3 above).

19. McDonald, North American Bison (see note 14 above).

20. Geist, Buffalo Nation (see note 3 above).

21. Ibid.

22. J. K. Townsend, Narrative of a Journey Across the Rocky Mountains to the Columbia River (Philadelphia: Henry Perkins, 1839; reprint, Lincoln: University of Nebraska Press, 1978).

23. D. Van Vuren, "Bison West of the Rocky Mountains: An Alternative Explanation," Northwest Science 61 (1987): 65–69; id., "Group Dynamics and Summer Home Range of Bison in Southern Utah," Journal of Mammalogy 64 (1983): 329–332.

24. J. E. Norland, "Habitat Use and Distribution of Bison in Theodore Roosevelt National Park" (master's thesis, Montana State University, Bozeman, 1984).

25. Van Vuren, "Bison West of the Rocky Mountains" (see note 23 above); id., "Group Dynamics" (see note 23 above).

26. Norland, "Habitat Use" (see note 24 above).

27. D. G. Peden et al., "The Trophic Ecology of Bison bison on Shortgrass Plains," Journal of Applied Ecology 11 (1974): 489–497.

28. W. E. Pinchak et al., "Beef Cattle Distribution Patterns on Foothill Ranges," Journal of Range Management 44, no. 3 (1991): 267–275.

29. T. Goodman et al., "Cattle Behavior with Emphasis on Time and Activity Allocations Between Upland and Riparian Habitats," in Riparian Resource Management, edited by R. E. Gresswell, B. A. Barton, and J. K. Kershner (USDI Bureau of Land Management, 1989); M. A. Smith et al., "Habitat Selection by Cattle Along an Ephemeral Channel," Journal of Range Management 45, no. 4 (1992): 385–390.

30. J. L. Holecheck, "Financial Benefits of Range Management Practices in the Chihuahuan Desert," Rangelands 14, no. 5 (1992): 279–284.

31. L. N. Carbyn, S. M. Oosenbrug, and D. W. Anions, Wolves, Bison . . . and the Dynamics Related to the Peace-Athabasca Delta in Canada's Wood Buffalo National Park, Circumpolar Research Series No. 4 (Edmonton: University of Alberta, Canadian Circumpolar Institute, 1993); Meagher, Bison of Yellowstone (see note 4 above).

32. H. F. Peters and S. B. Slen, "Hair Coat Characteristics of Bison, Domestic Bison Hybrids, Cattalo and Certain Domestic Breeds of Cattle," Canadian Journal of Animal Science 44 (1964): 48–57.

33. Geist, Buffalo Nation (see note 3 above).

34. R. N. Mack and J. N. Thompson, "Evolution in Steppe with Few Large, Hooved Mammals," American Midland Naturalist 119 (1982): 157–173; R. F. Miller, T. J. Svejcar, and N. E. West, "Implications of Livestock Grazing in the Intermountain Sagebrush Region: Plant Composition," in Ecological Implications of Livestock Herbivory in the West, edited by M. Vavra, W. A. Laycock, and R. D. Pieper (Denver: Society for Range Management, 1994).

35. G. P. Davis Jr., Man and Wildlife in Arizona: The American Exploration Period 1824–1865, edited by N. B. Carmony and D. E. Brown (Phoenix: Arizona Fish

and Game Department, 1982); P. S. Ogden, "Journal of Peter Skene Ogden: Snake River Expedition, 1827–1828," Oregon Historical Quarterly 11 (1910): 361–379; J. H. Simpson, Report of Explorations Across the Great Basin of the Territory of Utah for a Direct Wagon-Route from Camp Floyd to Genoa in the Carson Valley in 1859 (Reno: University of Nevada Press, 1983).

36. McDonald, North American Bison (see note 14 above).

37. R. F. Daubenmire, "The Western Limits of the Range of the American Bison," Ecology 66 (1985): 622–624.

38. Peden et al., "Trophic Ecology" (see note 27 above).

39. Mack and Thompson, "Evolution" (see note 34 above).

40. Van Vuren, "Bison" (see note 23 above).

41. Mack and Thompson, "Evolution" (see note 34 above).

42. R. N. Mack, "Alien Plant Invasion into the Intermountain West: A Case History," in Ecology of Biological Invasions in North America and Hawaii, edited by H. A. Mooney and J. A. Brake, Ecological Studies, vol. 58 (New York: Springer-Verlag, 1986); Mack and Thompson, "Evolution" (see note 34 above).

43. T. A. Jones, D. C. Nelson, and J. R. Carlson, "Developing a Grazing-Tolerant Native Grass for Bluebunch Wheatgrass Sites," Rangelands 13, no. 3 (1991): 147–150; Miller, Svejcar, and West, "Implications of Livestock Grazing" (see note 34 above).

COWS OR CONDOS

1. U.S. Department of Agriculture, National Resources Inventory (Washington, D.C.: Natural Resources Conservation Service, 1997; revised December 2000), www.nhq.nrcs.usda.gov/NRI/1997.

2. U.S. Department of Agriculture, America's Private Land: A Geography of Hope (Washington, D.C.: Natural Resources Conservation Service, 1997).

3. Ibid.

4. California Department of Fish and Game, Natural Heritage Division, GAP Analysis of Mainland California: An Interactive Atlas of Terrestrial Biodiversity and Land Management, CD-ROM (1995). A link to ordering this CD-ROM may be found at www.biogeog.ucsb.edu.

5. California Department of Conservation, California Farmland Conversion Report 1996–1998, Farmland Mapping and Monitoring Program Publication No. FM 2000-01 (Sacramento, 2000); California Department of Food and Agriculture, California Agricultural Resource Directory (Sacramento, 1998), www.cdfa.ca.gov.

6. USDA, National Resources Inventory.

7. Based on U.S. Census Bureau data, released 21 March 2001 and compiled by the Census and Economic Information Center, Montana Department of Commerce, in the table "Census 2000, Public Law 94-171 File: Total Population, Population Density, and Land Areas for Montana Counties," http://ceic.commerce.state.mt.us/C2000/PL2000/Plcountyarea.htm.

8. G. Wuerthner, "Subdivisions and Extractive Industries," Wild Earth 7, no. 3 (1997): 57–62.

9. L. Kolankiewicz and R. Beck, Weighing Sprawl Factors in Large U.S. Cities (Arlington, Va.: NumbersUSA, 2001).

10. Ibid.

11. J. L. Holecheck, "Western Ranching at the Crossroads," Rangelands 23, no.1 (2001): 17–21.

12. T. M. Power, Lost Landscapes and Failed Economies: The Search for a Value of Place (Washington, D.C.: Island Press, 1996).

13. Holecheck, "Western Ranching."

14. R. Petersen and D. L. Coppock, "Economies and Demographics Constrain Investment in Utah Private Grazing Lands," Journal of Range Management 54, no. 2 (2001): 106–114.

15. Ibid.

16. Ibid.

17. R. H. Liffmann, L. Hunsinger, and L. C. Forego, "To Ranch or Not to Ranch: Home on the Urban Range? Journal of Range Management 53, no. 4 (2000): 362–379.

18. Holecheck, "Western Ranching" (see note 11 above); Liffmann, Hunsinger, and Forego, "To Ranch" (see note

17 above); Petersen and Coppock, "Economies" (see note 14 above).

19. Petersen and Coppock, ibid.
20. H. I. Rowe, M. Shinderman, and E. T. Bartlett, "Change on the Range," *Rangelands* 23, no. 2 (2001): 6–9.
21. Ibid.
22. Florida Department of Environmental Protection, Greenways & Trails Program (2001), www.dep.state.fl.us./gwt/.

PART VII

THE LAST ROUNDUP

1. A. Leopold, "The Virgin Southwest," in *The River of the Mother of God and Other Essays by Aldo Leopold*, edited by S. L. Flader and J. B. Callicott (Madison: University of Wisconsin Press, 1991), pp. 173–180.
2. L. McMurtry, "Death of the Cowboy," *New York Review of Books*, 4 November 1999.

UPHOLDING THE LAW

1. 43 U.S.C. § 1732(b) (1994). Note that although the Wilderness Act, 16 U.S.C. §§ 1131–1136 (1994), "grandfathered" pre-1964 grazing use, it is still "subject to reasonable regulations," 16 U.S.C. § 1133(d)(4)(2) (1994).
2. 43 U.S.C. §§ 1712(c)(1), 1702(c) (1994). See *National Wildlife Federation v. BLM*, 140 IBLA 85 (1997) (the "Comb Wash" decision).
3. 43 C.F.R. § 4180.1 (2000).
4. See *Idaho Watersheds Project v. Hahn*, 187 F.3d 1035 (9th Cir. 1999).
5. The Multiple-Use Sustained-Yield Act of 1960, 16 U.S.C. §§ 528–531 (1994).
6. 16 U.S.C. § 1604(g)(2)(A) (1994).
7. 36 C.F.R. § 219.3 (2000). See also 36 C.F.R. § 219.20 (2000) (addressing suitability and capability of forest lands for livestock grazing). But see 65 Fed. Reg. 67514 (9 November 2000) (final rule revising the Forest Service's planning regulations at 36 C.F.R. part 219) and 66 Fed. Reg. 27552 (17 May 2001) (delaying implementation of the revised rules for one year).
8. See *The Wilderness Society* v. Thomas et al., 188 F.3d 1130 (9th Cir. 1999).
9. See *Appeal Decision of the Rio Grande National Forest Land and Resource Management Plan* (# 97-13-00-0057) (19 January 2001); *Discretionary Review Decision on the Chief's Appeal Decision Regarding the Rio Grande National Forest Land and Resource Management Plan* (29 March 2001).
10. 16 U.S.C. §§ 1531–1540 (1994).
11. 16 U.S.C. § 1533 (1994).
12. 16 U.S.C. § 1536 (1994).
13. 16 U.S.C. §§ 1271–1284 (1994).
14. 16 USC § 1274(d) (1994); see *National Wildlife Federation v. Cosgriffe*, 21 F.Supp.2d 1211 (D. Or. 1998).
15. See, e.g., *Oregon Natural Desert Association v. Green*, 953 F. Supp. 1133 (D. Or. 1997) (holding BLM plan for Donner und Blitzen River unlawfully allowed grazing to continue along river despite adverse impacts to vegetation, water quality, and fish habitat).
16. 33 U.S.C. §§ 1251–1387 (1994).
17. See 33 U.S.C. § 1323 (1994); 43 U.S.C. § 1712(c)(8) (1994) (FLPMA land use planning must provide for compliance with water quality standards); 43 C.F.R. § 4180.1(c) (2000) (BLM regulation requiring water quality compliance); 36 C.F.R. 219.1 (2001) (listing CWA as a principle authority for management of the national forests).
18. 42 U.S.C. §§ 4321 et seq. (1994).

INDEX

Kahn